Solitons in Mathematics and Physics

ALAN C. NEWELL

University of Arizona

SOCIETY for INDUSTRIAL and
APPLIED MATHEMATICS • 1985

PHILADELPHIA PENNSYLVANIA 19103

Copyright © 1985 by Society for Industrial and Applied Mathematics.

Library of Congress Catalog Card Number: 84-71051
ISBN: 0-89871-196-7

Contents

Introduction

There has been a revolution in nonlinear physics over the past twenty years. Two great discoveries, each of which, incidentally, was made with the aid of the computer experiment, have radically changed the thinking of scientists about the nature of nonlinearity and introduced two new theoretical constructs into the field of dynamics. The first of these is the soliton and the second is the strange attractor. Before this period of discovery, our understanding of nonlinear behavior in systems with many degrees of freedom was limited to situations which were capable of being described by a purely hyperbolic system of equations (compressible flows, shock waves) or which were small perturbations away from linear states. While there are still many nonlinear processes like fully developed turbulence and quantum systems with large fluctuations about which little is known, there are also several other types of nonlinear behavior which occur widely in nature and which can now be categorized, predicted and understood. No longer does the term nonlinear, which literally means not linear, connote a domain of understanding beyond the pale.

This is a book about solitons as they appear in mathematics and physics. It is an outgrowth of a set of lectures I gave in June 1982 as part of a series supported by the National Science Foundation through the Conference Board on the Mathematical Sciences (CBMS). In writing it, I have tried to keep the serious student who is a nonexpert uppermost in mind from the point of view of both style and price. This is not an encyclopaedia of information on solitons in which every sentence is interrupted by either a caveat or a reference. Rather, I have tried to tell the story of the soliton as I would like to have heard it as a graduate student, with some historical development, lots of motivation, frequent attempts to relate the topic in hand to the big picture and a clear indication of the direction or directions in which the subject is going. Often, the important ideas are repeated several times, sometimes in slightly different contexts. As a consequence of the style, the book at times is woefully inadequate in assigning proper credit to the many colleagues who have contributed so much to this fascinating subject. For these omissions, I apologize.

On the other hand, the book is not easy and, except for the opening chapter which tells the story of the soliton's discovery, is not meant for armchair reading. It demands a sharp pencil and even sharper wits. There are five chapters, the purpose of the first of which I have already mentioned. The second chapter introduces the reader to the origin and physics of the Korteweg–deVries (KdV) and nonlinear Schrödinger (NLS) equations and lays great emphasis on their universality and ubiquity, on which matter I will have more to say in the second half of this introduction. This chapter also discusses their derivation as asymptotic solvability conditions. In trying to stress what is of universal importance, the chapter focuses (no pun intended) a lot of attention on the Benjamin–Feir or envelope modulational instability which plays a major role in many physical contexts. It is a manifestation of the fact that monochromatic wavetrains are often unstable and develop local pulselike or soliton behavior. In one space dimension, the pulse develops until the envelope soliton shape is formed. In higher dimensions, the effect is more dramatic and the solution can develop singularities in finite time, the manifestation of which behavior is seen in nonlinear optics (as filamentation) and in plasma physics (collapse of Langmuir waves). The last section is devoted to a comprehensive discussion of the connection between what is generally known as Whitham theory and the nonlinear Schrödinger equation. On the face of it, it would seem that the latter is a simple, small amplitude limit of the former. That is not the case and a much more subtle limit process is involved. It turns out to be directly analogous to the problem of relating the behaviors of a continuous system away from and near to a phase transition. In the former circumstance, the amplitude of the order parameter is slaved to the gradient of the phase (analogous to solutions in Whitham theory) whereas in the latter circumstance the amplitude develops a life of its own.

The third chapter introduces soliton mathematics in the standard way. First, we show how one derives the families of integrable equations associated with a given eigenvalue problem and how to endow the equations with a Hamiltonian structure. Whereas the chapter mainly concentrates on the two simplest families, the KdV and NLS, the exercises at the end of Sections 3b and 3c involve headier stuff. The reader should eventually familiarize himself with these exercises; in particular, Exercise 3b(5) introduces the inverse scattering framework for evolution equations with more than one spatial dimension. Following these sections, I introduce the method of inverse scattering and show how to solve the initial-boundary problem for the Korteweg–deVries equation on the infinite line. I also give a lengthy discussion of how to use the ideas of inverse scattering theory to analyze situations which can be described by a perturbed Korteweg–deVries equation. In particular, the problem of the propagation of a solitary wave in a channel of slowly changing depth is discussed in great detail and a method is developed, which has general application, to compute the distorted flow field, including the wave of reflection. As you will see, this is a nontrivial problem because the perturbation not only changes the solitary wave but it also introduces new flow components. The last section of this chapter deals with ways of constructing special classes of solutions which are often of most interest in applications; here you will meet multisolitons, the rational solutions and finally the multiphase periodic solutions. Certain features of the new perspective on soliton equations, which will be introduced in Chapter 5, begin to make their appearance in the latter part of Chapter 3. The notion that one is finding solutions not just for a single equation, but for a family of equations, is continually stressed.

In Chapter 4, a new hero emerges. It is the τ-function. In the earlier sections of this chapter, I show how it is introduced as a potential function and as a natural consequence of the form of the conservation laws and symmetries. By this stage it should be absolutely clear to the reader that one is dealing with an infinite sequence of commuting flows and that τ is to be considered as a function of all the flow times $\{t_k\}$ as independent variables. The middle sections deal with the Hirota formalism for constructing multisoliton solutions and pay particular attention to the algebraic structure of the bilinear Hirota equations which admit N-soliton solutions for arbitrary N. In particular, we show how the existence of the N-soliton solution for arbitrary N of a particular Hirota equation is equivalent to the existence of an infinite family of Hirota equations of ascending degree to which the particular equation belongs, and which share the same phase-shift function. The role of the phase-shift function in constructing the infinite family is emphasized. Some of these ideas are quite new. Following the discussion on the Hirota formalism, I introduce the reader in the next section to the Painlevé property which all integrable systems appear to enjoy. The connection of this important property, which gives an-easy-to-apply test of exactly solvable systems, with the Hirota condition (the condition which a given Hirota polynomial must satisfy in order that the corresponding bilinear equation admits N-soliton solutions) is stressed. The last section in this chapter introduces Bäcklund transformations, by which means more complicated solutions can be built from simpler ones. It turns out to be particularly helpful to express a Bäcklund transformation in the form of $\tau_{\text{new}} = e^{\beta Y} \cdot \tau_{\text{old}}$. The operator Y is very important. It is called a *vertex operator*.

Throughout the first four chapters, there has been a gradual change in perspective. At first, one tends to think of the soliton equation as a nonlinear evolution equation, a prescription which describes how a given function of a space-like variable x evolves with respect to a time-like variable t. This is certainly the point of view one takes when one applies the inverse scattering transform, in which the evolution equation is clearly considered to be a Cauchy initial-boundary value problem. However, as the various miracles of soliton equations unfold, it becomes more clear that a given equation is best thought of as a *local* relation between a function (or functions) of an infinite number of independent variables and its various derivatives with respect to the independent variables, a relation which is special because of some underlying algebraic structure. Because the equation is local, there is no need to think of any one variable as space-like and therefore particularly distinguished.

Based on these ideas, a new framework for viewing soliton equations is introduced in Chapter 5. This is the longest chapter and deals with material which should be new to all but a few experts in the field. I have avoided expressing the new ideas in an excessively mathematical language or writing style so I expect the reader to be able to hang tough. I start by pointing out the role which the Wahlquist–Estabrook method plays in identifying the underlying algebraic structure of a given equation. In the cases of KdV and NLS, it turns out that the phase space on which the soliton flows live is an infinite-dimensional graded Lie algebra $G = \widetilde{sl}(2, C)$, the loop algebra $\sum_{+\infty}^{-N} X_i \zeta^{-i}$ of $sl(2, C)$. The expression $\sum_{\infty}^{-N} X_i \zeta^{-i}$ is simply a power series in which each of the coefficients X_j belongs to $sl(2, C)$ and can be represented in matrix form by 2×2 matrices of zero trace. The physicist will recognize that the Pauli spin matrices can be used as a basis for this vector space. The algebra G can be decomposed into two subalgebras and on the orthogonal complement of one of them, which can be identified with the dual of the other and which is therefore a Poisson manifold, there are natural Hamiltonian vector fields or flows. When generated by a special sequence of functions, *these vector fields are the soliton equations.* They are an overdetermined, infinite system of ordinary differential equations in infinitely many independent variables $\{t_k\}_0^\infty$. When one wishes to distinguish one of the independent variables, say t_1, which we then call x, the equations can be used to express the infinite number of dependent variables as higher and higher x derivatives of the first members in the sequence. The remaining equations then give the well-known AKNS hierarchy of soliton equations, the first nontrivial member of which is the NLS equation. However, it is equally permissible to choose t_2 as the distinguished variable x, in which case one expresses the infinite number of dependent variables as x derivatives of the first and second members of the sequence. The remaining equations then give a new hierarchy of soliton equations; in this case it is known as the DNLS (derivative nonlinear Schrödinger) hierarchy. In the sections in which these matters are discussed, we also discuss the connections between the new Hamiltonian structure which is naturally introduced with the algebra and the old variational Hamiltonian structure familiar from previous chapters. At the end of the section which outlines these ideas, I invite the reader to try (and help him through) several exercises in which the equations for the harmonic oscillator and finite Toda lattice with free ends are derived from this point of view.

It also turns out that the form of the equations invites the introduction of potentials which replace the infinite number of dependent variables. These are the Hirota τ-functions (there are three of them for $\widetilde{sl}(2, C)$, one "main" one called τ, and two auxiliary ones σ and ρ although we will see that this triplet is best thought of as a sequential threesome ρ, τ, σ in an infinite sequence $\{\tau_n\}$) and in these new potentials, the evolution equations are the Hirota bilinear equations. In this section we also introduce the generalized fluxes $F_{jk} = (\partial^2/\partial t_j\, \partial t_k) \ln \tau$ which play a very important role in the whole theory.

The later sections of this chapter go on to discuss gauge, Bäcklund and Schlesinger transformations, the inverse scattering method and the Riemann–Hilbert problem from a purely algebraic viewpoint and various other topics which are linked under the umbrella of our new approach. I will wait till Chapter 5 to discuss these things in more detail. At this stage, I want to discuss in some more detail with you the impact of the soliton discovery and its relation with other branches of physics. But before I do, there is one message I want to leave indelibly imprinted on your mind. The message is that soliton equations are magic purely for algebraic reasons which have to do with the structure of the equation as a very special relation between a function and its various derivatives. No global properties are required to give it its special significance.

Further discussion. The soliton itself is a dramatic new concept in nonlinear science. Here at last, on the classical level, is the entity that field theorists had been postulating for years, a local travelling wave pulse, a lump-like, coherent structure, the solution of a field equation with remarkable stability and particle-like properties. It is intrinsically nonlinear and owes its existence to a balance of two forces; one is linear and acts to disperse the pulse, the other is nonlinear and acts to focus it. Before the soliton, physicists had often talked about wave packets and photons, which are solutions of the linear time-dependent Schrödinger equation. But such packets would always disperse on a time scale inversely proportional to the square of the spread of the packet in wavenumber space. Nonlinearity is essential for stopping and balancing the dispersion process. In one dimension, the interplay between the dispersion and focusing of wave packets is described by an equation,

$$2iq_t + q_{xx} + 2q^2 q^* = 0,   \tag{1}$$

the nonlinear Schrödinger (NLS) equation, which describes the evolution of the envelope $q(x, t)$ of a wavetrain as seen from a frame of reference moving with the group velocity of the underlying carrier wave. It is a universal equation of nonlinear physics and occurs in a huge variety of situations: in nonlinear optics [19], in deep water wave theory [59], in the description of energy transport along alpha-helix proteins [112]. Not only is it ubiquitous, but one can easily give the recipe for the circumstances under which it will obtain.

Whereas the NLS was the first-born among soliton equations [21], it was the celebrated Korteweg–deVries (KdV) equation

$$q_t + 6qq_x + q_{xxx} = 0,   \tag{2}$$

which fathered the soliton [12]. It, too, is universal. It describes how a Riemann invariant, which, without other effects present, would travel undistorted along the straight parallel characteristics of a linear hyperbolic system (think of the d'Alembert solution $u(x, t) = f(x - t) + g(x + t)$ to the linear wave equation), slowly evolves due to the combined and cumulative influences of nonlinearity and dispersion. In (2), x is measured with respect to a frame of reference moving with the characteristic speed of the linear wave. In KdV, the nonlinearity has the effect of inducing the wave to break, of introducing convergence in the family of characteristics and thereby tending to cause the formation of infinite spatial derivatives in a finite time. Dispersion, on the other hand, mitigates the process by splitting the steepened front into a train of pulses or solitons each of which, if standing alone, has the shape

$$q(x, t) = 2\eta^2 \operatorname{sech}^2 \eta(x - x_0 - 4\eta^2 t).   \tag{3}$$

The KdV equation is also ubiquitous and, just as in the case of the NLS equation, one can give the recipe for the circumstances under which it obtains. It describes the evolution of shallow water waves, ion acoustic waves, long waves in shear flows (see [120] for references) and a host of other situations which the reader can find listed in the various survey papers and proceedings which are referenced.

Both the KdV and NLS equations arise as asymptotic solvability conditions. Briefly stated, the phrase asymptotic solvability condition refers to a condition on the leading order approximation to the solution of a more complicated set of equations which ensures that the later iterates of the approximation remain uniformly bounded. Other universal equations, also derived by this process and which also admit soliton solutions, are the modified Korteweg–deVries (MKdV) equation, the derivative nonlinear Schrödinger (DNLS) equation, the three wave interaction (TWI) equations, the Boussinesq equation, the Kadomtsev–Petviashvili (the two-dimensional KdV or KP) equation, the Benjamin–Ono (BO) equation, the intermediate long wave (ILW) equation, the Benney–Roskes–Davey–Stewartson (the two-dimensional NLS) equation, the sine- and sinh-Gordon equations, the massive Thirring model, the Landau–Lifshitz equation, the Gross–Neveu, the Vaks–Larkin–Nambu–Jona Lasinio chiral field models.

What is remarkable, and still unexplained as far as I am concerned, is that so many of the equations, derived as asymptotic solvability conditions under very general and widely applicable premises, are also soliton equations. In other words, why should an equation which is universal in physics also have such marvellous mathematical properties? I will explain these properties in more detail both in the next paragraphs and throughout these lectures, but one of the key properties of a soliton equation is that it has an infinite number of conservation laws and associated symmetries. It is certainly clear that, in developing mathematical models for physical situations, one naturally includes certain symmetries, like translational invariance, which discard the unnecessary and focus on the essential features of the process under investigation. But why should the process of finding the asymptotic solvability condition introduce so many symmetries, most of which are hidden and not readily accessible to physical interpretation? To stress the point: if one is given a hatful of equations and asked to pick one at random from this hat, it is very unlikely that it would be completely integrable. Yet in the hatful of equations that physics provides as asymptotic solvability conditions, there would appear to be a disproportionate share of ones with soliton properties. Can this be simply coincidence?

What do we mean by a soliton equation? All I have said so far is that a soliton is a solitary, travelling wave pulse of a nonlinear partial differential equation with remarkable stability and particle-like properties. I have hinted that a true soliton, a solution to an equation with very special qualities, is much more than a solitary wave. It is. Many equations admit solitary waves, namely local, travelling wave solutions with nonlinear stability properties. For example, if we change the Kerr or cubic nonlinearity in (1) to the saturable nonlinearity $-iq(1+2qq^*)^{-1}$ or the $6qq_x$ term in (2) to $6q^3q_x$, there are still travelling wave solutions which are neutrally stable to small disturbances. The solitary wave solutions of soliton equations have additional properties, however. One property is that two such solitary waves pass through each other without any loss of identity. For example, observe that the solitary wave (3) has an amplitude-dependent velocity. Now, imagine that at some initial time, we start off two solitary waves very far apart with the one to the left having the larger amplitude and velocity. The larger one will eventually overtake the smaller. The interaction will be very nonlinear and will not at all resemble the interaction of two linear waves in which the composite solution is a linear sum of the two individual waves. Nevertheless, after the nonlinear interaction, two pulses again will emerge, with the larger one in front, and each will regain its former identity precisely. There will be no radiation, no other mode created by the scattering process. The only interaction memory will be a phase shift; each pulse will be centered at a location different from where it would have been had it travelled unimpeded. Whereas this interaction property is remarkable and indeed often used as the test of soliton equations, it is not, by itself, sufficient. There are equations which admit solutions which are a nonlinear superposition of two solitary waves but which do not have all the properties enjoyed by soliton equations. A soliton equation, when it admits solitary wave solutions, must admit a solution which is a nonlinear superposition of N solitary waves for arbitrary N.

It is also exactly integrable in the sense of the infinite-dimensional extension of a completely integrable Hamiltonian system. We say that a finite-dimensional ($2m$ variables) Hamiltonian system is completely integrable if it admits m constants of the motion F_i, $i = 1, \ldots, m$, which are independent and in involution under the Poisson bracket associated with the Hamiltonian structure and the level surface defined by the intersection of the surfaces $F_i = c_i$ is compact and connected. There is a theorem that says that such a system can be canonically transformed (thereby preserving the Hamiltonian structure) to a set of new coordinates, the action-angle variables in which the system is completely separable. The action variables J_i, $1 \leq i \leq m$ (which are functions of the motion constants F_i) are constant in time and the angle variables θ_i change linearly in time; i.e. $\theta_i = \omega_i t + a_i$, $1 \leq i \leq m$, a_i, ω_i constant. As a consequence, the motion must be quasiperiodic and take place on an m-torus, topologically equivalent to the direct product of m circles. To date, all known soliton equations have Hamiltonian structures and an infinite number of independent motion constants in involution. There is also a canonical transformation (the inverse scattering transform or IST, the nonlinear analogue of the Fourier transform) which converts the soliton equation into an infinite sequence of separated equations for the action-angle variables, each member of which can be trivially integrated. In this way, one can, in principle, solve the Cauchy initial value problem. It turns out that some of the action variables are the soliton parameters; this is the reason that a soliton's identity, namely the parameters giving its shape, speed, amplitude, internal frequency etc., is preserved under collision. Other of the action variables are connected with the energy in each of the nonlinear radiation modes, the nonlinear analogue of the continuum of Fourier modes of a linear system.

Contrast this behavior with what one would expect to be the behavior of a mechanical system with strong coupling between many degrees of freedom. In general, one would not expect such a system to be separable. Consequently one would also not expect to see the power spectrum of the time series of any one of the dependent variables to consist of m distinct frequencies, as it would for example if it were a completely integrable Hamiltonian system with a compact Hamiltonian. On the contrary, one would expect there to be at least some spectrum broadening indicating stochastic, although not necessarily ergodic, behavior. Indeed, the second great discovery of the last decade, referred to in the opening paragraph, is the realization that one can have stochastic time dependence in systems with few degrees of freedom. What counts is the qualitative nature of the system of equations and not the dimension of the system. If the equations are such that solutions depend sensitively on initial conditions, then small errors in initial data are exponentially magnified by the flow to the point where it becomes completely impossible to predict the future state of the system. Even dissipative systems, in which a given volume in state space contracts under the flow, are not immune. It turns out that for systems of ordinary differential equations of dimension three or greater (e.g. the Lorenz and Rössler equations), for two-dimensional invertible maps (e.g. the Henon and Ikeda maps) or for one-dimensional noninvertible maps (e.g. the logistic equation $x_{n+1} = \mu x_n (1 - x_n)$ whose fascinating self similar structure in μ, x space was uncovered in the pioneering work of Mitchell Feigenbaum), the motion can take place on a new kind of attractor (as contrasted to the old types which were either a fixed point or a limit cycle) called a strange attractor. The attractor is called strange not simply because of its structure (locally it can be thought of as a direct product between R^n, $n = 1, 2, 3, \ldots$ and a Cantor set) but because motion on it depends sensitively on initial conditions. Indeed, there is reasonable hope that in some cases the apparent stochastic time dependence of systems with many degrees of freedom may be explained by the motion of the state space point on a strange attractor whose dimension is much smaller. Experimental evidence for this hope can be found in reference [123].

This digression on nonintegrable systems was intended to emphasize as strongly as possible the point that in order for an equation to be completely integrable it must have very special properties. Note that solutions of integrable systems do not depend sensitively on initial conditions. Initial errors are magnified in time by at most a linear rate. Before the soliton equations, the number of completely integrable systems could be counted on one hand. The most quoted were the harmonic oscillator, the motion of a body under a central force field and rigid body motion. Indeed the only other infinite dimensional, exactly solvable model in physics was not derived from Newtonian mechanics at all. Rather, it was the two-dimensional, nearest neighbor Ising model of equilibrium statistical mechanics, a model suggested in order to study phase transitions. In a celebrated tour de force, Onsager calculated the partition function by a series of ingenious and apparently miraculous steps. In what surely must be considered as a totally unexpected development, there appears to be a deep connection between soliton equations and the exactly solvable models of equilibrium statistical mechanics and quantum field theory. More about this later.

Unfortunately, the exactly integrable feature is not easy to determine a priori from the equation itself. Therefore, it is useful to search for other properties characteristic of soliton equations, which may be more readily applied to a given equation as a test for integrability. It is also instructive to find out what happens to these properties as the soliton nature of an equation is destroyed either by the addition of terms or by changing some of the crucial coefficients. It is to be expected that such a perturbation will give rise to some stochastic regions in the phase space, particularly in the neighborhood of homoclinic or heteroclinic orbits. If the perturbation is small, one would expect that a Kolmogorov–Arnol'd–Moser (KAM) type result would hold (although, for infinite dimensional systems, this has not yet been proved). Recall that in a completely integrable Hamiltonian system with a bounded Hamiltonian, the motion takes place on an invariant m-dimensional torus parametrized by the m values of the action variables. The KAM theorem says that under small perturbations most of these tori are preserved. There are, however, thin stochastic regions between these tori. One might ask now this feature is manifested in the breakdown of other properties of soliton equations. Furthermore, one might also use these ideas to characterize turbulent or stochastic behavior in the other models in physics, models which when unperturbed are exactly solvable like the Ising model, and which are intimately connected with soliton equations.

While many of the special properties of soliton equations will be discussed in these lectures, there are two I want to mention in the present discussion. The first of these is the Hirota property and is due to Hirota, who discovered a very useful and important method for calculating multisoliton solutions. One requires that the equations be written in bilinear form, a step which is achieved for (2) by writing

$$q(x, t) = 2 \frac{\partial^2}{\partial x^2} \ln \tau \tag{4}$$

but for which (step) there is no general algorithm. The function $\tau(x, t)$ satisfies a quadratic equation

$$\tau\tau_{xt} - \tau_x\tau_t + \tau\tau_{xxxx} - 4\tau_x\tau_{xxx} + 3\tau_{xx}^2 = 0. \tag{5}$$

Hirota developed a new calculus for these equations in which the derivatives $\partial/\partial t$, $\partial/\partial x$ are replaced by derivative-like operators D_t, D_x; in this notation the quadratic equation (5) can be written

$$P(D_t, D_x)\tau \cdot \tau \equiv (D_t D_x + D_x^4)\tau \cdot \tau = 0, \tag{6}$$

with P a polynomial in its arguments. From this equation, it is fairly straightforward to determine the conditions (the Hirota conditions) which the polynomial P must satisfy in order that the equation admit N-soliton solutions for arbitrary N. For $P(D_t, D_x) = D_x D_t + D_x^4$ (the KdV equation), or $P(D_1, D_x) = D_x D_t + D_x^6$ (the Kotera–Sawada equation), the Hirota conditions are satisfied.

For $P(D_t, D_x) = D_t D_x + D_x^8$, they are not and only two-solitary wave solutions can be found. The same conditions on P admit a related class of solutions, the infinite sequence of rational solutions, for each of which the function τ is a polynomial in x and t and for which the corresponding solution q is a rational function. The first three nontrivial rational solutions of (5) are $\tau = x$, $\tau = x^3 + 12t$, $\tau = x^6 + 60x^3 t - 720t^2$. The corresponding solutions $q(x, t)$ have double poles of strength -2 at each of the zeros $x = x(t)$ of $\tau(x, t)$.

The existence of these rational solutions is equivalent to another property enjoyed by soliton equations, the Painlevé property. This property was originally introduced in connection with second order nonlinear ordinary differential equations. The goal was to classify all second order equations whose solutions had the property that the only moveable singular points were poles. This means that the only singularities whose positions depend on initial data are pole singularities. For example, the solution of $dy/dx = -y^2$, $y(0) = 1/c$ is $y(x) = 1/(x + c)$ and has a pole at $x = -c$. On the other hand the points $x = 0$, ∞ are fixed critical points for the equation $2x\, dy/dx = y$. Painlevé found that there were fifty equation types satisfying this requirement, consisting of forty-four which were reducible to known equations and six new equations whose solutions are called Painlevé transcendents. The second equation in the list of six is

$$y_{xx} = xy + 2y^3, \tag{7}$$

about which a lot more will be said in Chapters 4 and 5. For now, the salient point is that (7) admits the solution

$$y = \frac{1}{x - x_0} + \sum_0^\infty a_n (x - x_0)^n, \tag{8}$$

in which x_c and a_3 are arbitrary and all other a_n's are uniquely determined. The equation that determines a_3 reads $0 \cdot a_3 = 0$, the zero on the right-hand side arising as the result of just the right combination of terms in (7). If the xy were replaced by $x^2 y$ or the y^3 term by y^4 (which would require the leading order term to be a second order pole), there would be an incompatibility in the equations for the $\{a_n\}$ which would necessitate the introduction of a term proportional to $(x - x_0)^m \ln (x - x_0)$. The equation would not then have the Painlevé property, for the arbitrary initial point x_0 would no longer be a pole singularity.

What is remarkable is that all integrable equations appear to have the Painlevé property although the idea has to be modified somewhat when applied to partial differential equations. It also appears to be exactly equivalent to the existence of an infinite sequence of rational solutions. Note in our example that if τ has a Taylor expansion about a point on the surface on which it vanishes, then q has a polar expansion. The Painlevé property in the language of the τ-function then seems to demand that the function τ has no moveable critical points.

This observation is significant and has potentially important consequences not only in the context of evolution equations but also for other exactly solvable models. I have mentioned already that there appears to be a connection between the nearest neighbor, two-dimensional Ising model and soliton equations. This connection was first established by the work of Sato, Miwa and Jimbo [103] who showed that, in the scaling limit, the n-point correlation function satisfies a system of very special nonlinear deformation equations which express the fact that the monodromy group of an associated linear system is preserved. Concretely, consider the linear nth order system

$$\zeta \frac{dW}{d\zeta} = (\zeta A + \zeta^{-1} B + C) W$$

in which the n-point correlation function τ appears in the matrix coefficients A, B, C. This system has irregular singular points at $\zeta = 0, \infty$. The condition that the monodromy group of this system is independent of the arguments of the n-point correlation function forces the latter to satisfy a nonlinear deformation equation. The solution to this equation can be constructed, in principle, by using the isomonodromy property. In this way, closed form solutions for the n-point correlation can be found [103]. This remarkable construction closely parallels the construction of solitons and other special types of solutions of soliton equations. In particular, the two-point function (in the scaling limit) satisfies the same equation as the one phase self-similar solution of the sinh-Gordon equation. In addition, there are further suggestive connections between Ising models and integrable systems. For example, McCoy and Wu [109] have shown that at the critical temperature, the two-point correlation function as function of the discrete separations satisfies an exactly integrable, discrete version of the Toda lattice.

If indeed it turns out that one can establish the exact connection between soliton equations and other integrable models, it seems natural to ask about turbulent or stochastic behavior in the latter. One measure of the loss of integrability in the former is the loss of the Painlevé property. Indeed Greene and Percival [110] have shown how in differential equation models exhibiting stochastic behavior the poles pile up in the complex time plane along natural boundaries (which themselves appear to have interesting self similiar behavior). As additional evidence, Segur [111] showed that for the very special choice of parameters in the Lorenz model for which the Painlevé property holds, the model is integrable.

What is likely to happen, then, in a nonintegrable model of statistical physics? In principle, the partition function, the free energy and the correlation functions are all defined. The most natural conjecture to make is that the correlation functions as functions of their arguments (recall they are analogous to the Hirota τ-function) do not have the Painlevé property and instead have algebraic and essential singularities which depend on data. The functions would then behave in a wildly erratic manner near these singularities with the result that a small lack of precision in the knowledge of the input data point would lead to catastrophic error.

In summarizing this discussion, let me again emphasize two points. The first is that the algebraic properties of solvable models are the unifying element and the second is that there is much to be gained by understanding the connections between soliton equations and their solvable cousins of statistical and quantum physics.

Acknowledgments. I want to thank those who have read and made useful suggestions about parts or all of these lectures: Alejandro Aceves, Jerry Bona, John Greene, Erica Jen, Dave McLaughlin, Martin Kruskal, Bob Miura, Tudor Ratiu and Norman Zabusky. Hermann Flaschka's advice and criticism, often terse and merciless but always relevant, were invaluable. He is a great friend. I am also most grateful to Gail Dickerson who managed to type the manuscript and still preserve her humor and sanity. The book is dedicated to my teachers, Victor Graham, David Benney, Willem Malkus and Martin Kruskal, and my children Jamie, Shane, Matt and Pippa, who also have taught me a great deal.

In addition, I am grateful to the Mathematics Sections of the Army Research Office and the Air Force Office of Scientific Research for their generous and encouraging support.

University of Arizona,
December 1983

CHAPTER 1

The History of the Soliton

1a. John Scott Russell's discovery. The discovery of the soliton, its remarkable properties and the incredible richness of structure involved in its mathematical description, occurred in two stages and over a period of almost one hundred and forty years. The story begins with the observation by John Scott Russell of "the great wave of translation". I shall let himself tell of the incident.

> I believe I shall best introduce this phenomenon by describing the circumstances of my own first acquaintance with it. I was observing the motion of a boat which was rapidly drawn along a narrow channel by a pair of horses, when the boat suddenly stopped—not so the mass of water in the channel which it had put in motion; it accumulated round the prow of the vessel in a state of violent agitation, then suddenly leaving it behind, rolled forward with great velocity, assuming the form of a large solitary elevation, a rounded, smooth and well-defined heap of water, which continued its course along the channel apparently without change of form or diminution of speed. I followed it on horseback, and overtook it still rolling on at a rate of some eight or nine miles an hour, preserving its original figure some thirty feet long and a foot to a foot and a half in height. Its height gradually diminished, and after a chase of one or two miles I lost it in the windings of the channel. Such, in the month of August 1834, was my first chance interview with that singular and beautiful phenomenon which I have called the Wave of Translation, a name which it now very generally bears [1].

If the hallmark of a great scientist is his or her ability to recognize what is essentially new, and it surely must be one of the key qualities, then Russell must be accorded this appellation. From the very first sighting, he was convinced that what he had observed was a new scientific phenomenon and thereafter he spent a major portion of his professional life carrying out experiments to determine the properties of the great wave.

> This is a most beautiful and extraordinary phenomenon: the first day I saw it was the happiest day of my life. Nobody had ever had the good fortune to see it before or, at all events, to know what it meant. It is now known as the solitary wave of translation. No one before had fancied a solitary wave as a possible thing. When I described this to Sir John Herschel, he said "It is merely half of a common wave that has been cut off". But it is not so, because the common waves go partly above and partly below the surface level; and not only that but its shape is different. Instead of being half a wave it is clearly a whole wave, with this difference, that the whole wave is not above and below the surface alternately but always above it. So much for what a heap of water does: it will not stay where it is but travels to a distance [2].

He knew that it was a fundamental mode of propagation in the sense that an arbitrary heap of water would disintegrate and resolve itself into a primary and residual wave. He knew that its velocity was proportional to its height and proposed after much experimental work the law $c^2 = g(h + \eta)$ where g, h and η are gravity, the undisturbed depth and the maximum height of the wave, as measured from the undisturbed level, respectively. He knew that waves of

1

depression behaved differently from waves of elevation and produced no permanent travelling forms. He knew about the interaction of solitary waves but does not appear to have noticed their soliton quality, a property I will discuss shortly. Indeed, if he had applied the reversibility property of the Euler equations and the fact that two waves would become infinitely separated as $t \to \pm\infty$, he could have inferred this truly remarkable feature. He also knew about the strange and uniquely nonlinear reflection properties of slightly oblique waves.

He also knew how to create them! I was recently privileged to attend a most enjoyable and well-organized conference at Heriot–Watt University celebrating this great man's work on the hundredth anniversary of his death. It was a grand occasion, filled with lively talks and stimulating discussions and involving an international array of scientists from at least a dozen different disciplines. The highlight of the meeting was to be the recreation of Russell's experience at the same spot on the Union canal where the original incident took place. I am not proud to admit that we failed, that with all our combined knowledge and experience, we were not able to improvise the means to create the wave when the engine of the powerful motorboat, which had worked so well in trial runs, failed the day before the event. The man himself used to do it on a regular basis, with a couple of horses, a couple of ropes, an old barge and a great intuitive understanding of how to transfer the momentum of the boat to the water. Standing on the bank, watching the efforts of swarms of eager young scientists charging along the canal in the role of large horses, one could see many a glass raised in silent respect.

Nevertheless, despite the setback of that day, there is not a soul now living who does not believe in the great wave, for the experiments of Russell have been repeated under carefully controlled circumstances and his predictions verified. It was not always that way, however. At first, Russell's ideas faced great hostility and scepticism from the leading lights in the scientific community of his day. Both Airy and Stokes questioned whether a wave which travelled without change in shape could be totally above the water and cited the diminution of amplitude as an indication that the wave was inherently nonpermanent. Russell had suggested (correctly) that this failure was due to friction. In fact, Stokes "proved" in his 1849 paper through the use of a small amplitude expansion of a sinusoidal wave, that the only permanent wave is basically sinusoidal with the nonlinear terms modifying the shape (the second and higher harmonics) and the speed (it becomes weakly dependent on amplitude). What he had discovered, of course, was the other limit (the modulus of the elliptic function tending to zero rather than one) of the general cnoidal wave solution to the equations of motion. Later on, Stokes was to recognize and admit his errors. It is somewhat ironic that the wave which Stokes discovered (the Stokes wave) is itself unstable when the ratio of depth to wavelength is approximately one (this ratio is small for the solitary wave). In deeper water, an almost monochromatic wave train of the type he described breaks up into a series of wavegroups.

It was not till the 1870's that Russell's work was finally vindicated and its scientific importance can be measured by the eminence of the men who did the job. Independently, Boussinesq [3] (1872) and Rayleigh (1876) found the hyperbolic secant squared solution for the free surface. Boussinesq's 1872 paper in fact did a lot more and introduced many of the ideas used nowadays by modern analysts. In particular, he found the conserved density of the third conservation law, a quantity he called the moment of instability. He derived his solution from the approximation to the water wave equations that now bear his name. In this approximation, the motion can be still bidirectional[1] but the basic idea of the balance between nonlinearity and dispersion is present. It was left to Korteweg and deVries in 1895, who apparently did not know the work of Boussinesq and Rayleigh and who were still trying to answer the objections of Airy and Stokes, to write down the unidirectional equation which now bears their names. (It would appear to have been the thesis project of deVries.)

In this first stage of discovery, the primary thrust was to establish the existence and resilience of the wave. The discovery of its universal nature and its additional properties was to await a new day and an unexpected result from another experiment designed to answer a totally different question.

1b. Fermi–Pasta–Ulam. The scene now changes. It is almost sixty years later and six thousand miles away. The place is Los Alamos and the principals are Enrico Fermi, John Pasta and Stan Ulam. The question of interest was: why do solids have finite heat conductivity? The solid is modelled by a one-dimensional lattice, a set of masses coupled by springs. In 1914, Debye had suggested that the finiteness of the thermal conductivity of a lattice is due to the anharmonicity of the nonlinear forces in the springs. If the force is linear (Hooke's law), energy is carried unhindered by the independent fundamental or normal modes of propagation. The effective thermal conductivity is infinite; no thermal gradient is required to push the heat through the lattice from one end to another and no diffusion equation obtains. Debye thought that if the lattice were weakly nonlinear, the normal modes (calculated from the linearized spring) would interact due to the nonlinearity and thereby hinder the propagation of energy. The net effect of many such nonlinear interactions (phonon collisions) would manifest itself in a diffusion equation with a finite transport coefficient. This suggestion motivated Fermi, Pasta and Ulam (FPU) [5] to undertake a numerical study of the one-dimensional anharmonic lattice on the Maniac I computer at Los Alamos. They argued that a smooth initial state in which all the energy was in the lowest mode or the first few lowest modes would eventually relax to a state of statistical equilibrium due to nonlinear couplings. In that state, the energy would be equidistributed among

---

[1] The equation I refer to is (2.14) with $D = 1$ which is bidirectional. In actual fact, Boussinesq simplified the RHS by replacing F_t by $-F_x$ and F_{xt} by $-F_{xx}$ which imposes the unidirectional approximation. His name, therefore, is associated with (2.26), which is bidirectional but does not describe two directional water waves, rather than (2.14).

all modes on the average. The relaxation time would then provide a measure of diffusion coefficient.

The model used by FPU to describe their one-dimensional lattice of length L consists of a row of $N-1$ identical masses each connected to the next and the end ones to fixed boundaries by N nonlinear springs of length h. Those springs when compressed or extended by an amount Δ exert a force

$$F = k(\Delta + \alpha \, \Delta^2) \tag{1.1}$$

where k is the linear spring constant and α, taken positive, measures the strength of the nonlinearity. The equations governing the dynamics of this lattice are

$$my_{itt} = k(y_{i+1} - 2y_i + y_{i-1})(1 + \alpha(y_{i+1} - y_{i-1})), \qquad i = 1, 2, \ldots, N-1,$$
$$y_0 = y_N = 0, \tag{1.2}$$

y_i being the displacement of the ith mass from its equilibrium position.

FPU usually took the energy to be in the few lowest modes of the corresponding linear problem. In the linear problem, the energy in each mode would persist unchanged forever and no new mode would be excited. In the nonlinear problem, the energy flows from the low modes to higher ones, and FPU expected this to continue until the energy became equidistributed over all modes accommodated in their numerical scheme. With 64 points in x-space, they had 64 different modes over which they hoped to see the energy distributed. The observed evolution could then serve as a model of thermalization for more complicated physical systems.

Now a great surprise was encountered—at least it seemed to surprise everyone who was involved in this problem or heard of it. The energy did not thermalize! In fact, after being initially contained in the lowest mode and then flowing back and forth among several low-order modes, the energy eventually recollected into the lowest mode to within an accuracy of one or two percent and from there on the process approximately repeated itself. FPU knew the phenomenon was not an example of Poincaré recurrence, the time for which, in a system of 63 independently moving masses, would be enormous. Rather, the system seemed to behave like a system of linearly coupled harmonic oscillators whose motion on a torus is quasiperiodic. (If the two fundamental frequencies were ω_1, ω_2 and $\omega_1/\omega_2 \approx m/n$ with m, n integers and relatively prime, then the initial state would recur approximately after a time $2\pi n/\omega_2$.) But how could this be? Why didn't the nonlinearity excite all the Fourier modes? Could the answer be that the system, when viewed in the right coordinates, was equivalent to a separable system of harmonic oscillators?

The FPU experiment had failed to produce the expected result and indeed the results it did produce challenged, as did the Michelson–Morley of the previous century, the basic thinking of physicists of the day. Nevertheless, since it was not connected with what was regarded at that time as frontier physics (little has changed in this regard today), it could easily have been dismissed, as

it was by many, as an anomalous curiosity. Other promising starts on unexpected nonlinear behavior sputtered out. For example, in 1962 Perring and
Skyrme found a two-soliton (two-particle elastic collision) solution for the
sine-Gordon equation which they were using as a model for a nonlinear meson
field theory. These exact solutions, which display a nonlinear superposition
principle, could have made contact with the work of Bäcklund and Bianchi who
invented in the latter part of the nineteenth century a general scheme for
constructing multisoliton solutions for the sine-Gordon equation which
emerges in the theory of surfaces of constant negative curvature. Clearly the
equation had very special properties, but Perring and Skyrme did not follow
the lead.

Fortunately, the curious results of the FPU experiment were not ignored by
all. The moment and opportunity were seized by two applied mathematicians
at Princeton University, Martin Kruskal and Norman Zabusky. They set out to
understand the abnormal, and in doing so discovered the soliton and a
wonderful new world of nonlinear behavior which today has captured the
imagination of scientists from every physical discipline and has given a renewed
life and richness to many previously discovered mathematical structures.

1c. Kruskal, Zabusky and discovery of the soliton. Kruskal and Zabusky
(KZ) [6]–[10] approached the FPU problem from the continuum viewpoint.
They argued that since the energy was contained in the lowest modes of the
system, the displacement of neighboring masses differed by $O(h/L)$ and therefore one could define a continuum displacement $y(x, t)$ where $y(ih, t) \approx y_i$.
Expanding the displacements y_{i+1}, y_{i-1} in a Taylor series, setting $kh^2/m = c^2$,
$2\alpha h = \varepsilon$, $h^2/12\varepsilon = \delta^2$, one finds that (1.2) becomes

$$y_{tt} - c^2 y_{xx} = \varepsilon c^2 y_x y_{xx} + \varepsilon c^2 \delta^2 y_{xxxx}, \tag{1.3}$$

where terms of order ε^2 have been ignored. The parameter ε is small. As we
shall see, it is crucial to retain the second term on the right-hand side, the
fourth-order derivative, in the approximation to the second-order difference.
Since FPU solved the equations by a centered time difference, it is also strictly
necessary to include a y_{tttt} term, but since to a good approximation $y_{tttt} = c^4 y_{xxxx}$
this inclusion will only change the size of δ^2. Because the numerical scheme
had to satisfy the Courant–Friedrichs–Lewy condition, the sign of δ^2 is not
changed.

How does one analyze (1.3)? It is clear that for times and distances of order
one, the solution behaves as if it satisfies the linear wave equation. An initial
profile decomposes into right and left going components, each of which would
travel undisturbed if it were not for the cumulative effects of the nonlinear and
dispersive terms on the RHS of (1.3). How does each of these components
evolve under these new influences? To see this, KZ looked for solutions of
(1.3) in the form

$$y(x, t) = f(\xi, T) + \varepsilon y^{(1)}(x, t) + \cdots, \tag{1.4}$$

where $\xi = x - ct$, $T = \varepsilon t$, and the dependence of f on T describes how the profile $f(\xi, T)$ evolves over long distances and times of order $1/\varepsilon$. The equation for $y^{(1)}$ reads

$$y_{tt}^{(1)} - c^2 y_{xx}^{(1)} = 2cf_{\xi T} + c^2 f_\xi f_{\xi\xi} + c^2 \delta^2 f_{\xi\xi\xi\xi}. \tag{1.5}$$

The solution $y^{(1)}$ will grow linearly with $\xi^- = x + ct$ and the asymptotic series (1.4) will be rendered nonuniform over long times unless the dependence of f on T is chosen to make the right-hand side of (1.5) zero. Setting $6q = f_\xi$, $\tau = cT/2$, one finds

$$q_\tau + 6qq_\xi + \delta^2 q_{\xi\xi\xi} = 0, \tag{1.6}$$

the Korteweg–deVries (KdV) equation. The hyperbolic secant squared solution ($\delta^2 = 1$) with parameter η,

$$q = 2\eta^2 \operatorname{sech}^2 \eta(\xi - v\tau), \tag{1.7}$$

the infinite period limit of the cnoidal wave periodic solutions, corresponds to the solitary wave seen by Russell.

At this point, it should be clear why it was necessary to introduce the second approximation to the finite difference $y_{i+1} - 2y_i + y_{i-1}$. If $\delta^2 = 0$, equation (1.6) has solutions which develop discontinuities in a finite time. For example, take as an initial condition $q(\xi, 0) = \frac{1}{6}\pi a \cos 2\pi\xi$ which corresponds to the initial conditions $y(x, 0) = a \sin 2\pi x$, $y_t(x, 0) = 0$ (remember, since $y_t(x, 0) = 0$, only half the initial profile goes right). The maximum negative slope q_ξ increases monotonically from $-\pi^2 a/3$ at $t = 0$ to $-\infty$ at $t = 1/\pi^2 a\varepsilon c$. Thus the naive continuum approximation to (1.2) breaks down. For δ^2 finite, but small, a different picture emerges. Figure 1, taken from the famous 1965 paper [6] of Zabusky and Kruskal announcing the soliton, shows the results of the KZ numerical experiment in which they use a centered difference, mass and (almost) energy conserving scheme to solve the KdV equation (1.6). They used

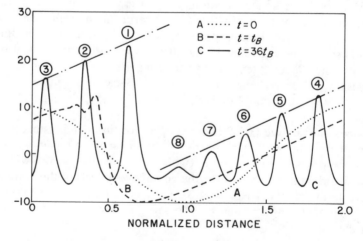

FIG. 1. *The temporal development of the wave form $q(x)$ (from [6]).*

periodic boundary conditions and their starting profile was sinusoidal. Initially the negative slope steepens, then the third derivative term causes fine structure wiggles of wavelength δ to appear near and to the left of the maximum of q (see Fig. 1, profile B). In time the wiggles separate, forming a train of pulses travelling to the right, with the largest on the right, each pulse seeming to take on a life and identity of its own and having a velocity proportional to its amplitude. These pulses each may be approximately described by the solitary wave solution (1.7) although strictly this is a solution valid for an isolated pulse on the infinite line. Because of the periodic boundary conditions, the solitary pulses eventually reappear on the left boundary and due to their higher velocity, the larger pulses overtake the smaller ones. At this point, the investigators noticed a remarkable phenomenon. Whereas during the interaction two pulses behaved in a most nonlinear way, afterwards they reappeared with the larger one in front, each bearing precisely its former identity (height, width and velocity). The only evidence of a collision at all was a phase shift whereby the larger one appeared to be ahead of the position it would have been had it travelled alone and the smaller one behind. If the two pulses were almost equal, the interaction seemed to take place by an exchange of identities in which the forward and smaller soliton became taller and narrower when it felt the leading edge of the larger one which then, in turn, took on the identity of the smaller one. If the two pulses were of greatly different amplitudes, the larger one rode over the smaller one in an adiabatic fashion. For amplitude differences in the in-between range, the interaction was more complicated. In a later analysis of the interaction Lax [14] (1968) verified these observations rigorously.

The pulses were very special indeed. They deserved and got a special name, the soliton, a name intended to connote particle-like qualities. After many passes through the grid, the solitons arrive once again at the same relative spatial positions, approximately produce the fine structure wiggles and then a gradually decreasing negative slope until the initial sinusoidal shape is almost recovered. This process is the mirror image (in both time and space) of the original break up of the initial shape. The time at which the pulses coalesce is called the recurrence time. The reason for the "almost recurrence" in such a short time is that the initial profile decomposes into relatively few soliton shapes. The approximate recurrence time is the minimum time for pulses with different constant velocities (one can improve the calculation by including the phase shifts) to arrive once again at a common point on a circle of length L.

This picture is only an approximation to the exact solution for two reasons. First, as found by later work (ca. 1976), for an initial profile which is analytic in ξ, the solution of (1.6) under periodic boundary conditions may be approximated by the so called finite-gap potential solution which is given by the second logarithmic derivative of the Riemann Θ function of vector arguments $k_j\xi + \omega_j\tau$, $j = 1, \ldots, N$, where N is the number of degrees of freedom. This is equivalent to saying that for a smooth initial profile most of the energy goes into relatively few soliton states. For large N, the gap widths are exponentially small. If $N = 1$, the Θ function becomes a periodic elliptic function

$$q(\xi, \tau) = \beta + (\alpha - \beta)\,\mathrm{cn}^2 \left\{ \sqrt{\frac{\alpha - \gamma}{2}}\,(\xi - 2(\alpha + \beta + \gamma)\tau);\, m^2 \right\}$$

where $m^2 = (\alpha - \beta)/(\alpha - \gamma)$, $\alpha > \beta > \gamma$ and we have taken $\delta^2 = 1$. The infinite period limit, $m^2 \to 1$, is (1.7) with $\alpha = 2\eta^2$ when we impose the condition that $q \to 0$ at $x = \pm\infty$. Second, whereas the solution is periodic in ξ (because the initial profile is), it is only quasiperiodic in the time τ since the ω_j's are not in general commensurate. A measure of the "recurrence" time, therefore, is found by taking rational approximations to the frequencies corresponding to these modes (gaps) containing significant energy. The exactness of the recurrence is then a function of the number of modes included and the chosen accuracy of the rational approximations to the frequencies.

In any event, the strange interaction properties of the solitons together with the almost recurrence property seemed more and more to indicate that, in some sense, the KdV equation was integrable. If this were the case, there should be lots of conserved quantities. However, the connection with Hamiltonian systems was far from uppermost in the minds of the investigators and the motivation for the search for conservation laws came from a different direction. Before I describe that, it is worth noting, as Zabusky has often stressed, the important and seminal role that numerical experimentation played in these discoveries. (I strongly recommend his article [10].) This was the first time in scientific history that investigators had this tremendous new power available to them. Indeed in the last few years we have seen continuing evidence that this mode of research, a combination of analysis and numerical experiment, is becoming increasingly more important in scientific discovery. The strange attractor, a new and fundamental concept in dynamics, and central to our current understanding of certain kinds of turbulence, was also discovered this way.

Now a caveat. The fact that two solitary waves of an equation preserve their form through nonlinear interaction is often taken to be both the acid test for and the definition of the soliton. I want to warn the reader that this condition is only necessary. There are equations (for example, in (1.54) replace D_x^4 by D_x^8) which admit two-phase solitary wave solutions, and therefore the asymptotic form of each individual solitary wave is preserved through collision, which do not possess all the ingredients for admission to the soliton class. The proper definition of a soliton involves its identification with certain of the scattering data of an eigenvalue problem. This we discuss in the section after next. Nevertheless, the computational test of colliding two solitary waves together to check their interaction behavior is still a very useful one. A better test is not only to collide a solitary wave with another one but also to test its resiliency under interaction with other special but local solutions of the equation. In the case of KdV, one might, for example, collide a solitary wave with a wave of depression.

1d. The conservation laws and the Miura transformation. The next step forward in the series of discoveries came about as the result of an attempt to describe the solution to (1.6) when δ^2 is small by averaging over the wiggly fine structure parts of the solution. This procedure would not be valid for all time, of course, and certainly things would be modified when the fine structure separated out into the well-defined soliton train of Fig. 1(C). Nevertheless, it might allow one to investigate the fascinating reversible-shock-like nature of the portion of the solution where q_ξ is large. By analogy with gas dynamics, therefore, it was important to find conservation laws in order that the jump conditions across the wiggly regions could be calculated. One would need four such pieces of information (the number of characteristic velocities entering or leaving the shock plus the position of the shock itself) as opposed to the usual three of gas dynamics. Two conservation laws

$$q_t + (3q^2 + \delta^2 q_{xx})_x = 0, \tag{1.8}$$

$$(\tfrac{1}{2}q^2)_t + (2q^3 + \delta^2(qq_{xx} - \tfrac{1}{2}q_x^2))_x = 0, \tag{1.9}$$

corresponding to conservation of mass and momentum (1.8) and energy (1.9) for water waves and to conservation of momentum and energy for the nonlinear spring, were already known; Whitham, who had developed a powerful theory for investigating modulated periodic waves about this time, had found a third, the conserved density, which corresponded to Boussinesq's famous moment of instability. Zabusky and Kruskal searched for and found a fourth, and their method of search (finding equations for the coefficients of all terms of degree 4, 5, 6, etc., degree $q = 1$, degree $\partial/\partial x = \tfrac{1}{2}$) indicated that the equations for the coefficients at stage 6 were overdetermined (as they are thereafter) and, therefore, they were not particularly surprised when they did not find a conservation law at this level. However, they had made an algebraic mistake, and more than a year was to go by before they continued again on this track.

The next surge of momentum came with the arrival of Robert Miura who was asked by Kruskal to get his feet wet by searching for a conservation law at level seven. He found one and then quickly filled in the missing sixth. Eight and nine fell quickly and Kruskal and Miura were fairly certain that there was an infinite number. However, rumors originated from the Courant Institute that nine was the limit. (In fact, what investigators there had discovered was what they perceived to be a change in algebraic structure.) Miura was therefore challenged to find the tenth. He did it during a two week vacation in Canada in the summer of 1966. (There is also a rumor that he was seen about this time in Mt. Sinai, carrying all ten.) It was now clear that there was a conservation law at every level. (Each conservation law has the form $(\partial U/\partial t) + (\partial F/\partial x) = 0$; U, called the conserved density and F, the corresponding flux, can be assigned a weight by adding the power of q to half the number of $\delta(\partial/\partial x)$ operations in each term in these quantities. For example, q has weight one, $3q^2$ and $\delta^2 q_{xx}$ both have weight two, $2q^3$, $\delta^2 qq_{xx}$ and $\delta^2 q_x^2$ all have weight three and so on.

Level then refers to the weight of the conserved density.) How to find them and what kind of constraints they would impose on the solution of the KdV equation were thoughts that ran through the minds of the investigators. The original motivation for looking for conservation laws was temporarily put aside. There were suddenly too many independent laws and the jump conditions (Rankine–Hugoniot relations) derived from these laws would have to be consistent with each other. Somehow, with all the new discoveries, the question did not seem to be quite as important as before.

For that matter, it is still not understood why a solid has finite heat conductivity!

Miura provided the next key [11]. He had found that the modified Korteweg–deVries equation (MKdV)

$$v_t + 6v^2 v_x + v_{xxx} = 0 \qquad (1.10)$$

also had an apparently infinite string of conservation laws. He observed that they could be matched with their corresponding counterparts in the KdV series by the transformation (which now bears his name)

$$q = v^2 - iv_x. \qquad (1.11)$$

From this point on, we will write x, t for ξ, τ and take $\delta^2 = 1$. In fact, he showed that

$$q_t + 6qq_x + q_{xxx} = \left(2v - i\frac{\partial}{\partial x}\right)(v_t + 6v^2 v_x + v_{xxx}), \qquad (1.12)$$

and therefore if $v(x, t)$ is a solution of (1.10), $q(x, t)$ is a solution of (1.6). Next, since (1.11) is a Riccati equation, the Miura transformation can be linearized by

$$v(x, t) = -i\frac{\phi_x}{\phi} \qquad (1.13)$$

to the Schrödinger equation with zero energy

$$q = -\frac{\phi_{xx}}{\phi}. \qquad (1.14)$$

The Galilean invariance of (1.6) suggests that nothing is changed by adding a constant velocity λ to q, whence (1.14) becomes

$$\phi_{xx} + (\lambda + q(x, t))\phi = 0, \qquad (1.15)$$

the stationary Schrödinger equation with potential $V(x) = -q(x; t)$ and energy $E = \lambda$. (Remark: The time variable in the time-dependent Schrödinger equation has nothing, repeat nothing, to do with the time t in the KdV equation.) The ingredients for what was to become the inverse scattering transform (IST) were now all there. By this stage, both Gardner and Greene had joined in the effort.

Before I relate this part of the story, it is worth describing Gardner's modification of the Miura transformation as it automatically includes λ and gives rise to the important idea that an infinite number of conserved quantities of one equation can be inferred from one conserved quantity of another if their solutions are related through a special kind of transformation (an example of a Bäcklund transformation). Gardner took

$$q = w + i\varepsilon w_x + \varepsilon^2 w^2 \qquad (1.16)$$

and found the equivalent relation to (1.12) to be

$$q_t + 6qq_x + q_{xxx} = \left(1 + 2\varepsilon^2 w + i\varepsilon\frac{\partial}{\partial x}\right)(w_t + 6(w + \varepsilon^2 w^2)w_x + w_{xxx}). \quad (1.17)$$

Now, linearize (1.16) with

$$w + \frac{1}{2\varepsilon^2} = \frac{i}{\varepsilon}\frac{\phi_x}{\phi} \qquad (1.18)$$

whence (1.16) becomes

$$\phi_{xx} + \left(q + \frac{1}{4\varepsilon^2}\right)\phi = 0. \qquad (1.19)$$

Also solve w as function of q and its derivatives as an asymptotic series in ε for small ε (we shall see later that the asymptotic expansion of ϕ for large λ is very important in the theory) and find

$$w = q - i\varepsilon q_x - \varepsilon^2(q_{xx} + q^2) + \cdots. \qquad (1.20)$$

Since $\int w\,dx$ is a constant in time when the integral is taken over the infinite line (assuming q and its derivatives approach zero as $x \to \pm\infty$) or the periodic interval, so are $\int q\,dx$, $\int q^2\,dx$ $\int(q^3 - \frac{1}{2}q_x^2)\,dx$, etc. One can also capture the infinite set of conservation laws for KdV from the first one for the equation $w_t + 6(w + \varepsilon^2 w^2)w_x + w_{xxx} = 0$.

1e. The inverse scattering transform [12]. Given (1.15), it was now natural to ask: if the potential $-q(x, t)$ evolves according to the KdV equation

$$q_t + 6qq_x + q_{xxx} = 0, \qquad (1.21)$$

how do $\lambda(t)$ and $\phi(x, t)$ evolve? We consider the whole real line $-\infty < x < \infty$, with q and its derivatives vanishing at $\pm\infty$ This can be done by substituting q from (1.15) into (1.21). One finds by direct calculation that

$$\lambda_t\phi^2 + (\phi Q_x - \phi_x Q)_x = 0 \qquad (1.22)$$

where $Q = \phi_t + \phi_{xxx} - 3(\lambda - q)\phi_x$. If ϕ vanishes as $x \to \pm\infty$ and is square integrable, then $\lambda_t = 0$. Thus the discrete eigenvalues $\lambda_n < 0$, $n = 1, 2, \ldots, N$ of (1.15) are constants of the motion. From the remaining part of (1.22) we find

$$\phi_t + \phi_{xxx} - 3(\lambda - q)\phi_x = C\phi + D\phi\int\frac{dx}{\phi^2} \qquad (1.23)$$

and since ϕ vanishes at $\pm\infty$, $D = 0$. We choose to normalize the bound state eigenfunction ϕ_n so that $\phi_n \sim \exp(\sqrt{-\lambda_n} x)$ as $x \to -\infty$. Thus $C_n = 4(-\lambda_n)^{3/2}$, and if $\phi_n \sim b_n(t) \exp(-\sqrt{-\lambda_n}\, x)$ as $x \to +\infty$,

$$b_{nt} = 8(-\lambda_n)^{3/2} b_n \tag{1.24}$$

and $b_n(t) = b_n(0) \exp(8(-\lambda_n)^{3/2} t)$. For $\lambda = \zeta^2 > 0$, a solution of (1.15) for large $|x|$ is a linear combination of $e^{\pm i\zeta x}$. We impose on ϕ the boundary conditions

$$\phi \sim e^{-i\zeta x} + R(\zeta, t) e^{i\zeta x} \quad \text{as } x \to \infty, \tag{1.25a}$$

$$\sim T(\zeta, t) e^{-i\zeta x}, \qquad x \to -\infty. \tag{1.25b}$$

In the usual quantum mechanical interpretation of (1.15), the coefficients of unity in (1.25a) and (implied) zero in (1.25b) indicate prescribed steady radiation coming from $x = \pm\infty$ only. The coefficients of transmission $T(\zeta, t)$ and reflection $R(\zeta, t)$ will be shown in Chapter 3 to satisfy $|T|^2 + |R|^2 = 1$. The spectrum for $\lambda > 0$ is continuous and we may choose λ constant, so that again (1.23) is valid provided $D = 0$. Because of the way ϕ is normalized at $+\infty$, $C = 4i\zeta^3$. Inserting (1.25a,b) we find in addition

$$T_t(\zeta, t) = 0, \qquad R_t(\zeta, t) = 8i\zeta^3 R(\zeta, t), \tag{1.26}$$

which means that the transmission coefficient as function of ζ is a constant of the motion and the reflection coefficient $R(\zeta, t)$ evolves by simply changing its phase linearly with time.

It was known from the early 1950's that the potential $-q(x)$ of Schrödinger's equation can be completely recovered from a knowledge of what is called the scattering data,

$$S = \{(\lambda_n, b_n)_1^N; R(\zeta), \zeta \text{ real}\}. \tag{1.27}$$

From S, the transmission coefficient $T(\zeta)$ can also be found. But if S is known from $q(x, 0)$ at $t = 0$, then (1.24) and (1.26) allow us to calculate $S(t)$ in a very simple manner. Hence $q(x, t)$ can be constructed at any arbitrary time. The prescription for the reconstruction involves a linear integral equation, the Gel'fand–Levitan–Marchenko equation. This and many other details will be derived in Chapter 3. For now, let us just remark that the general solution of (1.21) involves several components. The solitons, which propagate with positive velocity, are the physical manifestation of the discrete spectrum, one soliton for each eigenvalue. In isolation (at $t = \pm\infty$), each soliton has a height, width and speed proportional to $-\lambda_n$, $\sqrt{-\lambda_n}$ and $-\lambda_n$ respectively. Its position at any time can be calculated from b_n. The continuous spectrum gives rise to a component of the solution which, although nonlinear, bears a close resemblance to the solution of the linearized equation (1.32). The amplitude of the wave group associated with wavenumber ζ is measured by $|R(\zeta)|$, its position by Arg $R(\zeta)$. The region around $x = 0$ joining these two solution components involves among other things the self similar solution of (1.21) and is complicated, but it basically plays the role of a nonlinear Airy function.

(John Greene liked to attach a moral–religious categorization to the two solution components; the solitons were the soul of the solution while the continuous spectrum gave rise to the mortal flesh. I suppose it depends on your point of view which component deserves to be called good.) If one thinks of the various solution components as normal modes of the nonlinear system and this thinking is useful, then one can single out the soliton mode as special because it is totally new and has no linear analogue.

The essence, then, of the application of IST is as follows. The equation of interest

$$q_t + 6qq_x + q_{xxx} = 0 \qquad (1.28)$$

is written as the integrability condition of two linear equations

$$(-L + \lambda)\phi \equiv \phi_{xx} + (\lambda + q(x, t))\phi = 0 \qquad (1.29)$$

and

$$\phi_t = B\phi = -4\phi_{xxx} - 3q_x\phi - 6q\phi_x + C\phi \qquad (1.30a)$$
$$= (q_x + C)\phi + 4(\lambda - q/2)\phi_x \qquad (1.30b)$$

(where C is determined once a normalization is chosen for $\phi(x, t; \zeta)$). Then $q(x, 0)$ is mapped into the scattering data $S(0)$ of (1.29). The evolution of $S(t)$ is simple and linear. From a knowledge of $S(t)$, we reconstruct $q(x, t)$. Schematically,

$$q(x, 0) \xrightarrow{\text{direct transform}} S(0)$$
$$\qquad\qquad\qquad\qquad \Bigg\downarrow \begin{array}{l} \text{time evolution} \\ \text{of sc. data} \end{array} \qquad (1.31)$$
$$q(x, t) \xleftarrow{\text{inverse transform}} S(t).$$

The procedure is completely analogous to the way in which one would solve the linearized version of (1.28),

$$q_t + q_{xxx} = 0, \qquad (1.32)$$

by the Fourier transform. Here the direct transform is

$$b(k, t) = \frac{1}{2\pi} \int_{-\infty}^{\infty} q(x, t)e^{-ikx} \, dx, \qquad (1.33)$$

and $b(k, 0)$ is known once $q(x, 0)$ is given; the time evolution of $b(k, t)$ is

$$b_t(k, t) = ik^3 b(k, t); \qquad (1.34)$$

the inverse transform is

$$q(x, t) = \int_{-\infty}^{\infty} b(k, t)e^{ikx} \, dk. \qquad (1.35)$$

Indeed, we will show that IST, in the linear limit, reduces to the Fourier transform.

We also know that we can interpret both equations (1.28) and (1.32) as infinite dimensional Hamiltonian systems; we may write each formally as

$$q_t = \frac{\partial}{\partial x} \frac{\delta H}{\delta q} \tag{1.36}$$

where $\delta/\delta q$ is the variational derivative of the Hamiltonian functional $H[q]$; i.e.,

$$\lim_{\varepsilon \to 0} \frac{1}{\varepsilon} (H[q + \varepsilon \, \delta q] - H[q]) = \int_{-\infty}^{\infty} \frac{\delta H}{\delta q} \, \delta q \, dx.$$

Equation (1.36) is analogous to the expession

$$z^{\cdot} = J \, \nabla H(z) \tag{1.37}$$

valid for finite dimensional systems. Here z is a $2N$-vector (e.g., $q_1, \ldots, q_N$, $p_1, \ldots, p_N$), J a skew-symmetric matrix (e.g., $\begin{pmatrix} 0 & I_N \\ -I_N & 0 \end{pmatrix}$) and ∇ the gradient $(\partial/\partial q_1, \ldots, \partial/\partial p_N)$. In (1.36), $q(x)$ should be considered as an infinite dimensional vector, $\partial/\partial x$ a skew-symmetric operator instead of a matrix and $\delta/\delta q$ a variational derivative replacing the gradient. The corresponding two-form $\sum \delta q_i \wedge \delta p_i$ which is preserved under the flow is

$$\frac{1}{2} \int_{-\infty}^{\infty} \delta q(x) \wedge \left(\int_{-\infty}^{x} \delta q(y) \, dy \right) dx. \tag{1.38a}$$

The integral $\int^x$ is the inverse operation to J, where J is $\partial/\partial x$. The Poisson bracket of two functions F and G is

$$\{F, G\} = \int_{-\infty}^{\infty} \frac{\delta F}{\delta q} \frac{\partial}{\partial x} \frac{\delta G}{\delta q} \, dx. \tag{1.38b}$$

For (1.21), the Hamiltonian $H = \int_{-\infty}^{\infty} (\frac{1}{2}q_x^2 - q^3) \, dx$; for (1.32), $H = \int_{-\infty}^{\infty} \frac{1}{2}q_x^2 \, dx$. The Fourier transform (1.33) is a canonical transformation which carries the old coordinates $(q(x), -\infty < x < \infty)$ to new ones $A = 2\pi |b|^2/k$, $\theta = \text{Arg } b(k, t)$ in which the two-form (1.38)

$$\frac{1}{2} \int_{-\infty}^{\infty} \left(\delta q \wedge \int_{-\infty}^{x} dy \, \delta q \right) dx = \int_{0}^{\infty} \delta A(k) \wedge \delta \theta(k) \, dk \tag{1.39}$$

is preserved. (The reader should prove (1.39) for himself; for those unfamiliar with the wedge notation, $\delta q \wedge \delta w$ means $\delta_1 q \, \delta_2 w - \delta_2 q \, \delta_1 w$, where δ_1 and δ_2 are independent variations.) In (1.39), since $w_x = q$, one should not think of $q(x)$ and $w(x) = \int_{-\infty}^{x} q(y) \, dy$ as conjugate variables; rather one should read (1.39) as the continuous limit of $\frac{1}{2} \sum_i \delta q_i \wedge \sum_{j<i} \delta q_j$. On the other hand, the new coordinates A and θ are conjugate variables. From (1.34),

$$A_t = 0, \qquad \theta_t = k^3 \tag{1.40}$$

which are Hamilton's equations

$$A_t = -\frac{\delta H}{\delta \theta}, \qquad \theta_t = \frac{\delta H}{\delta A},$$

where

$$H = \frac{1}{2} \int_{-\infty}^{\infty} q_x^2 \, dx = \frac{1}{2} \int_0^{\infty} k^3 A^2(k) \, dk.$$

In an exactly analogous way, IST is a canonical transformation which carries the old coordinates $(q(x), -\infty < x < \infty)$ to the new ones which are the scattering data S given by (1.27).

Gardner [13] was the first to note that the KdV equation could be written in a Hamiltonian framework. Later, Zakharov and Faddeev [13] showed how it could be interpreted as a completely integrable Hamiltonian system. For a finite dimensional system of $2N$ dimensions, the term *completely integrable* means that the system possesses N independent constants of the motion $F_j(p, q)$ $(p = (p_1, \ldots, p_N), q = (q_1, \ldots, q_N)), j = 1, \ldots, N$ which are in involution with respect to the Poisson bracket. From this beginning, one can define N action variables (as functions of the F_j's) and N corresponding angle variables. For infinite dimensional systems, things are more formal. In this context, we will use the term completely integrable to mean that an infinite number of new coordinates can be found analogous to action-angle variables such that the former are constants of the motion and the latter vary linearly with time. As the reader may have already guessed, the action coordinates are functions of the infinite set of conserved densities.

Let me make a further remark on the time dependence of the transformed variables and point out the sense in which the infinite line problem for (1.28) $(q(x, t) \to 0, x \to \pm\infty,$ all $t)$ is simpler than the periodic problem $(q(x, t) = q(x + P, t)$ all x and $t)$. In the former, we know q at two points of the interval, namely $x = \pm\infty$, for all time. Indeed, the scattering data are a measure of how the asymptotic solutions of (1.29) $(\exp(\pm i\sqrt{\lambda} x))$ change as x traverses the potential between $-\infty$ and $+\infty$. From (1.30), one can see that at $\pm\infty$, the time dependence of $\phi(\infty, t; \zeta)$ is independent of q as q and its derivatives are zero there. In the latter problem, the periodic problem, q is not known for all time at any point on the interval $[0, P]$. Consequently, the time dependence of the scattering data is much more complicated.

Both these cases contrast with the way in which Burgers' equation $u_t = u_{xx} + 2uu_x$ is linearized. There the equation can be written as the integrability condition of $\phi_x = u\phi$ and $\phi_t = (u^2 + u_x)\phi$. Whereas it seems that the evolution of $\phi(x, t)$ depends on a knowledge of $u(x, t)$, this is not really the case as $(u^2 + u_x)\phi = \phi_{xx}$. Therefore, under the transformation $\phi_x = u\phi$, ϕ satisfies the linear heat equation. In the case of KdV on the infinite line, the equation for $\phi(x, t; \lambda)$ does not linearize; instead there is a free parameter λ and (1.28) is the integrability condition for (1.29) and (1.30) for all λ. So instead of knowing $\phi(x, t; \lambda)$ for all x and t, we know it for all t and λ at $x = \pm\infty$.

1f. The Lax equation [14]. Despite the fact that the time dependence of the angle variable is more complicated, both the infinite line and periodic problems show the key property that as $q(x, t)$ evolves according to (1.28), the spectrum of L of (1.29) considered as an operator on $L^2(R)$ ($R = (-\infty, \infty)$ or $(0, P)$), does not change. This property was expressed in an elegant manner by Lax (1968) in the same paper [14] in which he examined the two-soliton interaction. He noted that if $L(t)$ (here self-adjoint) and $L(0)$ share the same spectrum, they are unitarily equivalent; that is, there exists a unitary operator U ($UU^* = U^*U = I$, the identity) such that

$$L(t)U(t) = U(t)L(0). \tag{1.41}$$

Thus if $\phi(x, 0; \lambda)$ is an eigenfunction of $L = -(d^2/dx^2) - q(x)$ at $t = 0$ with eigenvalue λ, then $\phi(x, t; \lambda) = U(t)\phi(x, 0; \lambda)$ is the eigenfunction of $L(t)$ with the same eigenvalue. We see this directly from (1.41) since $L(t)U(t)\phi(x, 0; \lambda) = \lambda U(t)\phi(x, 0; \lambda)$. Differentiating (1.41) with respect to time gives us that

$$L_t = BL - LB = [B, L] \tag{1.42}$$

where $B = U_t U^*$ is skew adjoint. Equation (1.42) is called Lax's equation and L and B are called a Lax pair Note that B can be obtained by recognizing that $\phi_t(x, t; \lambda) = U_t\phi(x, 0; \lambda) = U_t U^*\phi(x, t; \lambda)$ so in the case of KdV, from (1.30) we read off B as (recall $\lambda\phi_x = -\phi_{xxx} - (q\phi)_x$)

$$B = -4\frac{\partial^3}{\partial x^3} - 3\left(q\frac{\partial}{\partial x} + \frac{\partial}{\partial x}q\right) + C. \tag{1.43}$$

It turns out that all the solvable equations like KdV can be expressed in Lax form. Indeed Lax showed that there was an infinite sequence of B's, one connected with each odd order of $\partial/\partial x$, and therefore an infinite family of flows q_t under which the spectrum of L is preserved. We will find formulae for these in Chapter 3. The reader might verify directly that (1.42) is indeed (1.21).

1g. Simultaneous developments in nonlinear optics and Bäcklund transformations. At approximately the same time as these advances were being made, the soliton made its appearance in a totally new context, the propagation of ultra-short (10^{-12} sec) optical pulses in resonant media. In 1967, McCall and Hahn [15] discovered the phenomenon of self-induced transparency, an effect whereby the leading edge of a pulse is used to invert an atomic population while the trailing edge returns the population to its initial state by stimulated emission. This process is realizable if it takes place in a time short compared to the phase memory time of the medium and also if the pulse has sufficient intensity to cause the population inversion. If we assume a medium of nondegenerate two-level atoms and neglect the effects of inhomogeneous broadening (the mismatch, due to Doppler shifting, between the carrier frequency of the incoming pulse and the energy difference of the two levels), then the process can be described in terms of a single field equation, the sine-Gordon equation

$$u_{xt} = \sin u, \tag{1.44}$$

where x measures distance from the beginning of the medium, t is the (retarded) time and the electric field envelope $E(x, t)$ is proportional to $\partial u/\partial x$.

This equation had been known for a long time. It had been studied long ago in connection with the theory of surfaces of constant negative curvature. In particular, A. V. Bäcklund (see [16]) had discovered that a new solution (or surface) $u_1(x, t)$ could be found of an old one $u_0(x, t)$ by the transformation

$$u_{1x} - u_{0x} = 4i\zeta \sin \frac{u_1 + u_0}{2}, \qquad (1.45a)$$

$$u_{1t} + u_{0t} = \frac{1}{i\zeta} \sin \frac{u_1 - u_0}{2}. \qquad (1.45b)$$

Such transformations are known as Bäcklund transformations (we will give a definition later in Chapter 4) and allow multisoliton solutions to be constructed in a fairly simple manner. For example, take $u_0 = 0$; then integrating (1.45), we find

$$u_1 = \pm 4 \tan^{-1} \exp \left(-2\eta x - \frac{t}{2\eta}\right). \qquad (1.46)$$

which describes a pulse u for which the corresponding electric field envelope E is equal to $4\eta \operatorname{sech} (2\eta x + t/2\eta)$ and the area $\int_{-\infty}^{\infty} E \, dx = u(\infty) - u(-\infty)$ is equal to 2π. These pulses are known as 2π pulses, and are called kinks (antikinks) if u increases (decreases) by 2π as x goes from $-\infty$ to $+\infty$. More complicated solutions can also be found by a ladder process, a theorem of permutability attributed to Bianchi, which may be written

$$\tan \frac{u_3 - u_0}{4} = \frac{\zeta_1 + \zeta_2}{\zeta_1 - \zeta_2} \tan \frac{u_1 - u_2}{4}. \qquad (1.47)$$

(I will leave the derivations of this as an exercise to the reader.) The more complicated solutions are known as 0π (u goes from 0 at $-\infty$, to 2π and back to 0 at $+\infty$; this pulse is effectively a superposition of a kink and antikink) and 4π (superposition of two kinks) pulses.

Just as in the case of the Korteweg–deVries equation, the sine-Gordon equation arises in a variety of contexts:

(1) as a model for dislocation in crystals (in which context Seeger, Donth and Kochendorfer [17] used the Bäcklund transformation to obtain the 2π pulse as early as 1953);
(2) as a model field theory (we have already referred to the work of Perring and Skyrme [18]);
(3) and in superconductivity where u describes the difference in the phases of the wave functions across a Josephson junction (see [19]).

It also describes a readily visualized mechanical model proposed by Scott which consists of a string of pendula suspended from and free to turn about a horizontal torsion wire. In this context $u(X, T)$ is the angle of twist measured from the vertical, and with $x = (X + T)/2$, $t = (X - T)/2$, (1.44) takes the form

$$u_{TT} - u_{XX} + \sin u = 0. \tag{1.48}$$

A kink is a counterclockwise twist of angle 2π; an antikink has the opposite orientation.

I do not want to leave the topic of the sine-Gordon equation without noting the important contributions that George Lamb has made to the subject. In a sequence of several papers, culminating with his 1971 article [20] in *Reviews of Modern Physics*, he elaborated on and discussed the physical relevance of the multisoliton and self-similar solutions of the sine-Gordon equation. He foresaw that the sine-Gordon equation is a sister under the skin to the Korteweg–deVries equation. Indeed, he independently discovered the inverse method for its solution. It is fitting, I think, to pay tribute to a modest man, a rare breed, a man who even resisted the urge to recount his accomplishments in his own book. He only lists one of his own references.

1h. Soliton factories and later developments of the 1970's. With all the ingredients out on the table as early as 1967, it is somewhat surprising that it took five years to take the KdV equation out of its integrable isolation. Some felt that it was a fluke, a clever transformation somewhat akin to the Hopf–Cole transformation (see also Forsythe, Vol. 6, p. 100 where the linearization of Burgers' equation is given as an exercise). Then in 1972 (published in 1971 in the Soviet Union), Zakharov and Shabat [21] found the Lax pair for the nonlinear Schrödinger equation, another universal equation about which a lot more will be said in Chapter 2. The Zakharov–Shabat result, the Potsdam conference of 1972, and the lectures of Kruskal on the sine-Gordon equation unleashed a great tide of energy on these problems. Wadati [22] discovered the setting in which to solve the modified Korteweg–de Vries (MKdV) equation. Ablowitz, Kaup, Newell and Segur (AKNS) [23], motivated by several key observations by Kruskal, solved the sine-Gordon equation (independently solved by Lamb and, a little later, by Faddeev and Takhtadzhyan) and, soon thereafter, showed how to write down the full set of equations (the AKNS hierarchy) solvable through the use of the Zakharov–Shabat eigenvalue problem

$$v_{1x} + i\zeta v_1 = q(x, t)v_2,$$
$$v_{2x} - i\zeta v_2 = r(x, t)v_1. \tag{1.49}$$

By this time, it had become clear how to begin with any eigenvalue problem and write down the evolution equations which kept its spectrum invariant. Soliton factories sprang up all over the world.

Tougher problems were handled. The Toda lattice (see [24]),

$$u_{n_{tt}} = e^{u_{n+1} - u_n} - e^{u_n - u_{n-1}}, \tag{1.50}$$

was shown to be an integrable model (Flaschka, Henon). It played the same seminal role for other classes of nonlinear partial difference equations that KdV did for partial differential equations.

The periodic problem for KdV was solved in several stages by several authors about and after 1976. The first discovery was the so-called finite gap solution in which the periodic and antiperiodic spectrum of (1.29) with periodic potential $q(x)$ (Hill's equation) consists of $(2n+1)$ simple eigenvalues λ_0, $\lambda_1, \ldots, \lambda_{2n}$ with remaining ones all double (McKean and van Moerbeke [25], Novikov [26], Its and Matveev [27], Krichever [28]). The infinite gap limit was treated by McKean and Trubowitz [29]. Many of the discoveries turned out to be rediscoveries of earlier work of Baker, Drach, Burchnall and Chaundy [30].

Lax pairs were also found for problems with spatial dimension more than one. In particular we shall discuss some results concerning the Kadomtsev–Petviashvili (KP) equation

$$\pm q_{yy} + (q_t + 6qq_x + q_{xxx})_x = 0 \tag{1.51}$$

(a weakly two-dimensional Korteweg–deVries equation). The initial value problem for this equation is very complicated and has only recently been solved (Manakov [31], Ablowitz, Fokas and Segur [32]). The instanton, the soliton of the self-dual Yang–Mills equations, has also been found, and a construction of a k-parameter instanton solution has been given by Atiyah, Hitchin, Drinfeld and Manin [33]. Several other field equations, important in nonlinear physics have been found to be integrable.

Before I leave this section, I want to tell you a bit about a giant in the field, V. E. Zakharov. He has contributed in so many areas: the Zakharov equations of plasma physics, his papers with Shabat outlining for the first time a general prescription for handling Lax pairs for equations with more than one spatial dimension, his work on the self-focusing singularity, and his paper in the Bullough–Caudrey volume [113] (see references on the method of "dressing" (building hierarchies of solutions)). He seems to have the knack of getting to all of the good problems first. He is a genius, brilliant and intuitive. A wild bull of a man of great good humor and appetites, he has a deep and abiding love and need for poetry and literature as well as physics. On one of our few meetings, he recited and acted out with great relish the opening scene from Macbeth. You will often come across his name.

1i. Soliton miracles and the need for a unifying point of view. A whole new world of integrable systems had been discovered together with constructive methods (IST, Bäcklund transformations, Hirota's method) for obtaining solutions. Hirota's method [34], which I have mentioned in the introduction, is quite ingenious. For the KdV equation, one sets

$$q(x, t) = 2 \frac{\partial^2}{\partial x^2} \ln \tau(x, t) \tag{1.52}$$

and obtains a quadratic equation for $\tau(x, t)$

$$\tau\tau_{xt} - \tau_x\tau_t + \tau_{xxxx} - 4\tau_x\tau_{xxx} + 3\tau_{xx}^2 = 0 \tag{1.53}$$

which Hirota rewrote in terms of a new differential operator which he intro-duced as

$$(D_tD_x + D_x^4)\tau \cdot \tau = 0. \tag{1.54}$$

I will explain the notation in Chapter 4. From (1.53), (1.54) N-soliton (and rational) solutions could be found simply by letting $\tau(x, t)$ be the sum of exponentials with phases linear in x and t and depending on arbitrary constants (which turn out to be the phase-shifts mentioned previously). For $N > 2$, the equations for the constants are overdetermined but consistent. What is it that brings about this consistency? The equation $(D_xD_t + D_x^6)\tau \cdot \tau = 0$ has similar properties. But $(D_xD_t + D_x^8)\tau \cdot \tau = 0$ has only 2-soliton solutions.

The Hirota method was once thought to be merely an ingenious device for finding solutions to soliton equations. In fact, the working definition of a system's integrability was taken to be: send your equation to Hirota; if you get it back solved within three weeks, then it's integrable! However, recent connections with quantum field theory and statistical mechanics indicate that Hirota's method plays a much more central role in the theory than heretofore believed. I hope in these lectures to show you one way in which it ties in very naturally with the general theory. I believe it also clearly relates to the Painlevé property [35] enjoyed by integrable systems. This property, which I will discuss in Chapter 4, says that the only singularities of integrable systems which are not fixed but depend on initial data are poles. This is almost equivalent (not quite, since there are fixed singularities with ugly natures) to saying that the Hirota $\tau(x, t)$ function is analytic in each of its variables. Certainly, as we will show, for certain solution classes, it will be.

What, then, is the general theory? What is the unifying structure which ties all of the miracles that happen in soliton mathematics together? The miracles include: an infinite number of conservation laws, membership in an infinite family of commuting flows (I will explain this term in Chapter 3), a Hamilto-nian structure, the Hirota formulation and the τ-function, the Painlevé prop-erty, the association with a linear eigenvalue problem, inverse scattering, isospectral, iso-Riemann surface, isomonodromic (the last two yet to be ex-plained) deformations [36], Bäcklund transformations.

The connecting link, I believe, comes from asking the question: given an evolution equation, how does one determine whether it is integrable and possesses all these remarkable properties? The first investigators to give a (reasonably) rational way of answering this question were Wahlquist and Estabrook [37] and I will describe my version of what they did in Chapter 5. In essence, they try to force the nonlinear equation of interest to be an integrability condition of two linear equations containing the unknown variable and its x derivatives as coefficients. In doing so, they obtain an infinite dimensional algebra or, to put it another way, a set of commutation relations that are not closed.

It is our contention (and my colleagues in these endeavors are Hermann Flaschka and Tudor Ratiu) that the Wahlquist–Estabrook method is trying to tell us that the appropriate phase space on which all the flows live is an infinite dimensional Lie algebra, which for problems in one spatial dimension is isomorphic to a Kac–Moody algebra. This algebra can be written as a direct sum of two subalgebras; the orthogonal complement of one is the dual of the other. On this dual, there are natural dynamical structures, a Poisson bracket and a Hamiltonian vector field. A special class of Hamiltonians gives rise to a set of commuting flows and each flow is then an exactly integrable equation. It is important to stress that integrable evolution equations always arise as members of an infinite family. From this starting point, many of the features on the list of miracles fall out as natural consequences [38] and we answer from two points of view the question

"What does the Lie algebra $sl(2)$ have to do with KdV?"

All this material will be discussed in depth in Chapter 5. Our work is complementary to recent work by groups in Kyoto [39] (M. and Y. Sato, T. Miwa, M. Jimbo, M. Kashiwara, E. Date) and Oxford [40] (G. Wilson, G. Segal).

CHAPTER 2

Derivation of the Korteweg–deVries, Nonlinear Schrödinger and Other Important and Canonical Equations of Mathematical Physics

2a. An outline of what we are going to do. In this chapter, the goal is to convince you of the reasons for the ubiquity and therefore the importance of the Korteweg–deVries (KdV) and nonlinear Schrödinger (NLS) equations. Whereas discussion of these equations will occupy most of the chapter, in the last section I will mention briefly other canonical systems.

The KdV equation will arise in all situations which can, at leading order, be approximated by a first order linear hyperbolic system but which also contain weakly nonlinear and weakly dispersive terms. The equation describes how each of the Riemann invariants, which travel unchanged along corresponding characteristic directions if nonlinear and dispersive effects are absent, slowly and independently evolve due to these influences. We saw an example of this in our opening chapter. There, the mechanical system was described to first order by the linear wave equation and the weak nonlinearity and dispersion was due to the nonlinear spring potential and the discreteness of the lattice respectively. A disturbance, initially confined in an order one interval, will evolve on an order one time scale into left and right going components according to the prescription of the linear wave equation. However, over long times and distances inversely proportional to the strength of the nonlinearity and dispersion, the subsequent evolution of each component will be described by two separate KdV equations. In the following section, we will show how the KdV equation obtains in the context of low amplitude long water waves on narrow and shallow channels. I chose this example for two reasons. The first is historical, and the second is that it provides an intuitive and readily visualized context in which to examine the effects of other influences which spoil the exact integrability of the KdV equation. In particular, we shall examine what happens to long waves as the channel depth decreases or increases slowly. The response of the wave is not purely adiabatic. We will also derive an equation (or I will ask you to derive it in the exercises) which can model situations in which all the waves crests are not exactly parallel to the shore line or each other. This equation is called the Kadomtsev–Petviashvili (KP) equation, or sometimes the two-dimensional KdV equation. It, too, has miraculous properties. In the exercises, I will also ask you to derive the equations for the Toda lattice and to discuss in what limits the waves in this lattice can be described by the KdV equation. We will also meet the Boussinesq equation which, like the

KdV and KP equations, enjoys the property that it is exactly integrable and we will discuss the situations in which it obtains.

If you were impressed by all the places in which either the KdV or one of its close companion equations arise, you will be truly amazed at the ubiquity of the nonlinear Schrödinger equation (NLS) and its close relations. It is an equation for a complex scalar field $q(x, t)$,

$$q_t = iq_{xx} \pm 2iq^2q^* \quad (* \text{ is complex conjugate}). \tag{2.1}$$

It describes the evolution of the envelope of a wave train and, unlike its linear counterpart, contains within it the soliton solution embodying the concept of a wave packet. The circumstances necessary for its occurrence are that the underlying wave packet is strongly dispersive, almost monochromatic and weakly nonlinear. The x in (2.1) is position measured with respect to a frame of reference moving with the (linear) group velocity corresponding to the wave number of the carrier wave, and the equation itself represents a balance between linear dispersion which has a tendency to break wave packets up and the focusing effect of the cubic nonlinearity produced by the self interaction of the wave with itself. We shall also meet variants of this equation. The nonlinearity is not always as simple as q^2q^*, arising from a $q^2e^{2i\theta} \times q^*e^{-i\theta}$ ($\theta = kx - \omega t$) interaction, but may also involve a mean (nonoscillatory) component $p(x, t)$ in the form pq. In some instances, the mean field p is algebraically proportional to qq^*, in which case (2.1) obtains. In other cases, (2.1) is augmented by another equation relating the evolution of the mean field p to spatial derivatives of qq^*. Rather than drown the reader in the extensive calculations which arise in many of the physical applications, I will examine each of the occurrences in their simplest nontrivial contexts, emphasizing the ingredients which lead to the different cases and then point the reader to the relevant papers in the literature.

I also want to introduce some related ideas and show their connection with NLS. In particular, we will examine how to find the nonlinear Schrödinger limit of Whitham's theory, which is a prescription for following the evolution of fully nonlinear wave trains in a slowly varying environment. This limit is subtle and the point is not often addressed in the literature. We shall also look at the effects of more dimensions. Somewhat counter to one's intuition, the replacement of $\partial^2/\partial x^2$ with ∇^2 in (2.1), with the plus sign on the nonlinearity, leads to the effective strengthening of the nonlinear focusing property of the equation to the extent that the solution becomes locally unbounded in finite time. This focusing phenomenon is widespread in physics and is seen in plasmas in the form of Langmuir wave collapse and in optics as filamentation. Of course, as the amplitude and the inverse of the envelope pulse width become very large, the premises under which the equation is derived are no longer valid and a new description has to be found. Nevertheless, the equation does describe the beginning of the process by which waves focus locally.

2b. Small amplitude, long waves in a channel of slowly changing depth. Equations of the KdV type [41], [42], [43]. In this section, we return to the model which started the whole business. The plan is to give a careful derivation of the KdV equation for the case of water waves. Following Johnson's paper [44], we will also include the effects of a slowly varying depth and comment on how to attack the resulting perturbed KdV equation. Several remarks on the Boussinesq (bidirectional) and Kadomtsev–Petviashvili (weakly two-dimensional KdV) equations will close out the section.

Consider the following situation shown in Fig. 2. (The amplitude and horizontal length scales of the disturbance are greatly amplified and diminished respectively so that everything will fit in the picture.) Assume a two-dimensional, vorticity free fluid with velocity field $\mathbf{u}(x, y, t)$ in a simply connected domain bounded by a time independent lower boundary $y = -H(x)$ and a free surface $y = h_0 + N(x, t)$. The conditions at the ends $x = -\infty, +\infty$ will be left unspecified for the moment but we will imagine that at these points the bottom boundary levels out to a constant depth. We introduce a velocity potential $\phi(x, y, t)$ given by $\mathbf{u} = \nabla\phi$.

We also assume that the disturbance in which we are interested has the following features. Its horizontal length l is large when compared to the average depth h_0; in fact $h_0^2/l^2 = \varepsilon \ll 1$. Its amplitude a is small compared with the average depth h_0 i.e., $a/h_0 = \mu \ll 1$. The phenomena of interest to us occur when these two scales balance. The distance over which the bottom boundary changes significantly (by order one) is greater than l. With these rules in mind we scale the independent and dependent variables as follows:

$$x \rightarrow lx, \qquad y \rightarrow h_0 y, \qquad t \rightarrow \frac{l}{\sqrt{gh_0}} t,$$

$$H \rightarrow h_0 h, \qquad N = a\eta, \qquad \phi \rightarrow \frac{a}{h_0} l\sqrt{gh_0}\phi. \tag{2.2}$$

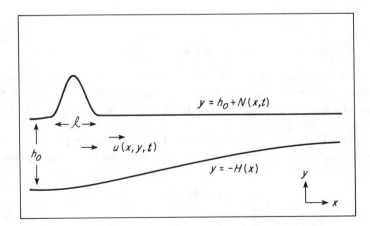

FIG. 2. *Solitary wave propagating in a channel of changing depth.*

Here g is gravity and, consistent with the last sentence, we assume $h_x = \varepsilon h_X$, $X = \varepsilon x$ and h_X is at most order one. With these scalings, the equations of continuity, the boundary condition on the normal velocity at $y = -h$, the continuity of normal stress (pressure) at the free surface (use Bernoulli's equation), and the free surface condition that equates the normal velocity of a particle at the surface with the normal velocity of the surface are (subscripts denote partial derivatives)

$$\phi_{yy} + \varepsilon \phi_{xx} = 0, \tag{2.3}$$

$$\phi_y = -\varepsilon^2 h_X \phi_x, \qquad y = -h, \tag{2.4}$$

$$\phi_t + \eta + \frac{1}{2}\mu\phi_x^2 + \frac{1}{2}\frac{\mu}{\varepsilon}\phi_y^2 = 0, \qquad y = 1 + \mu\eta, \tag{2.5}$$

$$\eta_t + \mu\phi_x\eta_x = \frac{1}{\varepsilon}\phi_y, \qquad y = 1 + \mu\eta. \tag{2.6}$$

The equations are solved by recognizing that (2.3) admits power series solutions in y (actually $y + h$ is more convenient) and we find, after using (2.4),

$$\phi(x, y, t) = F(x, t) - \frac{\varepsilon}{2}F_{xx}(y+h)^2 + \varepsilon^2\left(\frac{1}{24}F_{xxxx}(y+h)^4 - h_X F_x(y+h)\right) + \cdots. \tag{2.7}$$

Let us first derive the shallow water equations which are found by letting $\varepsilon \to 0$, μ finite. From (2.7)

$$\frac{1}{\varepsilon}\phi_y = -F_{xx}(1 + \mu\eta + h) - F_x h_x,$$

where we have written h_x for εh_X.

Now we can take the limit as $\varepsilon \to 0$ and setting $\phi_x = u$, we have for (2.5) and (2.6)

$$u_t + \mu u u_x + \eta_x = 0,$$
$$\eta_t + ((1 + h + \mu\eta)u)_x = 0, \tag{2.8}$$

the shallow water equations. (In the lattice model, this limit corresponds to taking the distance between the masses to be zero.) It is well known that, for most initial conditions, shockline structures, where η_x and u_x become infinite in finite time, are encountered.

However, the limit of interest to us is when nonlinearity measured by μ and dispersion measured by ε are both small and balance. Setting $\mu = \varepsilon$ and expanding (2.5), (2.6) about $y = 1$, we find

$$\phi_t + \eta + \tfrac{1}{2}\varepsilon\phi_x^2 = 0, \qquad y = 1, \tag{2.9}$$

$$\eta_t + \varepsilon(\phi_x\eta)_x = \frac{1}{\varepsilon}\phi_y, \qquad y = 1, \tag{2.10}$$

where now

$$\frac{1}{\varepsilon}\phi_y = -F_{xx}(1+h)+\frac{\varepsilon}{6}F_{xxxx}(1+h)^3 - \varepsilon F_x h_x.$$

The fundamental difference between the two limits is that dispersion in the form of higher order derivatives on F has been introduced at a level which can balance the tendency of the wave to break. Writing $u = F_x$ (u is only the horizontal component of velocity to leading order since $\phi_x = F_x - (\varepsilon/2(y+h)^2 F_{xxx} + O(\varepsilon^2))$, we find, after taking the x derivative of (2.9) and neglecting terms of order ε^2,

$$u_t + \eta_x = \varepsilon(\tfrac{1}{2}D^2 u_{xxt} - uu_x), \tag{2.11}$$

$$\eta_t + (Du)_x = \varepsilon(\tfrac{1}{6}D^3 u_{xxx} - (u\eta)_x), \tag{2.12}$$

where $D = 1 + h$. (Remember $\phi_x(y=1) = u - \varepsilon(h^2/2)u_{xx} + O(\varepsilon^2)$.) While these equations are the starting point for analyzing flows which have both left and right going components, one can obtain the simpler KdV limit by seeking solutions of (2.11) which are unidirectional waves, either right going or left going. It is vital that we use the correct characteristics which are

$$\Theta_\pm = \mp t + \frac{1}{\varepsilon}\int^X \frac{dX}{D^{1/2}}, \tag{2.13}$$

where $X = \varepsilon x$. We can make a single equation for F out of (2.11) and (2.12) by integrating (2.11) or solving (2.9),

$$\eta = -\phi_t - \tfrac{1}{2}\varepsilon\phi_x^2.$$

Substituting (2.7) into (2.12), we obtain

$$F_{tt} - (DF_x)_x = -2\varepsilon F_x F_{xt} - \varepsilon F_t F_{xx} + \varepsilon\frac{D^3}{3}F_{xxxx}. \tag{2.14}$$

It is now straightforward to derive the equation which describes the long distance behavior of F. As a first try, let us take $D = 1$. Then solve (2.14) iteratively in a manner akin to that used in Section 1c. Let $F = f + \varepsilon F_1 + \cdots$ where $f = f(\Theta_+ = -t + x, X = \varepsilon x)$. Then (2.14) is satisfied to leading order and at $O(\varepsilon)$ we obtain an equation for F_1 (let $\Theta_+ = \Theta$)

$$F_{1tt} - F_{1xx} = -4\frac{\partial^2 F_1}{\partial\Theta\,\partial\Theta} = 2f_{\Theta X} + 3f_\Theta f_{\Theta\Theta} + \tfrac{1}{3}f_{\Theta\Theta\Theta\Theta},$$

where $\Theta_- = t + x$. The term $2f_{\Theta X}$ comes from F_{xx} and takes account of the slow dependence of f on X. Since the RHS of this equation is independent of Θ_-, F_1 will grow linearly with Θ_- unless

$$2f_{\Theta X} + 3f_\Theta f_{\Theta\Theta} + \tfrac{1}{3}f_{\Theta\Theta\Theta\Theta} = 0. \tag{2.15}$$

Now if D is not constant but depends on $X = \varepsilon x$, we proceed exactly as before except that it is crucial to use the correct characteristic coordinates Θ_+ and Θ_-

as given by (2.13). I will leave it as an exercise to the reader to show that in order to suppress secular growth in F_1, where $F = f(\Theta_+, X) + \varepsilon F_1 + \cdots$, we must choose

$$q_\tau + 6qq_\Theta + q_{\Theta\Theta\Theta} = -\frac{9}{4}\frac{D_\tau}{D}q, \tag{2.16}$$

where we have written Θ_+ as Θ,

$$f_\Theta = \tfrac{2}{3}D^2q, \tag{2.17}$$

and τ is a rescaled distance coordinate

$$\tau = \frac{1}{6}\int^X D^{1/2}\,dX. \tag{2.18}$$

We call (2.16) the perturbed KdV equation or PKdV.

   Several remarks are in order. First, notice how we choose to write the equation as an evolution in x rather than t for a profile $q(\tau, \Theta)$ which depends on the (negative) retarded time

$$\Theta = -t + \frac{1}{\varepsilon}\int^X \frac{dx}{D^{1/2}}.$$

If D were constant, it would have been equally convenient to use either t (as we did in Section 1c.) or x in this role. Here, because the medium itself depends on x, it is necessary to use the latter. The particular problem we are going to analyze in some detail later is posed as follows: given q at $x = 0$ as function of t for all time and that $q \to 0$ as $t \to \pm\infty$, find $q(x, t)$ for all $x \geq 0$.

   Second, we point out that in order to treat the term on the right-hand side of (2.16) as a perturbation, we must take D_x/D to be of order $\sigma \ll 1$. But, recalling that we have already omitted terms of order ε^2 from the equations (2.11), (2.12), we insist that

$$\varepsilon \ll \sigma \ll 1. \tag{2.19}$$

   Third, let us look at the mass flux through a given station x for all time. From the equations themselves we know

$$\frac{\partial}{\partial x}\int_{-h}^{1+\varepsilon\eta} \tilde{u}\,dy = -\eta_t, \tag{2.20}$$

where $\tilde{u} = \phi_x$ is the true horizontal velocity. Thus if no net mass is added between $t = -\infty$ and $+\infty$, we have that

$$\frac{\partial}{\partial x}\int_{-\infty}^{\infty} dt\left(\int_{-h}^{1+\varepsilon\eta} \tilde{u}\,dy\right) = 0 \tag{2.21}$$

and, to leading order, this means that

$$\frac{\partial}{\partial x}\int_{-\infty}^{\infty} Du\,dt = 0, \tag{2.22}$$

with $u = F_x$. Now note from (2.16) that

$$\frac{\partial}{\partial \tau} \int_{-\infty}^{\infty} D^{9/4} q \, d\Theta = 0 \tag{2.23}$$

or converting to x, t coordinates,

$$\frac{\partial}{\partial x} \int_{-\infty}^{\infty} D^{3/4} u \, dt = 0. \tag{2.24}$$

This means that PKdV does not conserve the total mass flux through a given station. What happens is that some water is reflected and so in order to analyze correctly the influence of a change in depth, one must also look at the left going flow. This we will do when we investigate the propagation of a solitary wave towards a shore.

Fourth, let us derive a law that has been around for 150 years, Green's law. Suppose we were just considering a linear wave propagating in a region of slow depth change. Then the equation describing the evolution of the amplitude q would be given by (2.16) without the two last terms on the left-hand side. The result is that $D^{9/4}q$ or $D^{3/4}u$ or $D^{1/4}\eta$ (recall $\eta \sim -\phi_t \sim D^{-1/2}\phi_x$) is a function only of Θ and is therefore constant along right going characteristics.

We are now in a position to describe how to handle both (2.16) and the full bidirectional problem (2.11), (2.12). Imagine the following situation. A solitary wave q_s with amplitude η_0 arrives at $x = 0$ at time $t = 0$ at which point the unperturbed depth begins to change. The solitary wave will undergo an adiabatic change (its amplitude parameter will change slowly) to satisfy the conservation of energy law

$$\frac{\partial}{\partial \tau} \int q_s^2 \, d\Theta = -\frac{9}{2} \frac{D_\tau}{D} \int q_s^2 \, d\Theta.$$

However with $q = q_s$, the mass flux law

$$\frac{\partial}{\partial \tau} \int q \, d\Theta = -\frac{9}{4} \frac{D_\tau}{D} \int q \, d\Theta$$

then fails to be satisfied. We therefore must add to the right going flow a shelf of amplitude of order σ (relative to the solitary wave amplitude) which stretches between the point to which infinitesimal disturbances would have travelled $\Theta_+ = 0$, and the solitary wave, a distance of order σ^{-1} as measured in the units of the solitary wave width. Therefore it carries a mass flux of the same order as the solitary wave. Its amplitude in the immediate rear of the solitary wave is determined by the amount by which the local mass flux law fails to be satisfied. Its future evolution from that point is given by Green's law.

However, this is not enough because, as we have already noted, the mass flux law for the full bidirectional equation is still unsatisfied. Therefore, we must add another component, a wave of reflection. This has an amplitude (relative to the solitary wave) of order $\sigma \varepsilon$ and stretches over a distance $(\sigma \varepsilon)^{-1}$ times the width of the solitary wave so that it, too, carries a mass flux of equal order to each of the right going components. The velocity field of this reflected component is determined on the right going characteristic at $\Theta_+ = 0$ by the amount by which the mass flux law of the PKdV equation fails to satisfy the exact mass flux law of the full bidirectional equation. Its behavior elsewhere is determined by solving the Goursat problem: given u on $\Theta_+ = 0$ and $\eta = u = 0$ on $\Theta_- = 0$, find u, η which satisfies the linearized version of the equations (2.11), (2.12) in the quadrant $\Theta_+ < 0$, $\Theta_- > 0$. It turns out that Du and η are constant to within $O(\varepsilon\sigma)^2$ along the negative characteristics Θ_-. Green's law does not apply because the gradient of the field variables u and η is of the same order as the gradient in the undisturbed depth.

Using these ideas (which can probably be best characterized by the phrase "the judicious use of the conservation laws") we can find a completely consistent solution to the original problem. It is of interest to list the mass flux M (normalized by the factor $\frac{8}{3}\eta_0\rho\varepsilon^{1/2}h_0^2$) associated with the three solution components. ($\frac{4}{3}\eta_0^2\varepsilon h_0$ is the amplitude of the incoming solitary wave, ρ is density, and h_0, $h(x)$ and h_f are the initial, local and final undisturbed depths respectively.)

$$\text{solitary wave: } \frac{h}{h_0},$$

$$\text{right going shelf: } \left(\frac{h}{h_0}\right)^{1/4} - \left(\frac{h}{h_0}\right), \qquad (2.25)$$

$$\text{reflected wave: } \left(\frac{h_f}{h_0}\right)^{1/4} - \left(\frac{h}{h_0}\right)^{1/4}.$$

The sum of these three components is $(h_f/h_0)^{1/4}$, which is constant and equal to the right going mass flux after the pulse has reached the new constant undisturbed depth and no further reflection need be accounted for. Note in particular, the interesting result that if $h_f \ll h_0$, most of the water is reflected. We carry out some of the calculations in Chapter 3. Other details on these calculations and related results can be found in references [43], [45], [46]. Among some circles, the need for a reflected wave is still a point of controversy. The feeling is that a slowly changing slope should at best induce an adiabatic right going response and the wave of reflection should be exponentially small. However, the existence of a starting point at which the depth begins to change means that the response is not adiabatic. The solitary wave and the amplitude of the right going shelf are slowly varying but the range of the latter is not. It extends from the rear of the solitary wave to the point to which the longest linear wave would have travelled from the starting point. This means that the right going shelf is not infinite in length and therefore is not able to account for all the mass flux.

We close out this section with some comments on the roles of the exactly solvable "Boussinesq" [47] and Kadomtsev–Petviashvili [48] equations. The former is

$$v_{tt} - v_{xx} = v_x v_{xx} + v_{xxxx}. \tag{2.26}$$

We observe that the lattice equation (1.3) is exactly the same and, if we allow ourselves a little latitude (namely replace F_t by $-F_x$ in the nonlinear terms) (2.14) is too.[2] But in what sense is (2.26) a more accurate description of events than two uncoupled KdV equations? The answer is that it is no more accurate for, if the right-hand side in (1.3), the nonlinear and dispersion terms, is to be of equal magnitude with the left-hand side, then the terms we have ignored ($\varepsilon^2 y_{xxxxxx}$ and nonlinear terms such as $\varepsilon^2 y_{xx} y_{xxxx}$) are equally important. We may therefore ask: are there any circumstances in which (2.26) is the relevant canonical equation? The answer is yes. Recall that I pointed out that if in the first approximation the underlying system is hyperbolic with distinct characteristic velocities, then the slow distortion of the corresponding Riemann invariant about each characteristic direction is described by the Korteweg–deVries equation. However, if two of these characteristic velocities are close, then one cannot separate out the evolution along these two neighbors. It is a fairly simple exercise to show that the canonical equation describing the long time behavior of solutions to

$$\left(\frac{\partial}{\partial t} + (c - \sqrt{\varepsilon})\frac{\partial}{\partial x}\right)\left(\frac{\partial}{\partial t} + (c + \sqrt{\varepsilon})\frac{\partial}{\partial x}\right)u = \varepsilon\sqrt{\varepsilon}\,\frac{\partial u}{\partial x}\frac{\partial^2 u}{\partial x^2} + \varepsilon\frac{\partial^4 u}{\partial x^4}$$

is indeed the Boussinesq equation. In the frame of reference travelling with the average velocity c, the slow evolution of the field u as function of $X = x - ct$ and $T = \sqrt{\varepsilon}t$ is given by (2.26). The reason that the nonlinear term must be taken smaller than the dispersion term is that the initial resonance (the leading order approximation is $(\partial/\partial t + c\,\partial/\partial x)^2 u = 0$) causes the initial amplitude to grow by a factor of $1/\sqrt{\varepsilon}$ before nonlinearity and dispersion come into play.

As a final remark to this section, consider what happens if a weak dependence on the other horizontal coordinate, which we will call z, is included in the model for either water or lattice waves. This gives rise to another term on the left-hand sides in (2.14) and (1.3) proportional to $-\varepsilon F_{zz}$ or $-\varepsilon c^2 y_{zz}$. It is a trivial matter to show that the canonical long time equation corresponding to (1.6) will now be

$$f_{zz} + \frac{2}{c} f_{\xi T} + f_\xi f_{\xi\xi} + \delta^2 f_{\xi\xi\xi\xi} = 0,$$

which can also be written as (set $f_\xi = 6u$, $\tau = cT/2$ and take $\delta^2 = 1$)

$$u_{zz} + \frac{\partial}{\partial \xi}(u_\tau + 6uu_\xi + u_{\xi\xi\xi}) = 0. \tag{2.27}$$

---

[2] Notice that (2.14), with $D = 1$, becomes (2.26) if we only ask for its unidirectional solutions $F(-t+x)$ by setting $F_t = -F_x$, $F_{xt} = -F_{xx}$. Indeed it is this version of the equation which is associated with Boussinesq.

Equation (2.27) is known as the Kadomtsev–Petviashvili equation and it too, has many remarkable properties.

Exercises 2b.

1. Derive the equation for long lattice waves if the spring force is $F = k(\Delta + \alpha\Delta^3)$. You will find that the relevant equation is the modified Korteweg–deVries (MKdV) equation. Investigate its travelling wave solutions. Do they exist for both the soft ($\alpha < 0$) and hard ($\alpha > 0$) spring cases?

2. It turns out that a complexified version of the MKdV equation is also universal in the sense that it obtains as the asymptotic description in a variety of contexts. One particular application is to the lower hybrid waves in a plasma. The reader is directed to the references in [118] and in particular the article by G. J. Morales and Y. C. Lee, *Soliton-like structures in plasmas*, Rocky Mountain J. Math., 8, 1, 2, Winter, Spring, 1978.

3. Show that the long time-distance behavior of the field $u(x - ct, \sqrt{\varepsilon}t$ or $\sqrt{\varepsilon}x)$ described by the equation

$$\left(\frac{\partial}{\partial t} + (c - \sqrt{\varepsilon})\frac{\partial}{\partial x}\right)\left(\frac{\partial}{\partial t} + (c + \sqrt{\varepsilon})\frac{\partial}{\partial x}\right)u = \varepsilon\sqrt{\varepsilon}\frac{\partial u}{\partial x}\frac{\partial^2 u}{\partial x^2} + \varepsilon\frac{\partial^4 u}{\partial x^4}$$

is given by the Boussinesq equation. Can you find any concrete examples where this equation might be relevant[3]?. Find also the travelling waves of (2.26). In what respect do they differ from those of KdV?

4. Suppose the spring force in the lattice is $F = \exp(y_{n+1} - y_n)$. Write down the equation of motion and then go to the reference [50], in which the Toda lattice is discussed. In what limit are these waves adequately described by the KdV equation?

5. Start with a two-dimensional lattice model in which each mass is connected to two sets of two neighbors, east and west, and north and south. If the spring constant $k_\perp$ in the north-south springs is much weaker than the spring constant k in the east-west springs and of the same order of magnitude as the quadratic nonlinearity α in the latter, then, if $k_\perp \sim \alpha \sim h^2$ where h is the spacing of the lattice masses, show that the long time behavior of the slightly oblique right (or left) going waves in this lattice is described by the Kadomtsev–Petviashvili equation. Be accurate. Recall that the spring law is given in terms of the displacement in its length not just its horizontal or vertical component. Find the travelling wave solution for this model. How is it related to that of KdV?

2c. The nonlinear Schrödinger and other envelope equations.

It is best to start with a simple example. Let us consider the Scott model described in Chapter 1 consisting of a line of pendula very close together hanging vertically under gravity from a horizontal torsion wire about which each can twist. If the twist angle of the pendulum at x is $u(x, t)$, then its motion is given by the

---

[3] After I posed this question at the CBMS lectures, an example was found by C. H. Su [49].

sine-Gordon equation

$$u_{tt} - c^2 u_{xx} + \omega_p^2 \sin u = 0, \tag{2.28}$$

where the $-\omega_p^2 \sin u$ force is due to gravity and the $c^2 u_{xx}$ force models the effect of the twist. Now imagine that we wiggle one end of the pendulum chain with a very small amplitude motion of frequency ω. It is not unreasonable to expect that in order to follow this motion we can approximate $\sin u$ by its Taylor series about $u = 0$. Keeping the first two terms, we obtain

$$u_{tt} - c^2 u_{xx} + \omega_p^2 u = \frac{\omega_p^2}{6} u^3 + \cdots . \tag{2.29}$$

The linearized equation admits sinusoidal solutions $u = e^{-i\omega t + ikx}$ where k is real and given by the dispersion relation

$$\omega^2 = \omega_p^2 + c^2 k^2 \tag{2.30}$$

as long as $\omega > \omega_p$. For $\omega < \omega_p$, k is pure imaginary and the initial wiggle dies out exponentially in x. Let us suppose that $\omega > \omega_p$ so that real waves propagate down the string. Now it is also to be expected that eventually the nonlinear terms will cause some modulation of this motion as we well know that the period of a nonlinear spring (or indeed a single pendulum) depends on amplitude. To find this modulation, let us seek solutions to (2.29) in the form

$$u = \varepsilon(u_0 + \varepsilon u_1 + \varepsilon^2 u_2 + \cdots), \tag{2.31}$$

where

$$u_0 = a e^{ikx - i\omega t} + a^* e^{-ikx + i\omega t} \tag{2.32}$$

and where we will allow a to be a slowly varying function of time

$$a_t = \varepsilon A_1(a, a^*) + \varepsilon^2 A_2(a, a^*) + \cdots . \tag{2.33}$$

The coefficients $A_j(a, a^*)$ in (2.33) are chosen in order to suppress secular[4] behavior in $u_1, u_2, \ldots$. In the vernacular of multiple time scales we would write A_1 as $\partial a / \partial T_1$, $T_1 = \varepsilon t$ and A_2 as $\partial a / \partial T_2$, $T_2 = \varepsilon^2 t$, and so on. Solving (2.29) iteratively leads to $u_1 = 0$,

$$A_1 = 0, \qquad u_2 = -\frac{a^3}{48} e^{3i(kx - \omega t)} + (*),$$

$$A_2 = +\frac{i}{4} \frac{\omega_p^2}{\omega} a^2 a^*,$$

whence

$$u_0 = a_0 \exp\left(ikx - it\left(\omega - \frac{\omega_p^2}{4\omega} \varepsilon^2 a_0 a_0^*\right)\right) + (*) \tag{2.34}$$

---

[4] Secular behavior refers to a situation in which the iterates $u_1, u_2, \ldots$ grow algebraically in the fast time or space variables. If this were allowed to happen, the asymptotic series (2.31) would not be uniformly valid over long times and distances.

and the period of the motion increases. In the context of water waves, the
solution computed in this way is known as the *Stokes wave*. The reader might
wish to compute the period of a pendulum of maximum amplitude $2\varepsilon |a_0|$
exactly in terms of an elliptic integral and then take its low amplitude
expansion in order to check (2.34) (see Exercise 2c(1)). Note that in all this
calculation the spatial structure e^{ikx} plays a passive role. Once ω is fixed, then
so is k from (2.30). But no one is able to wiggle a chain at a perfectly tuned
frequency and so one might expect a small but finite band width μ of frequencies and wavenumbers to be excited. How can we incorporate these into the
description? One way is to look for solutions u_0 which are a finite sum of waves

$$u_0 = \sum_{k_j = k + \mu K_j} a_j e^{i(k_j x - \omega_j t)} + (*), \qquad \omega_j^2 = \omega_p^2 + c^2 k_j^2, \tag{2.35}$$

but this approach is clumsy and leads to a set of coupled nonlinear equations
for the amplitudes a_j which are not terribly enlightening. Another way, which
is suggested by (2.35), is to look for solutions of a form in which the amplitude
a is a slowly varying function of x as well as time, an idea originally introduced
in [53]. The most interesting balance between the various effects occurs when
$\mu = \varepsilon$.

Let us repeat the previous calculation where this time we allow A_1, A_2, to be
a function of a_X, a_X^*, a_{XX}, etc. $X = \varepsilon x$, as well as a and a^*. At $O(\varepsilon)$ we find

$$u_{1tt} - c^2 u_{1xx} + \omega_p^2 u_1 = (2i\omega a_{T_1} + 2ikc^2 a_X) e^{i(kx - \omega t)} + (*), \tag{2.36}$$

where $a_{T_1} = A_1$. In order to suppress secular growth in u_1, we must have that

$$a_{T_1} + \frac{c^2 k}{\omega} a_X = 0, \tag{2.37}$$

or that a moves with the group velocity $\omega' = d\omega/dk$, as calculated from (2.30),
of the wave packet. Then $u_1 = 0$. At order ε^2,

$$u_{2tt} - c^2 u_{2xx} + \omega_p^2 u_2 = \frac{\omega_p^2}{6} a^3 e^{3i(kx - \omega t)}$$
$$+ (2i\omega a_{T_2} - a_{T_1 T_1} + c^2 a_{XX} + \tfrac{1}{2}\omega_p^2 a^2 a^*) e^{i(kx - \omega t)} + (*),$$

and the suppression of secular terms requires that

$$a_{T_2} - \frac{i\omega''}{2} a_{\xi\xi} - \frac{i}{4} \frac{\omega_p^2}{\omega} a^2 a^* = 0. \tag{2.38}$$

In finding (2.38), we used (2.37) in order to write $a_{T_1 T_1}$ in terms of a_{XX}. Also,
$\xi = \varepsilon(x - \omega' t)$, $T_2 = \varepsilon^2 t$ and ω'' is the dispersion $d^2\omega/dk^2$ as calculated from
(2.30). Equation (2.38) is the nonlinear Schrödinger equation. Observe that it
contains as a special solution the x independent frequency modulation solution
(2.34) *but*, and this is a very important but, it is an unstable solution if the
product of the coefficients in front of the dispersion term $\omega''/2$ and the
nonlinear term $(\tfrac{1}{4}\omega_p^2/\omega)$ is positive, a situation which obtains in the present
example. This is the instability which was discovered by Benjamin and Feir

[51] when they tried to demonstrate experimentally the existence of the Stokes solution for surface water waves. It is a most important instability as it causes otherwise monochromatic wavetrains to evolve into a series of pulses. I will discuss the nature of this instability in more detail shortly.

For now, however, I want to return to the reasons for the ubiquity of the NLS equation and show how all the linear terms in the equation have a universal structure. Consider the equation

$$L\left(\frac{\partial}{\partial t}, \frac{\partial}{\partial x}\right)u = N(u^2, u^3, \ldots) \tag{2.39}$$

where Lu and $N(u^2, u^3, \ldots)$ are constant coefficient linear and nonlinear operators on u and its derivatives. Let the linear portion of (2.39) admit sinusoidal solutions of the form

$$u = ae^{i(kx-\omega t)}, \tag{2.40}$$

where

$$L(-i\omega, ik) = 0 \tag{2.41}$$

is the dispersion relation giving ω as function of k or vice versa. Since (2.41) holds for all k, we can deduce that

$$-i\omega'L_1 + iL_2 = 0, \tag{2.42}$$
$$-i\omega''L_1 - \omega'^2 L_{11} + 2\omega'L_{12} - L_{22} = 0 \tag{2.43}$$

by differentiating once and then twice with respect to k. L_j is the derivative of L with respect to the jth argument. Now, let us seek solutions of (2.39) in the form

$$u(x, t) = \varepsilon(u_0 + \varepsilon u_1 + \varepsilon^2 u_2 + \cdots) \tag{2.44}$$

with u_0 given by (2.32) with a a slowly varying function of position and time. We note that $L(\partial/\partial t, \partial/\partial x)$, under the multiple time scale algorithm, formally becomes

$$L\left(\frac{\partial}{\partial t} + \varepsilon\frac{\partial}{\partial T_1} + \varepsilon^2\frac{\partial}{\partial T_2}, \frac{\partial}{\partial x} + \varepsilon\frac{\partial}{\partial X}\right)$$

$$= L\left(\frac{\partial}{\partial t}, \frac{\partial}{\partial x}\right) + \varepsilon\left(L_1\frac{\partial}{\partial T_1} + L_2\frac{\partial}{\partial X}\right)$$

$$+ \varepsilon^2\left(L_1\frac{\partial}{\partial T_2} + \frac{1}{2}L_{11}\frac{\partial^2}{\partial T_1^2} + L_{12}\frac{\partial^2}{\partial T_1 \partial X} + \frac{1}{2}L_{22}\frac{\partial^2}{\partial X^2}\right) + \cdots \tag{2.45}$$

which we write as $L^{(0)} + \varepsilon L^{(1)} + \varepsilon^2 L^{(2)}$. Iteratively solving (2.39), we have

$$L^{(0)}\left(\frac{\partial}{\partial t}, \frac{\partial}{\partial x}\right)u_0 = 0 \tag{2.46}$$

with solution

$$u_0 = a(X, T_1, T_2, \ldots)e^{i(kx-\omega t)} + (*) \tag{2.47}$$

with (2.41) holding between ω and k. Next at order ε,

$$L^{(0)}u_1 = -L^{(1)}u_0 + N(u_0^2). \tag{2.48}$$

What are the secular terms? They are those terms on the RHS of (2.48) giving rise to a solution u_1 which grows algebraically in x or t. They can be recognized by the fact that they have an x, t structure which belongs to the null space of $L^{(0)}$. For example, $L^{(1)}u_0$ belongs to this class as $L^{(0)}L^{(1)}u_0 = L^{(1)}L^{(0)}u_0 = 0$. Which terms in $N(u_0^2)$ belong? The term u_0^2 itself contains second harmonics $a^2 e^{2i(kx-\omega t)}$ and mean terms aa^*. But, since the system is strongly dispersive, for almost all k, $\omega(2k) \neq 2\omega(k)$ and therefore $L^{(0)} e^{2i(kx-\omega t)} \neq 0$. Now, the constant solution aa^* may indeed belong to the null space of $L^{(0)}$. If it does and if $N(u_0^2)$ is nonzero, we must include in the zeroth order approximation u_0 a mean term which slowly varies with x and t. In this situation, the "wave" e^0 is the third leg of the triad in a triad resonance (see Section 2f).

What happens more often is that $N \cdot aa^*$ is zero to this order. This occurs because the equation has underlying symmetries like Galilean invariance which makes it impossible to force a mean flow directly. (For the sake of an example, think of N as $\partial/\partial x$.) On the other hand, due to the slow dependence of the envelope on x, a local mean flow, which has the form $\varepsilon^2(\partial/\partial X)aa^*$, can arise, and unless removed, cause a secular response at the ε^2 level in u_2. Such a term need not violate any overall conservation property. In one ε^{-1} patch, it may increase the mean level; in the next patch its parity can change so that no net "mass" is added to the system. I stress its potential presence, however, because sometimes it is easy to overlook it. In order to account for this effect we include a mean contribution as a homogeneous solution b in u_1 (or simply in u_0 at order ε, same difference). This mean term b then will contribute to potential secular behavior $e^{i(kx-\omega t)}$ at the $O(\varepsilon^2)$ level in u_2 through the quadratic product $N(u_0 u_1)$. Removing secular terms at $O(\varepsilon^2)$ then leads to a coupled system of equations in a, the wave envelope and b the slowly varying mean. Sometimes b is expressible as a constant times aa^*; sometimes it is not. We will meet both of these cases in the exercises and I will also point you to three concrete examples in physics where these effects are important.

For now let us assume that the mean flow is not in the null space of $L^{(0)}$ as, for example, would be the case if $L = \partial^2/\partial t^2 - c^2 \partial^2/\partial x^2 + \omega_p^2$. Then the only secular term in (2.48) is $L^{(1)}u_0$ and so we must choose the dependence of a on X and T_1 such that $L^{(1)}u_0 = 0$, namely

$$L_1 \frac{\partial a}{\partial T_1} + L_2 \frac{\partial a}{\partial X} = 0.$$

But from (2.42), $L_2 = +\omega' L_1$ and so if $L_1 \neq 0$ (which we assume)

$$a = a(X - \omega' T_1, T_2). \tag{2.49}$$

We then solve for u_1, which contains second harmonics and perhaps a mean term proportional to aa^*. At order ε^2, the secular terms which are nonlinear in

a and a^* arise from the quadratic product u_0u_1 and the cubic product u_0^3. These lead to a term which we write as $\beta L_1 a^2 a^*$. The linear terms which are secular are of the form $L^{(2)}u_0$ which, when we take account of (2.49), may be written as a product of $e^{i(kx-\omega t)}$ with

$$L_1\frac{\partial a}{\partial T_2}+\left(\frac{1}{2}\omega'^2 L_{11}-\omega' L_{12}+\frac{1}{2}L_{22}\right)\frac{\partial^2 a}{\partial X^2}.$$

But from (2.43), this is equal to

$$L_1\left(\frac{\partial a}{\partial T_2}-\frac{i\omega''}{2}\frac{\partial^2 a}{\partial X^2}\right),$$

and so the universal form of the NLS equation is

$$\frac{\partial a}{\partial T_2}=\frac{i\omega''}{2}\frac{\partial^2 a}{\partial \xi^2}+i\beta a^2 a^* \tag{2.50a}$$

where a is a function of $\xi=\varepsilon(x-\omega't)$ and $T_2=\varepsilon^2 t$.

In higher dimensions, I will leave it as an exercise for you to prove that (again assuming the mean is nonsecular)

$$\frac{\partial a}{\partial T_2}-\frac{i}{2}\sum_{r,s}\frac{\partial^2\omega}{\partial k_r\,\partial k_s}\frac{\partial^2 a}{\partial \xi_r\,\partial \xi_s}=i\beta a^2 a^*,$$

where $\partial^2\omega/\partial k_r\,\partial k_s$ is the dispersion tensor, $\xi_r=\varepsilon(x_r-\omega_r t)$ and $\omega_r=\partial\omega/\partial k_r$ is the component of the (vector) group velocity in the x_r direction.

Let me also say here and emphasize by way of an example in the exercises that the roles of x and t are interchangeable. We could equally well look for solutions in the form $a(\varepsilon(t-k'x),\varepsilon^2 x)$ with $k'=dk/d\omega=1/(d\omega/dk)$ and the coefficient of the dispersion is then $ik''/2$. This formulation is convenient when one or both of the parameters c^2 and ω_p^2, like the strength of the torsion wire or the length of the pendula in the original example, is a slowly varying function of x.

Equation (2.50a) belongs to the class of exactly solvable models. The transformation

$$\xi=X,\qquad \frac{\omega''T_2}{2}=\tau,\qquad q=\sqrt{\left|\frac{\beta}{\omega''}\right|}\,a$$

puts it into canonical form

$$q_\tau=iq_{XX}+2isq^2 q^* \tag{2.50b}$$

with $s=\operatorname{sgn}(\beta/\omega'')$. We show in Chapter 3 how (2.50b) can be incorporated into an inverse scattering framework. For $s=+1$, the asymptotic solution of the initial value problem for (2.50b) consists of a sequence of envelope solitons

$$q(X,\tau)=2\eta\operatorname{sech}2\eta(X+4v\tau-X_0)\exp(-2ivX-4i(v^2-\eta^2)\tau-i\phi_0) \tag{2.51a}$$

and radiation modes. For each soliton, the original field $u(x, t)$ has the form

$$u(x, t) = 2\sqrt{\left|\frac{\omega''}{\beta}\right|}\, \varepsilon\eta\, \text{sech}\, 2\varepsilon\eta(x - \omega'(k - 2v\varepsilon)t - x_0)$$

$$\cdot \exp\{i(k - 2v\varepsilon)x - i\omega(k - 2v\varepsilon)t + 2i\omega''\eta^2\varepsilon^2 t\}. \qquad (2.51b)$$

This expression points up an essential weakness in the NLS equation as a model for physical problems. Whereas the velocity of propagation of the oscillatory phase is amplitude dependent (depends on η), the velocity of the amplitude pulse is not. To be sure, the soliton parameters v, η are determined from the initial data (as are x_0, ϕ_0) $q(x, 0)$ (they are the analogue of the $\zeta_k = i\eta_k$ for the KdV equation), but it is clear that the velocity of the argument of the hyperbolic secant is the linear group velocity of $k - 2v\varepsilon$. The difficulty is that expansion has been carried only so far as to give correct phases to $O(\varepsilon^2)$; but there is an overall factor of ε on the phase of the hyperbolic secant. What we would really like is the expression for this argument to continue to order ε^3, that is, take the form

$$2\varepsilon\eta(x - \omega'(k - 2v\varepsilon)t - O(\varepsilon^2 t)).$$

The last term then would depend on η. This can of course be done, but if one does, one obtains a new governing equation, a perturbation of NLS which is no longer exactly integrable. Nevertheless, in certain circumstances, it is necessary to forego the mathematical convenience of exact integrability in order to capture the essential physics of the system one is modelling. This point can be illustrated dramatically if one studies the tunnelling properties of solitons. For a discussion, I refer the interested reader to reference [52].

Exercises 2c.

1. Show that the period of the pendulum which achieves maximum amplitude A is given by

$$T = \frac{4}{\omega_p}\int_0^{\pi/2} \frac{d\phi}{\sqrt{1 - m^2\sin^2\phi}}, \qquad m^2 = \sin^2\frac{A}{2}.$$

Show for A small that the period agrees with that predicted by (2.34).

2. Derive the NLS equation for the following examples.

(i) $u_{tt} - u_{xx} + \omega_p^2 u(1 - uu^*) = 0$, u a complex scalar field.

(ii) $u_t + u_{xxx} = -6\alpha\varepsilon u u_x$. In this example, note that $u = $ constant is a solution of the linear equation,

$$u_0 = ae^{i\theta} + (*), \qquad \theta = kx - \omega t, \qquad \omega = -k^3.$$

At order ε, find $a_{T_1} - 3k^2 a_X = 0$ and $u_1 = b(X, T_1, T_2) + (\alpha/k^2)a^2 e^{2i\theta} + (\alpha/k^2)a^{*2}e^{-2i\theta}$. At order ε^2,

$$u_{2t} + u_{2xxx} = -\frac{\partial u_0}{\partial T_2} - 3\frac{\partial^3 u_0}{\partial x\,\partial X^2} - \frac{\partial u_1}{\partial T_1} - 3\frac{\partial^2 u_1}{\partial x\,\partial X} - 6\alpha\frac{\partial}{\partial x}(u_0 u_1) - 6\alpha\frac{\partial}{\partial X}aa^*.$$

Eliminate terms on the RHS proportional to $e^{\pm i\theta}$, e^0 with the choice of a_{T_2} and b_{T_1}. Find $b_{T_1} = -6\alpha(aa^*)_X = (-2\alpha/k^2)(aa^*)_{T_1}$ which implies that $b = (-2\alpha/k^2)aa^*$. Also find $a_{T_2} + 3ika_{XX} - (6\alpha^2/k)ia^2a^* = 0$. Observe in particular how the nonlinear term comes from the sources, $a^2e^{2i\theta} \cdot a^*e^{-i\theta}$ and $ae^{i\theta} \cdot b$. Observe also that if you forgot the contribution from b, the sign of the nonlinear term would be the opposite, which would lead to completely erroneous conclusions.

(iii) $\quad u_{tt} - c^2u_{xx} + \gamma u_{xxxx} = \varepsilon\beta u_x u_{xx},$

$$u_0 = ae^{i\theta} + a^*e^{-i\theta} + b, \qquad \theta = kx - \omega t, \qquad \omega^2 = c^2k^2 + \gamma k^4,$$

$$u_1 = \frac{-i\beta}{12\gamma k}a^2e^{2i\theta} + \frac{i\beta}{12\gamma k}a^{*2}e^{-2i\theta}, \qquad a_{T_1} + \omega'a_X = 0.$$

At $O(\varepsilon^2)$, remove secular terms and find

$$a_T + \omega'a_X - \varepsilon\left(\frac{i\omega''}{2}a_{XX} - \frac{i\beta k^2}{2\omega}ab_X + \frac{i\beta^2k^2}{12\gamma\omega}a^2a^*\right) = 0,$$

$$b_{TT} - c^2b_{XX} = \beta k^2(aa^*)_X, \qquad T = \varepsilon t, \quad X = \varepsilon x.$$

Let $b_X = \rho$ and rewrite as

$$a_T + \omega'a_X - \varepsilon\left(\frac{i\omega''}{2}a_{XX} - \frac{i\beta k^2}{2\omega}a\rho + \frac{i\beta^2k^2}{12\gamma\omega}a^2a^*\right) = 0,$$

$$\rho_{TT} - c^2\rho_{XX} = \beta k^2(aa^*)_{XX}.$$

Compare this example to the interaction of Langmuir waves and ion acoustic waves of a plasma. In this context, they are called the Zakharov equations [58]. The RHS of the equation for the mean field $b(X, T)$ is due to what is called in the vernacular of plasma physics the pondermotive force. We can write this equation as

$$\rho_{TT} - c^2\rho_{XX} = \frac{\beta k^2((aa^*)_{XX} - (1/c^2)(aa^*)_{TT})}{1 - \omega'^2/c^2} + O(\varepsilon),$$

because $(aa^*)_T = -\omega'(aa^*)_X + O(\varepsilon)$. Therefore the part of the mean field induced by the slow gradients in the envelope of the fast field can be solved for:

$$\rho = \frac{\beta k^2 aa^*}{\omega'^2 - c^2}.$$

Note in particular the possibility of resonance, namely, when the group velocity of the fast field equals the phase velocity of the long wave or mean field. The reader should consult the papers by Benney (Stud. Appl. Math., 55 (1976), pp. 93ff.; 56 (1977), pp. 81–94) and Newell (SIAM J. Appl. Math., 35 (1978), pp. 650–664), for further discussion of the importance of this resonance. One now rewrites the equation for the envelope $a(\xi = X - \omega'T, \tau = \varepsilon T)$,

$$a_\tau - \frac{i\omega''}{2}a_{\xi\xi} + \frac{i\beta^2k^2}{12\gamma}\frac{\omega''}{\omega'^2 - c^2}a^2a^* = 0.$$

Finally, a comment about the Zakharov equations. If, in the present example, we had included a second spatial dimension

$$u_{tt} - c^2 \nabla^2 u + \gamma \nabla^4 u = \varepsilon \beta \nabla u \cdot \nabla(\nabla u),$$

$\nabla = (\partial/\partial x, \partial/\partial y)$, then one cannot simply eliminate the mean ρ in terms of aa^*. The reason for this is that in the two-dimensional case (see discussion in Section 2d), the "solitary wave" collapses and the argument of the envelope no longer is uniformly constant on the group velocity characteristic, except at the initial stage. As the envelope and mean field develop into collapsing spikes, the premises on which the equation is derived are no longer satisfied and it is necessary to include other terms which have been neglected in the analysis here. However, the behavior of the Zakharov equations are often taken as a guide as to what happens in real situations.

Examples (ii) and (iii) are also prototypes for the water wave problem in one horizontal space dimension. With the aid of references [54] and [55], derive the NLS equation

$$a_t + \omega' a_x - \frac{i\omega''}{2} a_{xx} + i\beta a^2 a^* = 0$$

for surface gravity waves. In this equation,

$$\omega^2 = gkT, \qquad T = \tanh kh, \qquad S = \operatorname{sech} kh,$$

$$\beta = \frac{1}{2}\omega k^2 \left\{ \frac{9T^4 - 10T^2 + 9}{8T^4} - \frac{1}{gh - \omega'^2}\left(\frac{\omega^2}{k^2 T^2} + \frac{\omega\omega' S^2}{kT^2} + \frac{ghS^4}{T^2} \right) \right\},$$

where $\eta(x, t)$, the surface displacement is $a/2e^{i(kx-\omega t)} + (*)$, g is gravity and h the undisturbed depth. Note β changes sign at $kh = 1.36$. This means that solitons form for deep water gravity waves $(kh > 1.36)$ but do not when the water gets shallower. For a discussion of the case of two horizontal space dimensions, the reader should consult Benney and Roskes [56] or Davey and Stewartson [57].

(iv) The following example is based on an experimental observation of Wu, Keolian and Rudnick (preprint) of the existence of what they call hydrodynamic polarons in a water trough resonator. The idea is roughly this. The linear dispersion relation for waves in a rectangular tank is $\omega^2 = gk \tanh kh$ where $k = (k_x^2 + k_y^2)^{1/2}$. Suppose that the tank is oblong with the y dimension L_y being small (2.54 cm) and the x dimension long (38 cm). In this experiment the mean depth is 2 cms. but the water is deep in the sense that $kh \approx 3$. The lowest mode in the y direction is $k_y = \pi/L_y$. Note that if the driving frequency ω is less than the natural frequency of the (0, 1) mode $k_y = 0$, $k_y = \pi/L_y$, $\omega_{01}^2 = gk_y \tan k_y h$, then there is no propagation in the x direction and the wave is trapped at an end wall with displacement

$$\eta(x, y, t) \propto e^{-(-k_x^2)^{1/2} x} \cos k_y y (A e^{-i\omega t} + A^* e^{-i\omega t}),$$

where $\omega_{01}^2 > \omega^2 = g(k_x^2 + k_y^2)^{1/2} \tanh (k_x^2 + k_y^2)^{1/2} h$, whence $k_x^2 < 0$. The above authors find that due to nonlinear effects balancing dispersive effects, one can have soliton shaped pulses which are located at arbitrary points along the channel. They mention that they are surprised to find local pulses when the excitation is uniform in x. This is not at all surprising for, as we shall see in the next section, the uniform response is unstable and the instabilities grow into solitons of the nonlinear Schrödinger equation. In examining this phenomenon, I shall use the model by Larraza and Putterman (preprint) who note that the NLS equation obtains and that therefore hyperbolic secant solutions are possible. Again, however, they fail to understand that they are inevitable. Consider

$$u_{tt} - c^2 \nabla^2 u + \gamma \nabla^4 u = \varepsilon \nabla^2 (\alpha u^2), \qquad 0 < \varepsilon \ll 1,$$

with

$$u = u(x, y, t), \qquad \nabla^2 = \frac{\partial^2}{\partial x^2} + \frac{\partial^2}{\partial y^2},$$

$$u_0 = (A(X, T)e^{-i\omega t} + (*)) \cos ky,$$

$$u_1 = B(X, T) + (A_2 e^{-2i\omega t} + B_2 + A_2^* e^{2i\omega t}) \cos 2ky,$$

where we choose $k(=k_y)$ and ω to be such that

$$\omega^2 = c^2 k^2 + \gamma k^4 - \varepsilon^2 \chi = \omega_{01}^2 - \varepsilon^2 \chi.$$

We choose the time dependence of $A(x, T)$, where $X = \varepsilon x$, $T = \varepsilon t$,

$$A_T = f_1 + \varepsilon f_2 + \cdots$$

and $B(X, T)$ in order to eliminate secular contributions proportional to $e^{i\theta}$ and e^0 in u_1, u_2 and u_3 and u_3 respectively. A little calculation shows $f_1 = 0$ (the group velocity $(\partial \omega / \partial k_x)(0, k)$ in the direction of slow modulation is zero)

$$A_2 = \frac{-\alpha}{6\gamma k^2} A^2, \qquad B_2 = -\frac{AA^*}{c^2 + 4\gamma k^2},$$

$$2i\omega A_T + \varepsilon \left(c^2 \frac{\partial^2 A}{\partial X^2} - \chi A - \alpha k^2 A \left(2B - \frac{\alpha AA^*}{6\gamma k^2} - \frac{k^2 AA^*}{c^2 k^2 + 4\gamma k^4} \right) \right) + O(\varepsilon^2) = 0,$$

$$B_{TT} - c^2 B_{XX} = \alpha \frac{\partial^2}{\partial X^2} AA^* + O(\varepsilon).$$

As in Example (iii), we can add $(-\alpha/c^2)(\partial^2/\partial T^2) AA^*$ to the RHS of the last equation as this term is $O(\varepsilon^2)$. Hence

$$B(X, T) = -\frac{\alpha}{c^2} AA^*,$$

and if $\tau = \varepsilon T$,

$$2i\omega A_\tau + c^2 A_{XX} + (\alpha^2 k^2 \beta AA^* - \chi)A = 0$$

with

$$\beta = \frac{2}{c^2} + \frac{1}{6\gamma k^2} + \frac{1}{c^2 + 4\gamma k^2},$$

positive if γ is. There are several points to note.

(a) Since the product of dispersion in the modulation direction x and β is positive, the asymptotic state consists of solitons with the form (2.51a).

(b) They will only move (i.e. $v \neq 0$) if the initial conditions are nonuniform in a special way or if external forces are applied. Their velocities do not depend on amplitude. Their locations and amplitudes are found from the initial conditions.

(c) The χ factor is removed simply by changing the phase of A.

And now a caveat. The experiment described by Wu, Keolian and Rudnick is not performed by simply shaking the trough (exciting the lowest sloshing mode) and allowing the surface to develop. Dissipation necessitates that the trough be continuously driven to overcome the loss. This means that one should add forcing and damping terms $E - \Gamma A$ to the right-hand side of the NLS equation. (If the frequency mismatch term $-\chi A$ is removed, E gains the phase factor $e^{i\Delta t}$, $\Delta = (-i\chi/2\omega)\varepsilon^2$.) The response of the forced, damped system can simply be the creation of a soliton whose amplitude is driven to a fixed value (see [45]). However, depending on the values of E and Γ, other responses are also possible. I do not plan to discuss them here but the interested reader should refer to the paper by Bishop, Fesser, Lomdahl and Trullinger in Physica D, 7 (1983), pp. 259 ff. that treats the forcing of breather solutions of the sine-Gordon equation. Low amplitude breathers of that equation are NLS solitons.

(v) $u_{tt} - c^2 u_{xx} + \omega_p^2(\varepsilon^2 x)u = \varepsilon^2 \gamma u^3$. γ, c^2 constant. This time we will look for the evolution of the wave packet in space. The relevant scales are

$$X_1 = \varepsilon x, \quad X_2 = \varepsilon^2 x, \quad t, \ T = \varepsilon t$$

$$u_0 = a e^{i\theta} + (*), \quad \theta = \int k \, dx - \omega t, \quad k^2 = \frac{\omega^2 - \omega_p^2(X_2)}{c^2}.$$

Note that

$$\frac{\partial^2 u_0}{\partial x^2} = e^{i\theta}(-k^2 a + 2ik\varepsilon a_{X_1} + \varepsilon^2(2ika_{X_2} + a_{X_1 X_1} + ik_{X_2}a)) + (*).$$

At order ε,

$$a_{T_1} + \omega' a_{X_1} = 0 \Rightarrow a = a\left(T - \frac{1}{\varepsilon}\int^{X_2} k' \, dX_2\right),$$

where $k' = 1/\omega' = dk/d\omega$. At order ε^2, find

$$a_{X_2} + \frac{ik''}{2} a_{TT} = \frac{3i\gamma}{2kc^2} a^2 a^* - \frac{1}{2k} k_{X_2} a.$$

As an additional exercise, use appropriate transformations to put this equation into canonical form $q_\tau - iq_{\theta\theta} - 2iq^2 q^* = \Gamma(k_\tau/k)q$. What are Γ, τ, θ, q? For a reference see [52].

(vi) *The envelope equation for a wave instability.* Suppose the L in (2.39) depends on a stress parameter R in such a way that the wave solution

$u(x_j, t) \propto e^{i(k_j \cdot x_j - \sigma t)}$, $\sigma = \omega - i\nu$ grows or decays depending on whether $R \gtreqless R_c$. The parameters ν and ω are determined as functions of k_j and R by the complex algebraic equation $L(-i\omega + \nu, k_j, R) = 0$. The critical surface is the surface in k_j, R space where $\nu = 0$. The critical values of k_j and R are those values on this surface where R is minimum. This is the lowest value of the stress parameter for which wave-like solutions grow. Use the ideas of this section to show that the (slowly varying) envelope $A(x_j, t)$ of the growing wave is governed by the complex envelope equation

$$\frac{\partial A}{\partial t} + \sum_{j=1}^{3} \frac{\partial \omega}{\partial k_j} \frac{\partial A}{\partial x_j} - \frac{1}{2} \sum_{j,l} \left(\frac{\partial \nu}{\partial R} \frac{\partial^2 R}{\partial k_j \partial k_l} + i \frac{\partial^2 \omega}{\partial k_j \partial k_l} \right) \frac{\partial^2 A}{\partial x_j \partial x_l}$$
$$= \left(\frac{\partial \nu}{\partial R} - i \frac{\partial \omega}{\partial R} \right) R_c \chi A - (\beta_r + i\beta_i) A^2 A^*.$$

The original field is written $u(\mathbf{x}, t) = \varepsilon A(\mathbf{x}, t) \exp(i\mathbf{k}_c \cdot \mathbf{x} - i\omega(R_c)t) + (*) + O(\varepsilon^2)$, $R = R_c(1 + \varepsilon^2 \chi)$, and $\mathbf{k}_c$ is one of the set of values on the critical surface which correspond to $R = R_c$. (Often, there is a degeneracy in the most critical wavevector due to a symmetry in the original system; e.g., convection in an infinite horizontal layer has a rotational symmetry.) The coefficients in the envelope equation are estimated at $\mathbf{k}_c$. Similar comments to those made earlier in this section about the necessity to include the excitation of the mean by slow gradients of AA^* may still be relevant.

After you have read the following section on the Benjamin–Feir instability, show that the x independent solution (take only one spatial dimension)

$$A = \sqrt{\frac{R_c \chi}{\beta_r}} \exp\left(-\frac{i\beta_i R_c \chi t}{\beta_r} \right)$$

is unstable in the Benjamin–Feir sense if

$$\beta_i \gamma_i + \beta_r \gamma_r < 0$$

where

$$\gamma_i = \left(\frac{\partial^2 \omega}{\partial k^2} \right)_c, \qquad \gamma_r = \left(\frac{\partial \nu}{\partial R} \frac{\partial^2 R}{\partial k^2} \right)_c.$$

Further discussion can be found in [127].

2d. The Benjamin–Feir instability. Recall that the underlying wave field $u(x, t)$ is given by

$$u(x, t) = ae^{i(k_0 x - \omega_0 t)} + (*), \qquad \omega_0 = \omega_0(k_0),$$

where a is a function of $\xi = \varepsilon(x - \omega_0' t)$ and $T = \varepsilon^2 t$ and satisfies

$$a_T = i \frac{\omega_0''}{2} a_{\xi\xi} + i\beta a^2 a^*. \tag{2.52}$$

Write $a = Ae^{i\phi}$. It is natural to define the local wave number k as the x derivative of the total phase and the local frequency as the negative of the t

derivative of the total phase $\theta = k_0 x - \omega_0 t + \phi(\xi, T)$

$$k = k_0 + \varepsilon \phi_\xi, \qquad \omega = \omega_0 + \varepsilon \omega_0' \phi_\xi - \varepsilon^2 \phi_T.$$

Note the relation

$$k_t + \omega_x = -\omega_0' \varepsilon^2 \phi_{\xi\xi} + \varepsilon^3 \phi_{\xi T} + \varepsilon^2 \omega_0' \phi_{\xi\xi} - \varepsilon^3 \phi_{T\xi} = 0, \tag{2.53}$$

which expresses the conservation of the number of waves. We will write the change in wave number $\varepsilon \phi_\xi$ as εK. Now the imaginary part of (2.52) gives

$$\phi_T = \frac{\omega_0''}{2}\left(\frac{A_{\xi\xi}}{A} - K^2\right) + \beta A^2,$$

which when differentiated with respect to ξ gives

$$K_T + \omega_0'' K K_\xi = \beta \rho_\xi + \frac{\omega_0''}{2}\left(\frac{A_{\xi\xi}}{A}\right)_\xi \tag{2.54a}$$

where $\rho = A^2$. Equation (2.54a) is the relation for conservation of waves (2.53), since

$$\omega = \omega_0 + \varepsilon \omega_0' K + \varepsilon^2\left(\frac{1}{2}\omega_0'' K^2 - \beta A^2 - \frac{\omega_0''}{2}\frac{A_{\xi\xi}}{A}\right).$$

(Recall $\partial/\partial t = \varepsilon^2 \, \partial/\partial T - \varepsilon \omega_0' \, \partial/\partial \xi$, $\partial/\partial x = \varepsilon \, \partial/\partial \xi$.) On the other hand the real part of (2.52),

$$(A^2)_T + \omega_0''(A^2 K)_\xi = 0, \tag{2.54b}$$

is the equation for conservation of wave action.

Next, look at the monochromatic solution

$$A = A_0, \qquad \phi = \beta A_0^2 T + \text{constant}, \tag{2.55}$$

which means that

$$k = k_0, \qquad \omega = \omega_0 - \beta A_0^2 \varepsilon^2.$$

This is the Stokes wave. Test its linear stability by setting $A = A_0 + \tilde{A}$, $K = \tilde{K}$ and find from (2.54)

$$\tilde{K}_T = 2\beta A_0 \tilde{A}_\xi + \frac{\omega_0''}{2}\frac{\tilde{A}_{\xi\xi\xi}}{A_0},$$

$$\tilde{A}_T = -\frac{\omega_0''}{2} A_0 \tilde{K}_\xi,$$

or

$$\tilde{A}_{TT} = -\beta \omega_0'' A_0^2 \tilde{A}_{\xi\xi} - \frac{\omega_0''^2}{4}\tilde{A}_{\xi\xi\xi\xi}. \tag{2.56a}$$

Therefore, if $\tilde{A} \propto e^{ik\xi + \sigma T}$,

$$\sigma^2 = \beta \omega_0'' A_0^2 K^2 - \frac{\omega_0''^2}{4} K^4, \tag{2.56b}$$

and so if $\beta\omega_0'' > 0$, the solution (2.55) is always unstable to long waves in the range $0 < K^2 < 4\beta A_0^2/\omega_0''$. The maximum growth rate occurs when $K^2 = 2\beta A_0^2/\omega_0''$ and is equal to $\beta^2 A_0^2$. The reader should read reference [59] by Lake, Yuen, Rungaldier and Ferguson who investigate this instability experimentally for water waves. He should also read the original paper of Benjamin and Feir [51].

One can also understand the reason for the instability using the following argument. Suppose one has a monochromatic wave with constant envelope and frequency $\omega_0 - \beta A_0^2\varepsilon^2$ which is perturbed at a point P so that the amplitude at P is less than A_0. Then if $\beta > 0$, ω at P is greater than ω to the left of P. Hence $\omega_x > 0$ in this region and by conservation of waves $k_t < 0$. Hence k decreases and if $\omega_0'' > 0$, ω_0' also decreases. To the right of P, ω_0' increases. Hence the regions to the left and right of P continue to separate and the amplitude perturbation deepens. For $\beta\omega_0'' < 0$, the perturbation is confined and smooths out.

So, what happens in practice to a wave train of surface gravity waves which is generated at a source S with a paddle oscillating at an almost constant frequency ω? If $hk > 1.36$, $\beta\omega'' > 0$ (see Exercise 2c(2)(ii)) and a monochromatic wave train is unstable. Because of the way the problem is posed, it is best to look at the evolution of a in $X = \varepsilon^2 x$ as function of $T = \varepsilon(t - x/\omega')$. If the duration of the paddle motion is finite, then the resulting wave packet breaks into a series of special pulses, the solitons of the nonlinear Schrödinger equation, the formula for which was given in (2.51). If on the other hand the paddle motion continues indefinitely at a constant amplitude or in some periodic fashion, then the relevant initial-boundary value problem to solve is one in which the amplitude $a(X, T)$ is periodic in T. Therefore, as the profile evolves in X, one would expect to see the very same type of recurrence previously seen for the KdV equation. The only difference is that here the field is periodic in time and quasiperiodic in space. What happens, then, is that first the wave train breaks into a number of separate pulses but then these recollect at some later position and reproduce the initial (at $x = 0$ as function of time) conditions. This recurrence has indeed been observed and I refer the reader to the Yuen–Ferguson paper [60].

If $\beta\omega'' < 0$ and one makes a small long wave perturbation in the envelope, then this perturbation propagates according to the KdV equation. The reader can observe the linear portion of the lattice equation (1.3) in (2.56a). As an exercise, I will ask you to include the nonlinear terms and derive the relevant KdV equation.

The Benjamin–Feir (or, as it is sometimes called, the modulational) instability is widespread and plays an important role in various nonlinear wave phenomena. Simply put, if dispersion and nonlinearity act against each other, monochromatic wave trains do not wish to remain monochromatic. The sidebands of the carrier wave can draw on its energy via a resonance mechanism with the result that the envelope becomes modulated. In one space dimension, this envelope modulation continues to grow until the soliton shape is reached. At this point, nonlinearity and dispersion are in exact balance and no further distortion occurs.

In two dimensions, if the product of β with the dispersion tensor $\partial^2\omega/\partial k_r\,\partial k_s$, r, $s = 1$, 2 is a positive definite matrix, the focusing process is never halted and continues until the pulse achieves locally an infinite amplitude which it does in finite time. In the context of nonlinear optics, such filamentation has been observed and the form of these filaments is discussed in Zakharov and Synakh [61]. Consider $q(\mathbf{r}, t)$ satisfying

$$2iq_t + \nabla^2 q + \beta(qq^*)^\sigma q = 0$$

which has the motion invariants

$$N(q, q^*) = \int qq^*\,d\mathbf{x},$$

$$P(q, q^*) = \int \frac{i}{2}(q\nabla q^* - q^*\nabla q)\,d\mathbf{x},$$

$$H(q, q^*) = \frac{1}{2}\int \left(|\nabla q|^2 - \frac{\beta}{\sigma+1}|q|^{2\sigma+2}\right)d\mathbf{x}.$$

The spatial dimension of the problem is n and the strength of the nonlinearity is measured by σ. For $\sigma < 2/n$, one can show the solution $q(\mathbf{r}, t)$ exists for all time. In the case of interest, $n = 2$, $\sigma = 1$ so that $\sigma = 2/n$ and this value of σ is a critical value. In one space dimension, the critical value of σ is 2. Now, if $N(q(\mathbf{r}, 0), q^*(\mathbf{r}, 0)) < N_0$, a value calculated by inserting into $N(q, q^*)$ the spherically symmetric solution $q(\mathbf{r}, t) = e^{it/2}R(|\mathbf{r}|)$, with $R(|\mathbf{r}|)$ everywhere positive, $R'(0) = 0$, $R(\infty) = 0$ and satisfying $\nabla^2 R - R + \beta R^{2\sigma+1} = 0$, then again $q(\mathbf{r}, t)$ exists for all time as long as $q(\mathbf{r}, 0)$ obeys certain weak conditions $\int (|q|^2 + |\nabla q|^2)\,d\mathbf{x} < \infty$.

The reader should prove for himself that

$$\frac{d^2}{dt^2}\int r^2 qq^*\,d\mathbf{x} = 4H. \tag{2.57}$$

Note that if $q(\mathbf{r}, 0)$ is such that H, a constant of the motion, is negative (H is exactly zero when $q(\mathbf{r}, t) = e^{it/2}R(|\mathbf{r}|)$), then the intrinsically positive quantity $\int r^2 qq^*\,d\mathbf{r}$ becomes negative in finite time. This cannot happen and the conclusion is that $q(\mathbf{r}, t)$ must have developed a singularity at $|\mathbf{r}| = 0$ before this. It is the nature of this singularity which is discussed in reference [61]. The thesis is that near the blow up time $t = t_0$, for $n = 2$, the amplitude of $q(\mathbf{r}, t)$ has an axially symmetric shape proportional to $\lambda R(\lambda |\mathbf{r}|)$ with $\lambda(t) = (t_0 - t)^{-2/3}$. In order to accommodate the extra number density, the difference between the initial number density and that carried by the solution $\lambda R(\lambda |\mathbf{r}|)$, namely N_0, one has to add to this central spike, a shelf which is almost constant in $|\mathbf{r}|$ for large distances and then drops off suddenly at some, as yet, uncalculated value. These last comments are based on observations of numerical experiments and some recent theoretical work. The exact structure of the blow-up solution at the critical value of $\sigma = 1$ is still unknown.

It is also not known whether the Zakharov equations in two spatial dimensions,

$$a_T - i\varepsilon\nabla^2 a - i\varepsilon\rho A = 0,$$

$$\rho_{TT} - c^2\nabla^2\rho = \nabla^2(AA^*), \qquad \nabla^2 = \frac{\partial^2}{\partial x^2} + \frac{\partial^2}{\partial y^2},$$ (2.58)

blow up in finite time. For these, there is no relation equivalent to (2.57).

Finally, if, as in the case of deep water gravity waves, the relevant NLS equation has an indefinite dispersion matrix, as in

$$a_T - ia_{xx} + ia_{yy} - 2ia^2a^* = 0,$$

then the soliton which would form in the x direction will also destabilize to a y-dependent disturbance, but this breakdown is much less dramatic than the process described in the previous paragraph. The reason for the instability is that the underlying carrier wave feeds some of its energy to neighboring oblique sidebands which together with the original wave form a quartet resonance. (See Whitham's book [55] for a discussion of the quartet resonance mechanism suggested by Phillips.)

Comment. It is remarkable that the instability of the Stokes wave was not discovered until the experiment of Benjamin and Feir. (The reader should also consult M. J. Lighthill, Proc. Roy. Soc. A, 299, pp. 28–53.) The formal method of constructing nonsinusoidal periodic solutions had been proposed by Stokes in 1849 (G. G. Stokes, *On the theory of oscillatory waves*, Trans. Cambridge Phil. Soc., 8, pp. 441–455) and a proof of convergence of the series for sufficiently small slopes was given by T. Levi–Cevita (Math. Ann., 93, pp. 264–314) in 1925. For a good summary of the role of the NLS equation in describing the instability and for comparisons with experiments, the reader should consult reference [59].

Exercise 2d.

Include the quadratic nonlinear terms in (2.56a) and show

$$\tilde{A}_{TT} + \beta\omega_0''A_0^2\tilde{A}_{\xi\xi} = -\frac{\omega_0''^2}{4}\tilde{A}_{\xi\xi\xi\xi} + Q,$$

where

$$Q = \frac{\omega_0''^2}{2} A_0(\tilde{K}\tilde{K}_\xi)_\xi - \beta\omega_0''A_0(\tilde{A}\tilde{A}_\xi)_\xi - \frac{1}{A_0}(\tilde{A}\tilde{A}_T)_T - \omega_0''(\tilde{A}\tilde{K})_{\xi T}.$$

Note that if $\beta\omega_0'' < 0$ and we look for unidirectional solutions,

$$\tilde{A} = \tilde{A}(X = \xi - cT, \varepsilon T), \qquad c^2 = -\beta\omega_0''A_0^2,$$

$$\tilde{K} = \frac{-2\beta}{c} A_0\tilde{A}, \qquad Q = -2\beta\omega_0''A_0(\tilde{A}^2)_{XX}$$

and the nonlinear version of (2.56a) is the Boussinesq equation. Notice that solitons of the resulting KdV equation are only possible when $\tilde{A} < 0$. Since these represent local reductions in the intensity of what was a monochromatic wave, they are called dark solitons. Also show directly that if $\beta \omega_0'' < 0$, (2.52) has solutions

$$a = \rho \exp\left(i\phi + \frac{iV}{2\xi} - \frac{i(\gamma + V^2/4)\omega_0''}{2T}\right),$$

$$\rho^2 = \rho_0^2 - \alpha^2 \operatorname{sech}^2 \alpha\left(\xi - \frac{V\omega_0''}{2} T\right),$$

$$\phi_x = \frac{h}{\rho^2}, \qquad h^2 = \rho_0^4(\rho_0^2 - \alpha^2),$$

$$\gamma + \frac{V^2}{4} = 3\rho_0^2 - \alpha^2.$$

Note as $\alpha^2 \to \rho_0^2$, $h \to 0$

$$\rho \to \rho_0 \tanh \rho_0\left(\xi - \frac{V\omega_0''}{2} T\right), \qquad \phi = \phi_0.$$

2e. Whitham theory [55]. In the mid 1960's another ingenious theory dealing with the propagation of fully nonlinear, almost periodic wave trains was developed. The theory is associated principally with the name of Whitham, although some of the ideas were developed independently by Kruskal in his efforts to understand the wiggly region which occurred in connection with the solutions of the Fermi–Pasta–Ulam lattice. The idea is quite simple. Suppose there exists a stable 2π-periodic travelling wave solution $f(\Theta, A)$, $\Theta = kX - \omega T$, A a constant amplitude, arising as a constant of integration, to a nonlinear partial differential equation. Then one can develop a wider class of solutions which describe the slow modulation of this wave train by allowing the former constants, wavenumber $\Theta_X = k$, frequency $\Theta_T = -\omega$ and amplitude A to be slowly varying functions of position X and time T. Equations for k, ω and A as functions of $x = \varepsilon X$, $t = \varepsilon T$, $0 < \varepsilon \ll 1$ are found as follows.

First, from the fact that ω and k are derived from a potential, the phase Θ, one has

$$k_t + \omega_x = 0, \tag{2.59}$$

a conservation law expressing the conservation of wave number. Second, when one substitutes the solution ansatz into the partial differential equation, the leading order part is a nonlinear ordinary differential equation in Θ for $f(\Theta)$. We know this has periodic solutions because of our assumption that the original partial differential equation admits periodic travelling wave solutions. The imposition of a fixed periodicity (2π is usually chosen) gives an algebraic relation between ω, k and A, called the dispersion relation. (It is crucial to impose a fixed periodicity. If one allows it to be a slowly varying function, it is impossible to control the growth of the later iterates.) Because these parameters vary slowly, there are $O(\varepsilon)$ terms, containing first derivatives of k, ω and A with respect to x and t, left over in the partial differential equation. The condition that the next iterate, which satisfies a linear ordinary differential equation in Θ with coefficients depending on f and its derivatives, is also 2π-periodic, imposes a solvability condition on the equation. This condition is a first order partial differential equation in k, A and ω and expresses the conservation of the wave action. Thus we have three equations, one algebraic and two differential, for the three unknowns k, ω and A.

Now we have already seen in Section 2c that the weakly nonlinear envelope of a carrier wave also contains three parameters A, k and ω (see equations (2.49)–(2.53)) and they are connected via the NLS equation. Are the two descriptions equivalent, the latter being merely Whitham's theory in the small amplitude limit? The answer is no. Clearly the NLS equation being a small amplitude theory cannot include Whitham theory which is valid for arbitrary amplitudes. On the other hand, in Whitham theory the amplitude is determined algebraically by the nonlinear dispersion relation whereas in NLS, it has a life of its own as it were, and satisfies a partial differential equation. Why is it that Whitham theory does not relax to the NLS equation? The difficulty is that when the amplitude is finite, the solvability condition at order ε changes in the sense that the null space of the linear operator acting on the first iterate of the solution is only half the size (dimension one) of what it would be if A were small. This results in only one equation, the conservation of wave action, which is an equation for the phase of the wave and corresponds to the imaginary part of the NLS equation which gives the evolution of ϕ, the phase of the complex amplitude. The amplitude A is already fixed by the dispersion relation. What happens when A is small? In that case it turns out that the dispersion relation must be augmented in order to satisfy the extra solvability requirements. The extra terms contain derivatives of A and so what was originally an algebraic relation giving A as function of ω and k now becomes a differential equation for A. This equation corresponds to the amplitude part of the NLS equation. We now present a version of Whitham theory which is uniformly valid as the amplitude becomes small and which unifies it with the NLS equation. The ideas I use were motivated by the expansions used by Ablowitz and Benney [62] in their investigations of a multiphase Whitham theory. Indeed both Ablowitz and Chu and Mei [63] noted the area of potential breakdown of Whitham theory and identified the problem terms. Moreover, Whitham himself shows how to include these higher order dispersion terms working from the averaged Lagrangian formulation (see [55, p. 522]).

It is interesting to note that exactly the same types of difficulties occur in the macroscopic behavior of systems far from equilibrium which can be described in terms of an order parameter. Away from the phase transition, the amplitude of the order parameter is given algebraically by the modulus of the phase gradient (an analogous expression to the nonlinear dispersion relation or eikonal equation) whereas near critical values of the bifurcation parameter (which measures the stress—think of it as the Rayleigh number in the context of fluid convection or as the temperature in magnetism—imposed on the system), the amplitude becomes small and satisfies a partial differential equation. This equation is analogous to the NLS equation and, when combined with the corresponding equation for the phase of the order parameter, is known as the Ginzburg–Landau equation. The reader interested in these remarks can consult [64] for further discussion. We now illustrate these comments by using two concrete examples.

The choice of the first example is particularly simple as it admits solutions of the form $f(\Theta) = Ae^{i\Theta}$ and this makes all the computations explicit. Consider a complex scalar field $u(X, T)$ described by the equation

$$u_{TT} - u_{XX} + \omega_p^2(1 - \beta uu^*)u = 0. \tag{2.60}$$

We look for solutions with the form

$$u(X, T) = f(\Theta, A) + \varepsilon u_1 + \varepsilon^2 u_2 + \cdots, \tag{2.61}$$

where $f(\Theta) = Ae^{i\Theta}$ and $\Theta_X = k$, $\Theta_T = -\omega$ and A are functions of $x = \varepsilon X$, $t = \varepsilon T$. The derivatives $\partial/\partial T$, $\partial/\partial X$ can be written $-\omega \,\partial/\partial\Theta + \varepsilon \,\partial/\partial t$, $k \,\partial/\partial\Theta + \varepsilon \,\partial/\partial x$ respectively, and when substituted into (2.60) give

$$\left\{ (\omega^2 - k^2)\frac{\partial^2}{\partial\Theta^2} - i\varepsilon\left(2\omega\frac{\partial^2}{\partial t\,\partial\Theta} + \omega_t\frac{\partial}{\partial\Theta} + 2k\frac{\partial^2}{\partial x\,\partial\Theta} + k_x\frac{\partial}{\partial\Theta} \right) \right.$$
$$\left. + \varepsilon^2\left(\frac{\partial^2}{\partial t^2} - \frac{\partial^2}{\partial x^2} \right) + \omega_p^2(1 - \beta u_0 u_0^* - \varepsilon\beta(u_0 u_1^* + u_0^* u_1) - \varepsilon^2\beta(u_0 u_2^* + u_1 u_1^* + u_0^* u_2)) \right\},$$
$$(u_0 + \varepsilon u_1 + \varepsilon^2 u_2 + \cdots) = 0. \tag{2.62}$$

The order one balance in this equation is

$$(\omega^2 - k^2)\frac{\partial^2 u_0}{\partial\Theta^2} + \omega_p^2(1 - \beta u_0 u_0^*)u_0 = 0, \tag{2.63}$$

whose solution

$$u_0 = Ae^{i\nu\Theta}, \qquad \nu^2 = \frac{\omega_p^2(1 - \beta A^2)}{\omega^2 - k^2}$$

is only 2π-periodic with no smaller period when $\nu^2 = 1$ or

$$\omega^2 - k^2 = \omega_p^2(1 - \beta A^2). \tag{2.64}$$

If we had allowed the period to be an arbitrary function of x and t, then when we calculated $\partial u_0/\partial x$ there would be terms of the form $i\nu_x \Theta$ appearing at the next order and it would be impossible to find a u_1 with the same period as u_0. This is analogous to the first step in the WKB method, in which it is very important to choose the correct fast time scale.

Equation (2.64) is the dispersion relation giving A as function of ω and k. However, in anticipation of what is to come we will, following the formulation suggested by Ablowitz and Benney [62], consider (2.64) to be the leading order term in the amplitude expansion

$$\omega^2 - k^2 - \omega_p^2(1 - \beta A^2) = \varepsilon g^{(1)} + \varepsilon^2 g^{(2)} + \cdots. \tag{2.65}$$

Now (2.62) becomes

$$\left\{ \omega_p^2(1 - \beta A^2)\left(\frac{\partial^2}{\partial \Theta^2} + 1\right) - i\varepsilon \left(2\omega \frac{\partial^2}{\partial t\, \partial \Theta} + \omega_t \frac{\partial}{\partial \Theta} + 2k \frac{\partial^2}{\partial x\, \partial \Theta} + k_x \frac{\partial}{\partial \Theta}\right) \right.$$

$$+ (\varepsilon g^{(1)} + \varepsilon^2 g^{(2)})\frac{\partial^2}{\partial \Theta^2} + \varepsilon^2\left(\frac{\partial^2}{\partial t^2} - \frac{\partial^2}{\partial x^2}\right) - \varepsilon \omega_p^2 \beta (u_0 u_1^* + u_0^* u_1)$$

$$\left. - \varepsilon^2 \omega_p^2 \beta (u_0 u_2^* + u_0^* u_2 + u_1 u_1^*) \right\}$$

$$\cdot (u_0 + \varepsilon u_1 + \varepsilon^2 u_2 + \cdots) = 0.$$

The order one balance gives $u_0 = A e^{i\Theta}$. The next order yields

$$\omega_p^2(1 - \beta A^2)\left(\frac{\partial^2}{\partial \Theta^2} + 1\right)u_1 - \omega_p^2 \beta (u_0 u_1^* + u_0^* u_1)\mu_0$$

$$= \frac{i e^{i\Theta}}{A}\left((\omega A^2)_t + (k A^2)_x\right) + g^{(1)} A e^{i\Theta},$$

from which we obtain $g^{(1)} = 0$, $u_1 = 0$ and

$$(\omega A^2)_t + (k A^2)_x = 0. \tag{2.66}$$

The choice (2.66) is necessary as otherwise $u_1 \propto \Theta e^{i\Theta}$; the choice $g^{(1)} = 0$ is convenient and then $u_1 = 0$ as a consequence. At order ε^2,

$$\omega_p^2(1 - \beta A^2)\left(\frac{\partial^2}{\partial \Theta^2} + 1\right)u_2 - \omega_p^2 \beta (u_0 u_2^* + u_0^* u_2)u_0 = e^{i\Theta}\left(\frac{\partial^2}{\partial x^2} - \frac{\partial^2}{\partial t^2}\right)A + e^{i\Theta} g^{(2)} A.$$

$$\tag{2.67}$$

It is here that the subtlety enters. Notice that any solution u_2 of the form $e^{i\Theta}$ annihilates the first term on the left-hand side of (2.67). As long as the right-hand side is $e^{i\Theta}$ times a real quantity (because A and $(u_0 u_2^* + u_0^* u_2)$ are real), (2.67) is solvable. For example if the RHS were $G e^{i\Theta}$, then $u_2 = (-G/2\omega_p^2 A\beta)e^{i\Theta}$ which is 2π-periodic. Notice however that the asymptotic series (2.61) is not well ordered as $A \to 0$. Therefore this limit is nonuniform. To say things in another way, the null space of the operator acting on u_2 is one-dimensional and spanned by $i e^{i\Theta}$ if A is not small, but is two-dimensional and spanned by $1 e^{i\Theta}$, $i e^{i\Theta}$ when A is. (It is actually four-dimensional when one includes the conjugate fields $e^{-i\Theta}$.) In order to facilitate the limit taking process, we will treat $1 e^{i\Theta}$ as if it were secular and choose $g^{(2)}$ to make the RHS of (2.67) zero. Then,

$$\omega^2 - k^2 - \omega_p^2(1 - \beta A^2) = \frac{\varepsilon^2}{A}\left(\frac{\partial^2}{\partial t^2} - \frac{\partial^2}{\partial x^2}\right)A. \tag{2.68}$$

The evolution of ω, k and A is determined by (2.59), (2.66) and (2.68) and, in fact, in this simple case, the solution $u = A e^{i\Theta}$ is exact. The RHS of (2.68) is only important when $A = O(\varepsilon)$ and $\omega^2 - k^2 - \omega_p^2 = O(\varepsilon^2)$ whence (2.68) is a differential equation of A.

Let me reiterate: the term $(\partial^2/\partial t^2 - \partial^2/\partial x^2)A$ is not secular in (2.67) for finite A. It is useful, however, if one plans to take the small A limit, to treat it as if it were secular and incorporate it in the dispersion relation.

The small A, almost monochromatic limit is taken as follows. Let $A \to \varepsilon A$, and choose $\omega \simeq \omega_0$, $k \simeq k_0$, $\omega_0^2 - k_0^2 = \omega_p^2$ by writing

$$\Theta = k_0 X - \omega_0 T + \phi(x, t). \tag{2.69}$$

Then $\omega = \omega_0 - \varepsilon\phi_t$, $k = k_0 + \varepsilon\phi_x$ and the conservation of waves equation is

$$((\omega_0 - \varepsilon\phi_t)A^2)_t + ((k_0 + \varepsilon\phi_x)A^2)_x = 0 \tag{2.70}$$

and the dispersion relation is

$$\varepsilon(-2\omega_0\phi_t - 2k_0\phi_x) + \varepsilon^2(\phi_t^2 - \phi_x^2) + \beta\omega_p^2\varepsilon^2 A^2 = \frac{\varepsilon^2}{A}(A_{tt} - A_{xx}). \tag{2.71}$$

The order one term in (2.70) and the order ε in (2.71) tell us that both A^2 and ϕ are functions of x and t in the combination $\xi = x - \omega_0't$, $\omega_0' = k_0/\omega_0$. Let $\tau = \varepsilon t$ and then (2.71) and (2.70) become respectively

$$\phi_\tau = \frac{1 - \omega_0'^2}{2\omega_0}\left(\frac{A_{\xi\xi}}{A} - \phi_\xi^2\right) + \frac{\beta\omega_p^2 A^2}{2\omega_0} \tag{2.72}$$

and

$$A_\tau = \frac{\omega_0'^2 - 1}{2\omega_0}(2A_\xi\phi_\xi + A\phi_{\xi\xi}), \tag{2.73}$$

or if $a = Ae^{i\phi}$

$$a_\tau = \frac{i\omega_0''}{2}a_{\xi\xi} + i\beta\frac{\omega_p^2}{2\omega_0}a^2a^*, \qquad \omega_0'' = \frac{1 - \omega_0'^2}{\omega_0}. \tag{2.74}$$

From Exercise 2c(2)(i), we saw that (2.74) is the NLS for (2.60). So, the term which Whitham theory neglects, namely $(\varepsilon^2/A)(A_{tt} - A_{xx})$, once included, gives back the NLS equation and corresponds to the first term in the RHS of (2.72).

Let me now also indicate what happens when one cannot solve for $f(\Theta)$ explicitly. I will use Whitham's original model

$$u_{TT} - u_{XX} + F(u) = 0 \tag{2.75}$$

where I take F to be odd in u and for small u, given by $u - \gamma u^3$. The reader may work out the NLS description as an exercise. If

$$u \simeq \varepsilon a(x - \omega_0't, \varepsilon t)e^{i(k_0 X - \omega_0 T)} + (*) + \cdots$$

with $\omega_0^2 - k_0^2 = 1$, $\omega_0' = k_0/\omega_0$, the group velocity, then

$$a_\tau = \frac{i\omega_0''}{2}a_{\xi\xi} + \frac{3i\gamma}{2\omega_0}a^2a^*, \tag{2.76}$$

where $\tau = \varepsilon t = \varepsilon^2 T$ and $\xi = x - \omega_0't = \varepsilon(X - \omega_0'T)$.

Next, applying Whitham theory to (2.75), we introduce

$$\Theta = \frac{\theta(x, t)}{\varepsilon}, \quad x = \varepsilon X, \quad t = \varepsilon T, \tag{2.77a}$$

$$\omega^2 - k^2 = g + \varepsilon g^{(1)} + \varepsilon^2 g^{(2)} + \cdots, \tag{2.77b}$$

$$u = f + \varepsilon u^{(1)} + \varepsilon^2 u^{(2)} + \cdots, \tag{2.77c}$$

and find

$$g \frac{d^2 f}{d\Theta^2} + F(f) = 0, \tag{2.78a}$$

$$g \frac{d^2 u^{(1)}}{d\Theta^2} + F'(f) u^{(1)} = R_1, \tag{2.78b}$$

$$g \frac{d^2 u^{(2)}}{d\Theta^2} + F'(f) u^{(2)} = R_2, \tag{2.78c}$$

where

$$R_1 = -g^{(1)} \frac{\partial^2 f}{\partial \Theta^2} + 2\omega \frac{\partial^2 f}{\partial \Theta \partial t} + 2k \frac{\partial^2 f}{\partial \Theta \partial x} + (\omega_t + k_x) \frac{\partial f}{\partial \Theta}$$

and

$$R_2 = -g^{(2)} \frac{\partial^2 f}{\partial \Theta^2} + 2\omega \frac{\partial^2 u^{(1)}}{\partial \Theta \partial t} + 2k \frac{\partial^2 u^{(1)}}{\partial \Theta \partial x} + (\omega_t + k_x) \frac{\partial u^{(1)}}{\partial \Theta}$$

$$- \frac{F''(f)}{2} u^{(1)2} - \left(\frac{\partial^2}{\partial t^2} - \frac{\partial^2}{\partial x^2} \right) f.$$

Multiplying (2.78a) by $f_\Theta (df/d\Theta)$ and integrating, we obtain

$$\tfrac{1}{2} g f_\Theta^2 + V(f) = E, \quad V' = F \tag{2.79}$$

from which we obtain

$$\sqrt{\frac{g}{2}} \int_{f_-}^{f} \frac{df}{\sqrt{E - V(f)}} = \Theta, \quad V(f_-) = E. \tag{2.80}$$

Without loss of generality, we take f to be odd in Θ. The dispersion relation is then

$$\sqrt{\frac{g}{2}} \int_{f_-}^{f_+} \frac{df}{\sqrt{E - V(f)}} = \pi, \quad V(f_+) = E \tag{2.81}$$

giving E as a function of g and vice versa.

The next task is to develop conditions on R_1, R_2 which allow $u^{(1)}$, $u^{(2)}$ to be 2π-periodic in Θ. Consider

$$g \frac{d^2 v}{d\Theta^2} + F'(f) v = R, \tag{2.82}$$

and write this as a system

$$\frac{dV}{d\Theta} = EV + \frac{1}{g}\begin{pmatrix} 0 \\ R \end{pmatrix}, \qquad (2.83)$$

where

$$V = \begin{pmatrix} v \\ v_\Theta \end{pmatrix}, \qquad E = \begin{pmatrix} 0 & 1 \\ -F\dfrac{1}{g} & 0 \end{pmatrix}. \qquad (2.84)$$

The matrix E is 2π-periodic and Floquet theory obtains. In particular, if U, a row vector, is a 2π-periodic solution of

$$\frac{dU}{d\Theta} = -UE, \qquad (2.85)$$

then, if V is to be 2π-periodic, it is necessary that (the proof is straightforward: just multiply (2.83) by the row vector U and integrate)

$$\int_0^{2\pi} U \cdot \begin{pmatrix} 0 \\ R \end{pmatrix} d\Theta = 0. \qquad (2.86)$$

Let Φ be a fundamental solution matrix of $dV/d\Theta = EV$; then $\Phi(\Theta+2\pi)$ is also a fundamental solution matrix (it satisfies the equation) and there exists a matrix $M = e^{2\pi R}$ independent of Θ and called the monodromy matrix, such that $\Phi(\Theta+2\pi) = \Phi(\Theta)M$. The eigenvalues of M are called the Floquet multipliers. If unity is a repeated eigenvalue then at least one of the associated eigenvectors gives rise to a 2π-periodic solution of the homogeneous equation (2.83). The rows of the inverse $\Phi^{-1}(\Theta)$ satisfy (2.85).

In the present case, we can construct $\Phi(\Theta)$ explicitly by noting that both $v_1 = f_\Theta$ and $v_2 = f_g + \Theta f_\Theta/2g$ satisfy the homogeneous version of (2.82). Here subscripts denote partial derivatives;

$$\Phi(\Theta) = \begin{pmatrix} f_\Theta & f_g + \dfrac{1}{2g}\Theta f_\Theta \\ f_{\Theta\Theta} & f_{g\Theta} + \dfrac{1}{2g}f_\Theta + \dfrac{1}{2g}\Theta f_{\Theta\Theta} \end{pmatrix},$$

$$\det \Phi(\Theta) = \frac{1}{g}E_g \quad \text{(differentiate (2.79))}$$

and

$$\Phi^{-1}(\Theta) = \frac{g}{E_g}\begin{pmatrix} f_{g\Theta} + \dfrac{1}{2g}f_\Theta + \dfrac{1}{2g}\Theta f_{\Theta\Theta} & -f_g - \dfrac{1}{2g}\Theta f_\Theta \\ -f_{\Theta\Theta} & f_\Theta \end{pmatrix}.$$

In this case M has a double eigenvalue unity and is equal to

$$\begin{pmatrix} 1 & \dfrac{\pi}{g} \\ 0 & 1 \end{pmatrix}.$$

The necessary condition (2.86) for V to be 2π-periodic is

$$\int_0^{2\pi} f_{\Theta} \cdot R \, d\Theta = 0. \tag{2.87}$$

Next, solve (2.83) by variation of parameters and obtain

$$v(\Theta) = \left(c_1 - \frac{1}{E_g} \int_0^{\Theta} \left(f_g + \frac{1}{2g} \Theta f_{\Theta}\right) \cdot R \, d\Theta\right) v_1(\Theta) + \left(c_2 + \frac{1}{E_g} \int_0^{\Theta} f_{\Theta} \cdot R d\Theta\right) v_2(\Theta). \tag{2.88}$$

Demanding that $v(\Theta)$ be 2π-periodic gives us that

$$c_2 = \frac{g}{\pi E_g} \int_0^{2\pi} \left(f_g + \frac{1}{2g} \Theta f_{\Theta}\right) R \, d\Theta. \tag{2.89}$$

Therefore, provided that with this choice of c_2, the asymptotic series (2.77c) remains well ordered (this means that the ratios of successive terms remain order ε for all values of the parameters), condition (2.87) is both necessary and sufficient. When applied to (2.78b), it gives

$$\frac{\partial}{\partial t} \omega \int_0^{2\pi} f_{\Theta}^2 \, d\Theta + \frac{\partial}{\partial x} k \int_0^{2\pi} f_{\Theta}^2 \, d\Theta = 0, \tag{2.90}$$

the conservation of wave action and analogue of (2.66). Because $f(\Theta)$ is odd, the term involving $g^{(1)}$ is zero.

What happens as the parameter E in f becomes small? A little calculation shows that in the small amplitude limit

$$f(\Theta) = A \sin \Theta + \tfrac{1}{32}\gamma A^3 \sin 3\Theta + \cdots, \tag{2.91a}$$

$$g = 1 - \tfrac{3}{4} \gamma A^2 + \cdots, \tag{2.91b}$$

$$E = \tfrac{1}{2}A^2 + \cdots. \tag{2.91c}$$

Now look at c_2, and in particular calculate its size when one takes for R the term $-(\partial^2/\partial t^2 - \partial^2/\partial x^2)f$ in R_2. We note that $f_g + \Theta f_{\Theta}/2g \approx A_g \sin \Theta$ is order A^{-1} since $g_{A^2} = O(1)$. Also $E_g = E_{A^2} \cdot (A^2)g = -\tfrac{2}{3}\gamma + \cdots$. Thus c_2 is order $1/A$ and the resulting solution $u^{(2)}$ is also. This means that (2.77c) is nonuniform when $A = O(\varepsilon)$.

In order to capture this behavior in another way, we do not use c_2 to impose 2π-periodicity on $v(\Theta)$ but rather use $g^{(2)}$. We demand

$$\int \left(f_g + \frac{\Theta f_{\Theta}}{2g}\right)\left(-g^{(2)} f_{\Theta\Theta} - \left(\frac{\partial^2}{\partial t^2} - \frac{\partial^2}{\partial x^2}\right)f\right) d\Theta = 0$$

(there is no contribution of comparable amplitude from the $u^{(1)}$ term) and find, using (2.91),

$$g^{(2)} = + \frac{1}{A} (A_{tt} - A_{xx}). \tag{2.92}$$

Now, the dispersion relation

$$\omega^2 - k^2 = 1 - \tfrac{3}{4}\gamma A^2 + \frac{\varepsilon^2}{A}(A_{tt} - A_{xx}) + \cdots \qquad (2.93)$$

has exactly the same form as in (2.68) and the NLS limit is recovered from (2.90) and (2.93) exactly as before.

Even though Whitham's theory does not contain the full NLS equation, it does contain enough of it to exhibit the Benjamin–Feir instability. With A^2 given by the dispersion relation, equations (2.59) and (2.66) are a first order system for k and ω. If elliptic, the Cauchy initial value problem is ill posed in the sense that any perturbation grows exponentially in time with the shortest waves growing the fastest. If we write Θ as θ/ε and solve (2.59) by setting $k = \theta_x$, $\omega = -\theta_t$, then (2.66) is $(-\theta_t A^2)_t + (\theta_x A^2)_x = 0$ where A^2 is a function of x and t through the combination $l = \theta_t^2 - \theta_x^2$. We find

$$\theta_{tt}\left(-A^2 - 2\theta_t^2 \frac{dA^2}{dl}\right) + 2\theta_{xt}\theta_x\theta_t \frac{dA^2}{dl} + \theta_{xx}\left(A^2 - 2\theta_x^2 \frac{dA^2}{dl}\right) = 0. \qquad (2.94)$$

The system is elliptic and unstable if

$$A^2\left(A^2 + 2l\frac{dA^2}{dl}\right) < 0, \qquad (2.95)$$

where

$$A^2 = -\frac{1}{\beta}\left(\frac{\omega^2 - k^2}{\omega_p^2} - 1\right).$$

For A small, this occurs when $\beta > 0$ or $\beta\omega_0'' > 0$ since $\omega_0'' = (\omega_0^2 - k_0^2)/\omega_0^3$ is always positive. Because Whitham's theory does not include the $A_{\xi\xi\xi\xi}$ term in (2.56), it does not predict the sideband of the underlying carrier wave which grows the fastest nor does it predict a finite band of unstable waves. On the other hand, however, its great power is that it is not a small amplitude theory.

We end this section by asking about some of the physical consequences of nonlinear dispersion relations. In particular we address the question: Can nonlinear waves tunnel? Imagine that we have initiated a train of waves of amplitude A with frequency ω into an environment where ω_p is slowly varying. Then $k(x)$ is given by

$$k^2 = \omega^2 - \omega_p^2(1 - \beta A^2).$$

Suppose initially $\omega^2 > \omega_p^2$ but that ω_p^2 grows (the pendula in the Scott model get shorter). No linear wave can propagate past the point where $\omega_p = \omega$. But if $\beta > 0$, then the point at which $\omega_p^2(1 - A^2) = \omega^2$ is further into the medium than the caustic for linear waves. Does this mean that nonlinear wave trains can tunnel without loss through barriers where linear waves are severely attenuated. The answer is both yes and no. It is no because we have just shown, at least in the weak nonlinear limit, that whenever $\beta > 0$, the monochromatic nonlinear wave train is unstable. However, what happens is that it breaks up into pulses, the solitons of the NLS equation, which do exhibit lossless tunnelling properties. The reader is referred to reference [52] for more details.

For large amplitudes, (2.95) becomes $3A^2 - 2/\beta < 0$. Thus even though $\beta > 0$, for waves of sufficiently large amplitude, stability is regained.

Many interesting questions remain open when the Whitham equations are hyperbolic, although some very beautiful work has been done. See Whitham's book [55], the beautiful and elegant paper of Flaschka, Forest and McLaughlin [65] (they discovered that the Riemann invariants are simply the spectrum of the periodic KdV problem) and the intriguing work of Lax and Levermore [66]. Principal among these questions are questions related to the long time behavior of the system. Do shocks evolve? If they do, what do they mean? How are new phases introduced? (The work of Lax and Levermore suggests that the (x, t) plane is split into regions in each of which a one phase or two or higher phase flow obtains.) What other long time behavior can there be? My Ph.D. student Krishna [67] integrated the Whitham equations numerically for a broad class of initial conditions in which the envelope of the fast oscillations decayed at infinity, in which circumstance inverse scattering theory obtains. He found that, in the case of one phase solutions, various members of two of the three families of characteristics became parallel and straight as time advanced. The equations then take on a parabolic rather than hyperbolic form. The lines in (x, t) space along which the characteristics become parallel correspond to the velocities of the solitary waves which emerge from doing the initial value problem.

Exercise 2e.
By setting $A \to 2\varepsilon A$, solve the corrected Whitham equations

$$\omega^2 - k^2 = 1 - 3\gamma A^2 \varepsilon^2 + \varepsilon^2 \frac{A_{tt} - A_{xx}}{A}, \quad k_t + \omega_x = 0, \quad (\omega A^2)_t + (kA^2)_x = 0$$

iteratively by setting $\omega = \omega_0 + \varepsilon \omega_1 + \varepsilon^2 \omega_2$, $k = k_0 + \varepsilon K$. Show that ω_0, k_0 should be constant (otherwise shocks occur), $\omega_0^2 = k_0^2 + 1$, $\omega_1 = k_0 K/\omega_0$, and both A^2 and K are functions of $\xi = x - (k_0/\omega_0)t$ and $\tau = \varepsilon t$. Introducing $K = \phi_\xi$ gives

$$\phi_\tau = \frac{\omega_0''}{2}\left(-K^2 + \frac{A_{\xi\xi}}{A}\right) + \frac{3\gamma}{2\omega_0}A^2, \quad (A^2)_\tau + \omega_0''(A^2)_\xi = 0,$$

which are the components of NLS for $a = Ae^{i\phi}$.

2f. Other canonical equations. There are many other equations of importance in physics which have the soliton property. The list of references contains many review articles, special issues and conference proceedings and the reader is invited to peruse this volume in search of his favorite equation. Nevertheless, there are two more equations which deserve special mention because they turn up so often.

One is the sine-Gordon equation. It can be found in nonlinear optics where it models the propagation of pulses in resonant media; in condensed matter physics and magnetization where it is used to model charge density waves in periodic pinning potentials and spin waves in liquid helium 3; in superconductivity where it describes propagation in a Josephson transmission line; in statistical mechanics, where it arises in the description of the critical region of Ising-like models. The reason for its ubiquity is that so many of these systems share an equivalence to a dynamical system which is derivable from a Lagrangian with a kinetic energy proportional to $\frac{1}{2}\int \phi_t^2 \, dx$ and a potential energy arising from two sources. The first source of potential energy is an elastic force which, in the continuum limit, may be modelled by

$$\frac{1}{2}c^2 \int \phi_x^2 \, dx;$$

the second source of potential is due to the influence of some background lattice structure and often the best model of this potential is

$$\omega_p^2 \int (1-\cos \phi) \, dx.$$

It is not hard to see that the Euler equation arising from this Lagrangian is

$$\phi_{tt} - c^2 \phi_{xx} + \omega_p^2 \sin \phi = 0. \tag{2.96}$$

For a discussion and a list of the many contexts in which (2.96) is found, I refer you to references [68], [69], [19], [71].

The second equation of note is not a single equation but a set of equations describing the resonant interaction of three waves. This set may be written [53]

$$\frac{\partial A_j}{\partial t} + \mathbf{c}_j \cdot \nabla A_j = \theta_{jkl} A_k^* A_l^* \tag{2.97}$$

for (j, k, l) cycled over $(1, 2, 3)$. Here $A_j(x, t)$ is the slowly varying envelope of a weakly nonlinear wave train with underlying carrier wave $\exp i(k_j x - \omega_j t)$, $\omega_j = \omega(k_j)$; $\mathbf{c}_j$ is the linear group velocity $\nabla \omega_j$. The left-hand side occurs for the very same reason that, at order ε (where ε is the bandwidth of waves about k), the envelope $a(X, T_1)$ of the NLS satisfies $a_{T_1} + \omega' a_X = 0$. The right-hand side arises from the quadratic nonlinear terms X, which in a weakly nonlinear system represent the strongest nonlinearity, which allow the resonance

$$\mathbf{k}_1 + \mathbf{k}_2 + \mathbf{k}_3 = 0, \qquad \omega(\mathbf{k}_1) + \omega(\mathbf{k}_2) + \omega(\mathbf{k}_3) = 0. \tag{2.98}$$

In the wave-particle picture, the resonance relation (2.98) represents the conservation of momentum and energy in a three-particle collision.

Considering its fundamental structure, it is not too surprising that equation (2.97) is of great importance in areas of physics where wave processes dominate. It describes, in essence, how energy is redistributed throughout the spectrum by nonlinear resonant interactions. It arises in plasma waves, in atmospheric waves and in ocean waves. (Sometimes the linear dispersion relation $\omega(\mathbf{k})$ does not allow (2.98) to be satisfied for any triad $\mathbf{k}_1$, $\mathbf{k}_2$, $\mathbf{k}_3$. In these cases, energy redistribution takes place at the next order through a resonant quartet interaction, for example, surface gravity waves.)

For a beautiful and detailed review of (2.97) and its integrability properties (it is exactly solvable for certain choices of θ_{jkl}), the best source is the article of Kaup, Riemann and Bers [72]. (See also [74] for its solution.)

CHAPTER 3

Soliton Equation Families and Solution Methods

3a.   Introduction. In this chapter, we follow four themes. In Sections 3b, c we use the traditional and somewhat pedestrian methods for identifying families of integrable nonlinear partial differential equations. In Sections 3d, e, f we discuss in some depth the inverse scattering method as used to solve the initial value problem on $-\infty < x < \infty$ for members of the KdV family. In Section 3g, we introduce methods for dealing with perturbations and discuss in considerable detail the problem of a solitary wave travelling in a channel of decreasing depth. This problem has many nontrivial features and has proved to be a very difficult nut to crack. Finally in Section 3h, we discuss methods for finding multisoliton, rational and multiphase periodic solutions to soliton equations.

3b.   The Korteweg–deVries equation family. The first goal of this section is to show how to derive soliton equation families. The second is to show in what sense each flow is Hamiltonian. The third goal is to introduce some important asymptotic expansions, the coefficients of which are constants of the motion and proportional to the Hamiltonians.

In order to derive the equation families, we will use the algorithm:
  (i) start with a chosen eigenvalue problem in x in which the coefficients depend on another variable t;
  (ii) assume a general form for how the eigenfunction will change as the coefficients evolve in t;
  (iii) write down the solvability condition for these two equations and determine which evolution equations are compatible with this condition.

Let us illustrate the method by identifying the KdV family. As we have discussed in Chapter 1, the appropriate eigenvalue problem is the Schrödinger equation

$$v_{xx} + (\lambda + q(x, t))v = 0, \tag{3.1}$$

which can also be written as a first order system

$$v_{1x} + i\zeta v_1 = q(x, t)v_2,$$
$$v_{2x} - i\zeta v_2 = -v_1, \qquad v_2 = v, \quad \lambda = \zeta^2. \tag{3.2}$$

We write the t dependence of v as

$$v_t = A(\lambda; q, q_x, \ldots)v - B(\lambda; q, q_x, \ldots)v_x. \tag{3.3}$$

Dependence on higher derivatives in v is incorporated since A and B depend on λ. Cross-differentiate (3.1) and (3.3), and find on setting the coefficients of v

61

and v_x to zero that

$$A = \tfrac{1}{2}B_x + \text{const.} \tag{3.4}$$

and

$$q_t = -\tfrac{1}{2}MB - 2\lambda NB, \tag{3.5}$$

where

$$M = D^3 + 4qD + 2q_x, \qquad N = D = \frac{\partial}{\partial x}. \tag{3.6}$$

Now we come to the second half of part (iii) in the algorithm. Recall that the goal is to find functions $B(\lambda; q, q_x, q_{xx}, \ldots)$ such that the solvability condition of (3.1) and (3.3), namely (3.5), is an evolution equation in the form

$$q_t = q_t(q, q_x, q_{xx}, \ldots).$$

A particularly simple class of such functions B are the polynomials

$$B = B_0 \lambda^n + \cdots + B_n \tag{3.7}$$

because, from (3.5), $B_{0x} = 0$ and, by comparing successive powers of λ, each B_{k+1} is given explicitly (up to a constant) in terms of its next lowest neighbor B_k. Without loss of generality, we take $B_0 = -1$ and find

$$NB_{k+1} = -\tfrac{1}{4}MB_k, \qquad k = 0, 1, \ldots, n-1. \tag{3.8}$$

Then, from the coefficient of λ^0,

$$q_t = -\tfrac{1}{2}MB_n = 2NB_{n+1} \tag{3.9}$$

where we define B_{n+1} by

$$NB_{n+1} = -\tfrac{1}{4}MB_n. \tag{3.10}$$

We can write

$$M = -4NL, \tag{3.11}$$

where

$$L = -\frac{1}{4}D^2 - q + \frac{1}{2}\int_{\infty}^{x} dx\, q_x \tag{3.12}$$

is nonlocal.

Let us calculate a few. We note that $B_{n+1} = LB_n + \text{const.}$; for the moment we ignore the constant. Then

$$B_0 = -1,$$

$$B_1 = LB_0 = \tfrac{1}{2}q,$$

$$B_2 = \tfrac{1}{2}Lq = -\tfrac{1}{8}(q_{xx} + 3q^2), \tag{3.13}$$

$$B_3 = \tfrac{1}{2}L^2 q = \tfrac{1}{32}(q_{xxxx} + 5q_x^2 + 10qq_{xx} + 10q^3),$$

where we have assumed q and all its derivatives are zero at the point we call $x = \infty$. The equations are respectively

$$q_{t_1} = q_x, \tag{3.14}$$

$$q_{t_3} = -\tfrac{1}{4}(q_{xx} + 3q^2)_x, \tag{3.15}$$

$$q_{t_5} = \tfrac{1}{16}(q_{xxxx} + 5q_x^2 + 10qq_{xx} + 10q^3)_x, \tag{3.16}$$

where we have labeled the $(2n+1)$th flow with linear dispersion $\omega(k) = -2(k/2)^{2n+1}$, $n = 0, 1, 2, \ldots$ with the time coordinate t_{2n+1}. The constant in B_k, $k < n$, merely adds a linear combination of the kth flow to the nth flow Setting all the constants to zero gives what we term the *pure* family. The constant in (3.4) is a different matter. Its choice allows us to normalize the eigenfunction $v(x, t, \zeta)$ in a convenient manner. This we do in Section 3c.

Each flow has a Hamiltonian structure. I next want to introduce a most important result, namely that each flow in the family has Hamiltonian form. It will be proved in Section 3f that one can write

$$2B_{n+1} = L^n q = \frac{\delta H_{2n+1}}{\delta q} \tag{3.17}$$

as the variational derivative of a functional of q. The latter is defined as

$$\sum_0^\infty \left(-\frac{d}{dx}\right)^n \frac{\partial \tilde{H}(q^{(n)})}{\partial q^{(n)}},$$

that is,

$$\lim_{\varepsilon \to 0} \frac{H[q + \varepsilon \eta] - H[q]}{\varepsilon} = \int_{-\infty}^\infty \frac{\delta H}{\delta q} \eta \, dx, \tag{3.18}$$

where $q^{(n)} = D^n q$, $D = d/dx$ and

$$H[q] = \int_{-\infty}^\infty \tilde{H}(q, q_x, q_{xx}, \ldots) \, dx. \tag{3.19}$$

For example,

$$H_1 = \frac{1}{2} \int_{-\infty}^\infty q^2 \, dx, \qquad H_3 = \frac{1}{8} \int_{-\infty}^\infty (q_x^2 - 2q^3) \, dx,$$

and their variational derivatives are respectively q, $-\tfrac{1}{4}(q_{xx} + 3q^2)$. Thus the $n = 0$ flow corresponding to a pure translation of the initial profile may be written as

$$q_{t_1} = q_x = \frac{\partial}{\partial x} \frac{\delta H_1}{\delta q}; \tag{3.20}$$

the KdV flow, $n = 1$, may be written

$$q_{t_3} = \frac{\partial}{\partial x} \frac{\delta H_3}{\delta q}, \tag{3.21}$$

and the $(2n+1)$th flow of the family

$$q_{t_{2n+1}} = \frac{\partial}{\partial x} \frac{\delta H_{2n+1}}{\delta q} = N \frac{\delta H_{2n+1}}{\delta q}. \tag{3.22}$$

Alternatively, using (3.11), we have

$$q_{t_{2n+1}} = M\left(-\frac{1}{4}\frac{\delta H_{2n-1}}{\delta q}\right), \tag{3.23}$$

since $\delta H_{2n+1}/\delta q = 2B_{n+1} = 2LB_n$ and $2NLB_n = -\frac{1}{2}MB_n = -\frac{1}{4}M(\delta H_{2n-1}/\delta q)$. Notice that each flow has the form

$$q_t = J\nabla H, \tag{3.24}$$

where J is a skew-symmetric operator, ∇ is a gradient and H is a Hamiltonian functional.

The remarkable fact is that every one of the functionals H_{2n+1}, $n = 0, 1, \ldots$, each of which generates a flow, is a constant of the motion for every other flow and they commute with each other under the natural Poisson bracket (Gardner [13])

$$\{F, G\} = \int_{-\infty}^{\infty} \frac{\delta G}{\delta q}\frac{\partial}{\partial x}\frac{\delta F}{\delta q}\, dx. \tag{3.25}$$

This bracket is natural in that if we think of q as evolving under the t_{2k+1} flow, then the rate of change of $I[q]$, a functional of q, i.e,

$$I[q] = \int_{-\infty}^{\infty} \tilde{I}(q, q_x, q_{xx}, \ldots)\, dx,$$

is given by

$$\frac{dI}{dt_{2k+1}} = \{H_{2k+1}, I\}. \tag{3.26}$$

What all this means is that q can be considered a function of an infinite number of independent variables x, t_{2k+1}, $k = 0, 1, \ldots$, and $(\partial/\partial t_{2j+1})(\partial q/\partial t_{2k+1}) = (\partial/\partial t_{2k+1})(\partial q/\partial t_{2j+1})$. The notion that solutions of soliton equations are functions of infinitely many variables is very important. As a consequence of the commutability property, if one begins with $q(x, \mathbf{0})$ and allows q to flow under $(KdV)_{2j+1}$ for a time t_{2j+1} and then under $(KdV)_{2k+1}$ for t_{2k+1}, one obtains the same result as would have been obtained had one allowed the evolutions to occur in the opposite order.

Observe that each flow can be given two Hamiltonian structures. We may derive $q_{t_{2n+1}}$ from the Hamiltonian functional H_{2n+1} using the Poisson operator N, or from H_{2n-1} using M. This dual structure is present in all integrable systems and it allows one to form a ladder process for identifying all the integrable flows in the family in exactly the same manner as we have already done and as (3.10) suggests. Given the two independent symplectic structures M, N and the Hamiltonian H_1 which generates translation given by NH_1; then the next flow is $-\frac{1}{4}MH_1$ (KdV) which is written NH_3; we continue, $-\frac{1}{4}MH_3$ is the next flow and so on.

I will now show how these functionals are obtained but delay till Section 3f the proof that they are constants of the motion. Let me remark before doing this that if we are dealing with flows under periodic boundary conditions in x, then the integral in (3.25) is taken over the period.

An asymptotic expansion for $v(x, \zeta)$ as $\zeta \to \infty$. The algorithm for constructing the constants of motion is as follows. Look for an asymptotic expansion as $\zeta \to \infty$ for the eigenfunction $v(x, t; \zeta)$ of (3.1) by setting $v = e^{i\zeta x + \Phi}$ whence

$$-2i\zeta\Phi_x = q + \Phi_{xx} + \Phi_x^2. \tag{3.27}$$

Solve iteratively for Φ_x,

$$\Phi_x = \sum_1^\infty \frac{1}{(2i\zeta)^n} R_n$$

and find $R_1 = -q$ and the recursion relation

$$R_{n+1} = -R_{nx} - \sum_{k=1}^{n-1} R_k R_{n-k}, \qquad n \geq 1. \tag{3.28}$$

The first five members of the list are $R_1 = -q$, $R_2 = q_x$, $R_3 = -q_{xx} - q^2$, $R_4 = q_{xxx} + 4qq_x$, $R_5 = -(q_{xx} + 3q^2)_{xx} + q_x^2 - 2q^3$. In particular, for reasons given in Sections 3e, f, we will be interested in the quantities

$$ve^{-i\zeta x} \sim 1 - \frac{1}{2i\zeta} \int_{-\infty}^x q(y)\, dy, \tag{3.29a}$$

and

$$\lim_{x \to -\infty} \left(v_2 - \frac{1}{2i\zeta}v_1\right)e^{-i\zeta x} \sim -\sum_1^\infty \frac{1}{(2i\zeta)^n} \int_{-\infty}^\infty R_n\, dx, \tag{3.29b}$$

as the former tells us how to find $q(x)$ from $v(x, \zeta)e^{-i\zeta x}$ and the latter turns out to be a constant of the motion (constant with respect to all times t_{2n+1}). Recall that $v_2 = v$ and $v_1 = -v_x + i\zeta v$ from (3.2). As an exercise I will also ask you to relate this result to that obtained using the Miura–Gardner transform (Section 1d). We will show that

$$H_{2n+1} = \frac{4i}{(2i)^{2n+3}} \int_{-\infty}^\infty R_{2n+3}\, dx, \qquad n = 0, 1, 2, \ldots. \tag{3.30}$$

Observe that R_1 is *not* a member of this family and that $\int_{-\infty}^\infty R_{2n}\, dx = 0$, $n = 1, \ldots$. Indeed the functional $-\frac{1}{2}H_{-1} = \int_{-\infty}^\infty q\, dx$ has a zero Poisson bracket with any other functional (put $F = H_{-1}$ in (3.25)); it is called a Casimir functional.

Exercises 3b.

1. Using the definition (3.25), show that this Poisson bracket satisfies the Jacobi identity $\{\{F, G\}, H\} + \{\{H, F\}, G\} + \{\{G, H\}, F\} = 0$.

2. Recall that in (1.18) we defined $w + 3/\varepsilon^2 = (6i/\varepsilon)v_x/v$ and found $v_{xx} + (q + 1/4\varepsilon^2)v = 0$. Identify ζ with $-1/2\varepsilon$, and show that $w = -12i\zeta((v_x/v) - i\zeta) = -12i\zeta\Phi_x$. Therefore the *conserved densities* of the KdV

equation introduced in Chapter 1 are $\int_{-\infty}^{\infty} R_n \, dx$ and are proportional to the Hamiltonians. The members of the infinite KdV family share the same conserved quantities.

3. At first sight, this exercise may not look meaningful. Take my word for it, it is! We shall meet the ideas again in Chapter 4. Let

$$q(x, t_3, \dots) = 2 \frac{\partial^2}{\partial x^2} \ln \tau(x, t_3, t_5, \dots)$$

and show up to $O(1/\zeta^5)$ that

$$e^{\Phi} = \exp\left(\ln \tau\left(x - \frac{1}{i\zeta}, t_3 - \frac{1}{3i\zeta^3}, t_5 - \frac{1}{5i\zeta^5}, \dots\right) - \ln \tau(x, t_3, t_5, \dots)\right),$$

whence v is asymptotic to

$$v(x, t) \sim e^{i\zeta x + i\zeta^3 t_3 + i\zeta^5 t_5} \frac{\tau(x - 1/i\zeta, t_3 - 1/3i\zeta^3, \dots)}{\tau(x, t_3, \dots)}.$$

Hint: you will need to write

$$\int_{\infty}^{x} R_3 \, dx \quad \text{as} \quad -\frac{1}{3} \int_{\infty}^{x} (q_{xx} + 3q^2) - \int_{\infty}^{x} \frac{2}{3} q_{xx}$$

and then use the fact that

$$\int_{\infty}^{x} \int_{\infty}^{x} q_{t_3} = 2 \frac{\partial}{\partial t_3} \ln \tau.$$

4. Show that by using (3.1), (3.3) with $A = \frac{1}{2}B_x$, $B = -\lambda + \frac{1}{2}q$, is

$$v_t = -v_{xxx} - \frac{3}{2}qv_x - \frac{3}{4}q_x v.$$

5. Do the following formal calculations. Let

$$L = D + a_1 D^{-1} + a_2 D^{-2} + \cdots,$$

where $D = \partial/\partial x$ and D^{-1} is the formal integral operator $\int^x dx$. Define $B_n = (L^n)_+$ by which notation we mean the differential operator part of L^n. For example $(L)_+ = D$, $(L^2)_+ = D^2 + 2a_1$, $(L^3)_+ = D^3 + 3a_1 D + 3Da_1 + 3a_2$. Consider the hierarchy of equations

$$v_{x_n} = B_n v, \quad n = 2, 3, 4, 5, \dots.$$

Their solvability condition is

$$(B_n)_{x_m} - (B_m)_{x_n} + [B_n, B_m] = 0. \tag{A}$$

Case (i). Suppose the coefficients a_1, a_2 are independent of x_2 in which case one can set $v \to e^{\lambda x_2} v$ and replace $v_{x_2} = B_2 v$ by $\lambda v = B_2 v$. Show that (A) then gives $a_2 = -\frac{1}{2}a_{1x}$, and

$$(2a_1)_{x_3} - \frac{3}{2}(2a_1)(2a_1)_x - \frac{1}{4}(2a_1)_{xxx} = 0.$$

Note that if we set $x_3 = -t$, $v_{x_3} = B_3 v$ is precisely the form of $v_t = Bv$ in the last exercise.

Case (ii). Let a_1, a_2 be independent of x_3. Then

$$a_2 = -\frac{1}{2} a_{1x} + \frac{1}{2} \int^x a_{1x_2} + C, \qquad -\frac{3}{2} a_{1x_2x_2} - \frac{1}{2} a_{1xxxx} - 6(a_1 a_{1x})_x = 0.$$

Let $a_1 = \frac{1}{2}(q - c^2/2)$ and find

$$q_{x_2x_2} - c^2 q_{xx} = -\frac{1}{3} q_{xxxx} - 2(qq_x)_x,$$

the Boussinesq equation (2.26).

Case (iii). Finally, let a_1, a_2 depend on x, x_2 and x_3. Now you will find

$$(2a_{1x_3} - \frac{1}{2} a_{1xxx} - 6a_1 a_{1x})_x - \frac{3}{2} a_{1x_2x_2} = 0$$

which, after making the appropriate transformations, is the KP equation (2.27).

Remark 1. The system $v_{x_n} = B_n v$, $n = 2, 3, \ldots$, gives rise to the KP hierarchy of equations. The interested reader should consult reference [39] for a discussion of the algebraic structure and the τ-function for the family.

Remark 2. For case (i), the KdV equation, the initial-boundary value problem is solved by focusing one's attention on the eigenvalue problem $\lambda v = B_2 v$ and using the equation $v_{x_3} = B_3 v$ to calculate the x_3 dependence of the scattering data. For case (ii), the Boussinesq equation, the eigenvalue problem is $\lambda v = B_3 v$, a third order system. The x_2 evolution of its scattering data is found from $v_{x_2} = B_2 v$. For case (iii), things are not that simple. There is no eigenvalue problem at all! One must solve the two-dimensional scattering problem $v_{x_2} = v_{xx} + q(x, x_2; x_3)v$. The reader is referred to [124] which is the proceedings of the 1983 meeting in Kiev. In particular, he should note the articles and references of Ablowitz, Fokas and Zakharov, Manakov.

3c. The AKNS hierarchy and its properties [23].

This time we start out with a different eigenvalue problem, a generalization of the one introduced by Zakharov and Shabat [21] for studying the NLS equation,

$$V_x = PV = \begin{pmatrix} -i\zeta & q(x, t) \\ r(x, t) & i\zeta \end{pmatrix} V, \qquad V = \begin{pmatrix} v_1 \\ v_2 \end{pmatrix}. \tag{3.31}$$

Let us determine the system of evolution equations which can be solved exactly via (3.31). If q and r evolve with some parameter t, then let

$$V_t = QV, \qquad Q = \begin{pmatrix} h & e \\ f & -h \end{pmatrix}, \tag{3.32}$$

and the solvability condition for (3.31) and (3.32) is

$$P_t - Q_x + [P, Q] = 0, \tag{3.33}$$

where $[P, Q]$ is the matrix commutator $PQ - QP$. In component form, (3.33) is

$$h_x = qf - re, \quad e_x + 2i\zeta e = q_t - 2hq, \quad f_x - 2i\zeta f = r_t + 2hr. \tag{3.34}$$

We look for polynomial solutions for Q of degree n (we will therefore designate this Q as $Q^{(n)}$ and its corresponding time variable as t_n). As an aid to the computations it is convenient to introduce the matrices

$$H = \begin{pmatrix} 1 & 0 \\ 0 & -1 \end{pmatrix}, \quad E = \begin{pmatrix} 0 & 1 \\ 0 & 0 \end{pmatrix}, \quad F = \begin{pmatrix} 0 & 0 \\ 1 & 0 \end{pmatrix} \tag{3.35}$$

which obey the commutation relations for the algebra $sl(2)$

$$[H, E] = 2E, \quad [H, F] = -2F, \quad [E, F] = H. \tag{3.36}$$

For the moment, the algebra is just a computational convenience; later on in Chapter 5 we will show that it is central to the whole theory.

We look for solutions to (3.33) in the form

$$Q^{(n)} = -iH\zeta^n + Q_1\zeta^{n-1} + \cdots + Q_n, \tag{3.37}$$

where $P = Q^{(1)} = -iH\zeta + qE + rF$, $Q_k = h_kH + e_kE + f_kF$. At ζ^{n-k+1}, the components e_k, f_k, of E and F in Q_k are determined; at ζ^{n-k}, the diagonal component h_k is found. The first few are

$$Q_0 = -iH,$$

$$Q_1 = \begin{pmatrix} 0 & q \\ r & 0 \end{pmatrix},$$

$$Q_2 = \begin{pmatrix} \dfrac{-i}{2}qr & \dfrac{i}{2}q_x \\[2mm] \dfrac{-i}{2}r_x & \dfrac{i}{2}qr \end{pmatrix},$$

$$Q_3 = \begin{pmatrix} -\tfrac{1}{4}(qr_x - rq_x) & -\tfrac{1}{4}(q_{xx} - 2q^2r) \\[2mm] -\tfrac{1}{4}(r_{xx} - 2qr^2) & \tfrac{1}{4}(qr_x - rq_x) \end{pmatrix}. \tag{3.38}$$

The equation

$$P_{t_n} - Q_{nx} + \left[\begin{pmatrix} 0 & q \\ r & 0 \end{pmatrix}, Q_n \right] = 0 \tag{3.39}$$

is then the evolution equation for q and r. The flows corresponding to $n = 0, 1, 2, 3$ are respectively

$$q_{t_0} = -2iq, \quad r_{t_0} = 2ir \tag{3.40a}$$

with

$$Q^{(0)} = -iH; \tag{3.40b}$$

$$q_{t_1} = q_x, \qquad r_{t_1} = r_x \tag{3.41a}$$

with

$$Q^{(1)} = -iH\zeta + qE + rF; \tag{3.41b}$$

$$q_{t_2} = \frac{i}{2}(q_{xx} - 2q^2 r), \qquad r_{t_2} = -\frac{i}{2}(r_{xx} - 2qr^2) \tag{3.42a}$$

with

$$Q^{(2)} = -iH\zeta^2 + \zeta(qE + rF) - \frac{iqr}{2}H + \frac{i}{2}q_x E - \frac{i}{2}r_x F; \tag{3.42b}$$

$$q_{t_3} = -\tfrac{1}{4}(q_{xxx} - 6qrq_x), \qquad r_{t_3} = -\tfrac{1}{4}(r_{xxx} - 6qrr_x) \tag{3.43a}$$

with

$$Q^{(3)} = -iH\zeta^3 + \zeta^2(qE + rF) + \zeta\left(-\frac{iqr}{2}H + \frac{i}{2}q_x E - \frac{i}{2}r_x F\right)$$

$$-\tfrac{1}{4}(qr_x - rq_x)H - \tfrac{1}{4}(q_{xx} - 2q^2 r)E - \tfrac{1}{4}(r_{xx} - 2qr^2)F. \tag{3.43b}$$

The zeroth (t_0) flow corresponds to a scaling of the coordinates in which all products qr are constant; the next (t_1) flow is translation; the second (t_2), with $r = \pm q^*$, a consistent approximation, is NLS; the third (t_3) flow is, with $r = \pm q$, again a consistent approximation, the MKdV equation. Notice that the strings $Q^{(k)}$ for all flows t_k, $k \leq n$ are congruent in the sense that $Q^{(k+1)} = \zeta Q^{(k)} + Q_{k+1}$, $k = 0, 1, \ldots, (n-1)$.

The following results, which are not easy to prove in the present formalism, hold.

(i) All the equations $P_{t_n} = Q_x^{(n)} + [Q^{(n)}, P]$ commute; i.e. $P_{t_n t_m} = P_{t_m t_n}$.

(ii) All these equations are Hamiltonian; there are certain functionals $\{F_n\}_0^\infty$, such that

$$q_{t_n} = \frac{\delta F_n}{\delta r}, \qquad r_{t_n} = -\frac{\delta F_n}{\delta q}; \tag{3.44}$$

e.g. for $n = 2$, $F_2 = -(i/2)\int_{-\infty}^{\infty}(q_x r_x + q^2 r^2)\, dx$.

(iii) The F_n are integrals of certain polynomials $\tilde{F}_n$ in q, r and their x derivatives and there are relations of the form

$$\frac{\partial \tilde{F}_n}{\partial t_j} = \frac{\partial G_{nj}}{\partial x}, \qquad n = 0, 1, 2, \ldots, \tag{3.45}$$

called *conservation laws*. Here $\tilde{F}_n$ is called a *conserved density* of the jth flow and G_{nj} is called the *flux*. The $\tilde{F}_n$'s, as in the case of the KdV family are usually found from asymptotic expansions in ζ of certain functionals. In Chapter 5, I will give you formulae, local in q, r and their derivatives, for the G_{nj}.

At this point I want to make several remarks and observations. First, we will derive all these results from a single starting point in Chapter 5. Second, observe that Hamilton's equations have the more familiar form $Z^{\cdot} = J\nabla H$

where $Z = \binom{q}{r}$ and J is $\left(\begin{smallmatrix} 0 & 1 \\ -1 & 0 \end{smallmatrix}\right)$. In these systems q and r are conjugate variables and the two-form which is preserved is $\int_{-\infty}^{\infty} \delta r \wedge \delta q \, dx$, where by the wedge product $\delta r \wedge \delta q$ I mean $\delta_1 r \, \delta_2 q - \delta_2 r \, \delta_1 q$ where δ_1 and δ_2 are independent variations. Third, in this formulation, the independent variable x has been given a special role; for example, all the coefficients in Q_r are x derivatives. Note, however, that in Q_3 the coefficient of E could be written as a t_2 derivative $(i/2)q_{t_2}$. Also, the conservation laws are all written in the form

$$\frac{\partial}{\partial t_j} (\text{conserved density}) = \frac{\partial}{\partial x} (\text{flux}).$$

There are in fact many more relations which take the form

$$\frac{\partial}{\partial t_j} F_{kl} = \frac{\partial}{\partial t_k} F_{lj}.$$

The reader might say at this stage; "x is special because of the fact that the range of integration in defining F_n is all x, whereas all the time evolutions are local in t." True, but remember that $\delta F_n / \delta q$ is really only a symbol for $\sum_0^\infty (-d/dx)^r (\partial F/\partial q^{(r)})$ where $q^{(r)}$ is $\partial^r q / \partial x^r$, and all these terms are purely local. Fourth, and this remark will lead us towards a new way of looking at things, we must have that

$$Q_{t_j}^{(k)} - Q_{t_k}^{(j)} + [Q^{(k)}, Q^{(j)}] = 0 \tag{3.46}$$

as the compatibility condition for all equations

$$V_{t_k} = Q^{(k)} V. \tag{3.47}$$

Now, recall that $Q^{(j)}$ is a polynomial in ζ of degree j. Let us divide (3.46) by ζ^j and take the limit of the resulting equation as $j \to \infty$. If we call

$$Q = \lim_{j \to \infty} \frac{Q^{(j)}}{\zeta^j} = \sum_0^\infty \frac{Q_r}{\zeta^r}, \tag{3.48}$$

then (3.46) becomes ($\lim_{j \to \infty} (Q_{t_j}^{(k)}/\zeta^j) = 0$; think of ζ as being large),

$$Q_{t_k} = [Q^{(k)}, Q]. \tag{3.49}$$

Now the equation for all the flows has a much more algebraic structure and is in Lax form (although here Q is not an operator in d/dx but rather a Laurent series in ζ). It admits the solution

$$Q = V \tilde{Q}_0 V^{-1} \tag{3.50}$$

where $\tilde{Q}_0$ is a fixed arbitrary matrix, independent of t_k and V satisfies (3.47). Often we will take the normalization of V to be such that $\tilde{Q}_0 = -iH$; in other circumstances it might be written as $Q(0)$, the value of Q at some fixed values of x and t_k. (Recall, in all we do, x and t_1 are interchangeable.)

Exercises 3c.

1. By direct computation, show that $q_{t_2 t_3} = q_{t_3 t_2}$.

2. Look for a solution for Q in the form $Q^{(-1)} = (1/\zeta)Q_{-1}$. Can you obtain from this the sine-Gordon equation

$$u_{xt_{-1}} = \sin u \quad ?$$

(Hint: take $-q = r = u_x/2$; what happens if you take $q = r = u_x/2$?)

3. Notice that when one makes restrictions like $r = q$, they are only compatible with certain of the flows. For example, if $r = \pm q$ initially, then the even flows (i.e. t_{2n}) will destroy this relation. However all the odd flows (i.e., t_{2n+1}) will preserve it. The opposite happens for $r = \pm q^*$.

4. Find that linear combination of the pure flows $Q^{(1)}$, $Q^{(3)}$ and $Q^{(-1)}$ which is appropriate for the evolution equations

a) $q_t = 7q_x + 4(q_{xxx} + 6q^2 q_x)$,

b) $u_{xt} = \sin u + u_{xxxx} + \frac{3}{2}u_x^2 u_{xx}$.

By this, I mean, find $Q = \alpha Q^{(1)} + \beta Q^{(3)} + \gamma Q^{(-1)}$ such that the integrability condition of $V_t = QV$ and (3.31) with $r = -q$ is (a) and (b).

5. By allowing the parameter ζ a time dependence $\zeta_t = \sum_0 \alpha_r \zeta^r$, one can include x dependent coefficients in the equations. Show that by taking $Q = Q^{(2)} - ix\alpha H$ (the term $i\zeta_t$, here taken to be $i\alpha$, α real, is added to the LHS of the first equation in (3.34)), the corresponding evolution equations are

$$q_t = \frac{i}{2}(q_{xx} - 2q^2 r) - 2i\alpha x q, \qquad r_t = -\frac{i}{2}(r_{xx} - 2qr^2) + 2i\alpha x r.$$

This allows us to study the effects of density gradients in contexts where the NLS equation obtains. The effect of these influences is to make the discrete spectrum (which was introduced in Chapter 1 and which will receive more attention in the following section), which up to now has always been a constant of the motion, move in a prescribed way. (See [75].)

6. By starting from the eigenvalue problem [76]

$$V_x = \begin{pmatrix} i\beta_1\zeta & p & q \\ \pm p^* & i\beta_2\zeta & r \\ \pm q^* & \pm r^* & i\beta_3\zeta \end{pmatrix} V, \qquad \sum_{j=1}^{3} \beta_j = 0,$$

show how to choose V_t such that the resulting equations are the three wave interaction equations in one space dimension, $p_t + c_1 p_x = d_1 q^* r^*$ plus two other equations obtained by cyclic permutation of p, q and r.

7. For the general $(n \times n)$ systems, you might want to consult references [76] and [77].

8. By taking the eigenvalue problem [78], [79]

$$V_x = \cdot \begin{pmatrix} -i\zeta^2 & \zeta q \\ \zeta r & i\zeta^2 \end{pmatrix} V$$

show how to choose V_t so as to obtain the derivative nonlinear Schrödinger equation (DNLS) and the massive Thirring model.

Remark. We will see that the set of equations which arise from this eigen-value problem are sisters under the skin to the AKNS hierarchy. See [38].

3d. The direct transform for the Schrödinger equation, or scattering on the infinite line [12], [80]. For reasons given in Chapter 1, it seems natural to associate the Schrödinger equation

$$v_{xx} + (\zeta^2 + q(x, t))v = 0 \qquad (3.51)$$

with solutions of the KdV equation. In Section 3b we found that if $q(x, t)$ evolves in the parameter t according to

$$q_t + 6qq_x + q_{xxx} = 0, \qquad (3.52)$$

then $B = -4\zeta^2 + 2q$, $A = q_x + C$, C a free constant, and

$$v_t = (q_x + C)v + (4\zeta^2 - 2q)v_x. \qquad (3.53)$$

(See (3.15), and its corresponding $B = B^{(1)}$. Note that we have written $t_3 = 4t$.) "So what?" you might ask. "How does this help?" The transformation from $q(x, t)$ to $v(x, t; \zeta)$ does not lead to an equation for $v(x, t; \zeta)$ which is any easier to solve. It does not, for example, linearize the KdV equation in the way a similar transformation does for Burgers' equation (Exercise 1d). Fortunately, the group (Gardner, Greene, Kruskal, Miura) who discovered the equation were familiar with quantum physics, and once Schrödinger's equation popped up it seemed natural to compute its *scattering data*. Furthermore, as I have told you in Chapter 1, it turns out that the map from the potential $q(x, t)$ to the scattering data (or a subset thereof) is precisely the right transformation to make in order to render the infinite dimensional, coupled mechanical system (3.52) separable and solvable.

Scattering theory, the main ideas. The word scattering connotes time, a "before" and an "after" and brings to mind a situation in which a given entering pulse or wave is (in one dimension) partially transmitted and partially reflected by some inhomogeneity represented by the potential $q(x, t)$. The first point to make is that *the time connoted by the word scattering and the parameter t (time) in the KdV equation have nothing to do with each other.* We will call the former variable τ and for the moment keep the latter time fixed. Consider, for example, the Scott model of a continuum of pendula hanging from a torsion wire, and imagine either that gravity disappears or the lengths of the pendula get very long as $x \to \pm\infty$. Then (2.28) can be written

$$u_{\tau\tau} - c^2 u_{xx} + \omega_p^2 u = 0, \qquad (3.54)$$

where the coefficients c^2 and ω_p^2 are functions of x with the property that $c^2 \to c_0^2$, $\omega_p^2 \to 0$ as $x \to \pm\infty$. For simplicity let us take $c^2 = 1$ for all x. Equation (3.54) would represent the propagation of waves in a string which is encased in an elastic medium over a portion of its length. In any event, because ω_p^2 is independent of τ, one can seek solutions to (3.54) in the form $u(x, \tau) = \int_{-\infty}^{\infty} v(x, \zeta)e^{-i\zeta\tau} d\zeta$, whence $v(x, \zeta)$ satisfies (3.51) with $-\omega_p^2(x) = q(x, t)$. Imagine that the t in $q(x, t)$ represents some parameter by which it, i.e. $q(x, t)$, can be changed continuously. But again I stress that as far as scattering theory is concerned, it is considered a constant.

Since $q(x)$ goes to zero at $x \to \pm\infty$, the asymptotic behavior in x of $v(x, \zeta)$ is given by a linear combination of $e^{\pm i\zeta x}$ with coefficients that can depend on ζ. The $e^{\pm i\zeta x}$ solutions will represent in $u(x, \tau)$ a function of $x \pm \tau$ and therefore a wave moving to the left (right). Consequently, if a Dirac delta function pulse is sent in towards the potential from $x = +\infty$, part of it will reflect and thus the solution $v(x, \zeta)$ at $x = +\infty$ will look like

$$v_\infty(x, \zeta) = e^{-i\zeta x} + R(\zeta)e^{i\zeta x}, \tag{3.55}$$

where $R(\zeta)$ is called the reflection coefficient, and part of it will be transmitted and so at $x = -\infty$, $v(x, \zeta)$ will look like

$$v_{-\infty}(x, \zeta) = T(\zeta)e^{-i\zeta x}, \tag{3.56}$$

where $T(\zeta)$ is called the transmission coefficient. It follows both from the theory of ordinary differential equations (we will say why shortly) and from intuitive considerations that

$$|R|^2 + |T|^2 = 1. \tag{3.57}$$

Now in the sine-Gordon model we have chosen a $q(x) = -\omega_p^2$ which is always negative. However, in the context of quantum physics, the potential $-q(x)$ in which the electron travels can be negative at some locations, and solutions of (3.51) which are other than wavelike are also possible. These are the so-called bound states and they arise from values of the energy $\lambda = E = \zeta^2$ which are negative whence ζ is purely imaginary. Whereas the wavelike solutions are admissible for all real values of ζ (all positive E), there are only a discrete and finite number of eigenvalues $\zeta_k (= i\eta_k)$ for which the corresponding eigenfunction is square integrable over $(-\infty, \infty)$. One can see this as follows. Imagine a solution which, as $x \to -\infty$, looks like $v_{-\infty}(x, \zeta) = e^{-i\zeta x}$ which decays exponentially there for $\zeta = i\eta$, $\eta > 0$. In general after interaction with the potential, one expects this solution to have both components $e^{-i\zeta x}$ and $e^{i\zeta x}$ as $x \to +\infty$. Whereas the latter behaviour is admissible ($e^{-\eta x}$, $\eta > 0$) the former is clearly not because the corresponding solution would not be square integrable. Therefore ζ has to be very special in order that the asymptotic behavior of $v(x, \zeta)$ as $x \to +\infty$ only contain the $e^{i\zeta x}$ term.

It is for this reason that it is convenient to normalize the solutions of (3.51) that we study as follows. For ζ real, we define $\phi(x, \zeta)$, $\psi(x, \zeta)$ to be those solutions with asymptotic behaviors

$$\phi(x, \zeta) \sim e^{-i\zeta x}, \qquad x \to -\infty \tag{3.58a}$$

and

$$\psi(x, \zeta) \sim e^{i\zeta x}, \qquad x \to +\infty. \tag{3.58b}$$

We define two pairs of linearly independent solutions ϕ, $\bar\phi$ and ψ, $\bar\psi$; $\bar\phi$ has asymptotic behavior $e^{i\zeta x}$ as $x \to -\infty$ and $\bar\psi \sim e^{-i\zeta x}$ as $x \to +\infty$. For $q(x)$ real, the functions $\bar{\phi}(x, \zeta) = \phi(x, -\zeta) = \phi^*(x, \zeta)$ and $\bar{\psi}(x, \zeta) = \psi(x, -\zeta) = \psi^*(x, \zeta)$ for ζ real. The asterisk denotes the complex conjugate. These two sets of linearly independent solutions are related,

$$\phi(x, \zeta) = a(\zeta)\psi(x, -\zeta) + b(\zeta)\psi(x, \zeta), \tag{3.59}$$

where from the symmetry conditions expressed above it is easy to show that $a^*(\zeta) = a(-\zeta)$ and $b^*(\zeta) = b(-\zeta)$. Also, the second order equation (3.51) has no first order term (if expressed as a system (3.2), the coefficient matrix is trace-free) and therefore the Wronskians of $(\phi, \bar{\phi})$ and $(\psi, \bar{\psi})$ are constants in x. Since $W(\phi, \bar{\phi}) = \phi\bar{\phi}_x - \phi_x\bar{\phi} = 2i\zeta$ and $W(\bar{\psi}, \psi) = 2i\zeta$, we must have (carry out this calculation!)

$$aa^* - bb^* = 1. \tag{3.60}$$

Now compare (3.59) and (3.55), (3.56). In order to get "one" in front of $e^{-i\zeta x}$ as $x \to +\infty$, divide (3.59) by $a(\zeta)$. Then it is clear that

$$T(\zeta) = \frac{1}{a(\zeta)}, \qquad R(\zeta) = \frac{b(\zeta)}{a(\zeta)}, \tag{3.61}$$

and that (3.60) is simply (3.57).

Remark. The "linear" limit. It is worth observing at this time how the reflection coefficient $R(\zeta)$ is related to the ordinary Fourier transform

$$\hat{q}(k) = \frac{1}{2\pi} \int_{-\infty}^{\infty} q(x)e^{-ikx}\,dx$$

of $q(x)$. To see this, use the formulation (3.2), to write the pair of integral equations for $\psi_1 = -\psi_x + i\zeta\psi$ and $\psi_2 = \psi$, where $\psi(x, \zeta)$ is a solution of (3.1) with asymptotic behavior (3.58b),

$$\psi_1 e^{i\zeta x} = \int_{\infty}^{x} q(y)\psi_2 e^{i\zeta y}\,dy,$$

$$\psi_2 e^{-i\zeta x} = 1 - \int_{\infty}^{x} \psi_1 e^{-i\zeta y}\,dy.$$

Now, solve these equations iteratively, successively obtaining terms which contain higher and higher powers of q. In fact, it is just these expansions which are used to prove the assertions (3.64) below. In this calculation, however, we think of q as being small and only keep terms up to linear in q. We find, then, that $\psi_1 e^{i\zeta x}$ is approximated by $\int_{\infty}^{x} q(y)e^{2i\zeta y}\,dy$. But from the inverse of the relation (3.59),

$$\psi(x, \zeta) = a(\zeta)\phi(x, -\zeta) - b(-\zeta)\phi(x, \zeta), \tag{3.62}$$

we note that $\psi_1 e^{i\zeta x} = (-\psi_x + i\zeta\psi)e^{i\zeta x}$ tends to $-2i\zeta b(-\zeta)$ as $x \to -\infty$. Thus,

$$-2i\zeta b(-\zeta) = \int_{\infty}^{-\infty} q(y)e^{2i\zeta y}\,dy$$

and therefore

$$\hat{q}(k) = -\frac{ik}{2\pi} b\left(\frac{k}{2}\right).$$

Also, in this limit $a \equiv 1$ and so the scattering data is the Fourier transform.

I want to emphasize that this calculation is just a paradigm, a useful way in which to see that the inverse scattering transform is a nonlinear analogue of the Fourier transforms. In reality, one has to be very careful about what one means by the "small q" limit. There are many potentials which are small everywhere in x but which, nevertheless, have bound states. For example, in Exercise 3d(3), the amplitude Q can be as small as we like; yet there is always one bound state. In this sense, the limit $a \to 1$ for all ζ, Im $\zeta > 0$, is nonuniform.

The scattering data and its properties. So far we have dealt only with solutions where ζ is real. It turns out that if a real $q(x)$ obeys the condition

$$\int_{-\infty}^{\infty} (1 + x^2) |q(x)| \, dx < \infty, \tag{3.63}$$

then the following results hold.

(i) $\phi(x, \zeta)e^{i\zeta x}$, $\psi(x, \zeta)e^{-i\zeta x}$ and $a(\zeta)$ (which is defined by $(1/2i\zeta)W(\phi, \psi))$ are analytic functions of ζ for Im $\zeta > 0$. $\qquad$ (3.64a)

(ii) $\phi(x, \zeta)e^{i\zeta x}$, $\psi(x, \zeta)e^{-i\zeta x}$ and their derivatives with respect to ζ exist and are continuous everywhere in Im $\zeta \geq 0$ including $\zeta = 0$. $\quad$ (3.64b)

The reader should consult Deift and Trubowitz [80] for details. The mathematical problem in inverse scattering is to characterize scattering data that correspond to a specified class of potentials. Originally Faddeev had studied the class $q(x)$ for which $\int_{-\infty}^{\infty} (1 + |x|) |q(x)| \, dx < \infty$, but Deift and Trubowitz pointed out that in order to control the ζ derivative of $\psi(x, \zeta)e^{-i\zeta x}$ at $\zeta = 0$, the slightly stronger condition (3.63) is necessary.

Next, consider (3.29) with $v = \psi$. Recall that $\psi_1 = -\psi_x + i\zeta\psi$ and, using (3.62), we see that the quantity $(\psi + (1/2i\zeta)(\psi_x - i\zeta\psi))e^{-i\zeta x}$ tends to $a(\zeta)$ as $x \to -\infty$. Hence (3.29) gives us that

$$\ln a(\zeta) \sim -\sum_1 \frac{1}{(2i\zeta)^n} \int_{-\infty}^{\infty} R_n \, dx. \tag{3.65}$$

In writing (3.65), we have assumed, of course, that all the integrals on the right-hand side still exist, a much stronger condition than (3.63) imposes. If we only have (3.63), then all we can infer is that $a(\zeta) \to 1$ and $\zeta \to \infty$, Im $\zeta \geq 0$. For the moment, however, this is all we need. We know that $a(\zeta)$ is analytic for Im $\zeta > 0$, exists when Im $\zeta = 0$ and tends to unity as $\zeta \to \infty$, Im $\zeta \geq 0$. Therefore, it can only have a finite number, N, of zeros in Im $\zeta > 0$ as otherwise the zeros of $a(\zeta)$ would have a finite accumulation point and then, by an analytic continuation argument, $a(\zeta)$ would be identically zero. (If the accumulation point is at $\zeta = 0$, a more detailed argument is required.) From (3.59), we observe that $a(\zeta)$ is proportional to the Wronskian of the solutions $\phi(x, \zeta)$, $\psi(x, \zeta)$

$$a(\zeta) = \frac{W(\phi, \psi)}{2i\zeta} = \frac{1}{2i\zeta}(\phi\psi_x - \phi_x\psi),$$

and, therefore, at each zero ζ_k, $k = 1, \ldots, N$, of $a(\zeta)$,

$$\phi(x, \zeta_k) = b_k\psi(x, \zeta_k), \qquad k = 1, \ldots, N. \tag{3.66}$$

Observe, also, that $a(\zeta)$ may have a pole at $\zeta = 0$. In fact, this is the rule. The exceptions occur when $|R(0)| < 1$. The reader might think of this property when he is working out exercises (ii), (iii), (iv), (v) at the end of the section.

The set

$$S = \left\{ R(\zeta) = \frac{b(\zeta)}{a(\zeta)}, (\zeta_k, b_k/a_k')_{k=1}^N \right\} \tag{3.67}$$

is called *the scattering data* (a_k' is $\partial a/\partial \zeta |_{\zeta_k}$). From it $a(\zeta)$ can be constructed as follows. The function

$$f(\zeta) = \prod_1^N \frac{\zeta + i\eta_k}{\zeta - i\eta_k} a(\zeta)$$

is analytic for Im $\zeta > 0$, tends to 1 at $\zeta = \infty$, Im $\zeta \geq 0$ and hence by Cauchy's theorem

$$\ln f(\zeta) = \frac{1}{2\pi i} \int_{-\infty}^{\infty} \frac{\ln f(\xi)}{\xi - \zeta} d\xi,$$

$$0 = \frac{1}{2\pi i} \int_{-\infty}^{\infty} \frac{\ln f(\xi)}{\xi + \zeta} d\xi, \qquad \text{Im } \zeta > 0.$$

Letting $\xi \to -\xi$ in the second equation, noting that for real ξ, $f(\xi) f(-\xi) = a(\xi)$ $a(-\xi) = |a(\xi)|^2$, and adding the two equations, we obtain

$$\ln a(\zeta) = \sum_{k=1}^N \ln \frac{\zeta - i\eta_k}{\zeta + i\eta_k} - \frac{1}{2\pi i} \int_{-\infty}^{\infty} \frac{\ln (1 - |R|^2)}{\xi - \zeta} d\xi,$$

from which it trivially follows that

$$\ln a(\zeta) = \sum_{k=1}^N \ln \frac{\zeta - i\eta_k}{\zeta + i\eta_k} - \frac{\zeta}{\pi i} \int_0^{\infty} \frac{\ln (1 - |R|^2)}{\xi^2 - \zeta^2} d\xi. \tag{3.68}$$

(It is easily seen that the possible simple pole at $\zeta = 0$ in $a(\zeta)$ does not affect the result.) Note that taking the asymptotic expansion as $\zeta \to \infty$ of both sides of (3.68) gives

$$H_{2n-1} = \frac{4i}{(2i)^{2n+1}} \int_{-\infty}^{\infty} R_{2n+1} dx = \frac{8i^{2n+2}}{2n+1} \sum_{k=1}^N \eta_k^{2n+1} - \frac{4}{\pi} \int_0^{\infty} \xi^{2n} \ln (1 - |R|^2) d\xi. \tag{3.69}$$

Equations (3.69) are the *trace formulae* giving the Hamiltonian functionals H_{2n-1}, $n = 1, 2, \ldots$ (the formula holds for $n = 0$, but H_{-1} is not a legitimate Hamiltonian)[5] as functions of the eigenvalues $\{\eta_k\}_{k=1}^N$ and the modulus of the reflection coefficient $|R(\xi)|$, ξ real and positive. They therefore express the constants of the motion in the old variables $(q, q_x, q_{xx}, \ldots)$ as functions of the constants of the motion in the new variables (the scattering data) and in particular tell us how the Hamiltonians map under the transformation from $q(x, t)$ to $S(t)$. The first few are worth writing down explicitly:

$$\int_{-\infty}^{\infty} q \, dx = 4 \sum_{1}^{N} \eta_k + \frac{2}{\pi} \int_{0}^{\infty} \ln(1 - |R|^2) \, d\xi, \tag{3.70a}$$

$$\int_{-\infty}^{\infty} q^2 \, dx = \frac{16}{3} \sum_{1}^{N} \eta_k^3 - \frac{8}{\pi} \int_{0}^{\infty} \xi^2 \ln(1 - |R|^2) \, d\xi, \tag{3.70b}$$

$$\int_{-\infty}^{\infty} (q_x^2 - 2q^3) \, dx = \frac{64}{5} \sum_{1}^{N} \eta_k^5 - \frac{32}{\pi} \int_{0}^{\infty} \xi^4 \ln(1 - |R|^2) \, d\xi. \tag{3.70c}$$

It is worth checking these formulae for the case when $q(x)$ is a one-soliton, reflectionless potential. By reflectionless, we mean that the reflection coefficient $R(\zeta)$ is identically zero. Then the nontrivial scattering data are simply the bound states $\{\zeta_k = i\eta_k, b_k\}_1^N$. In particular, if $N = 1$, $q(x) = 2\eta \operatorname{sech}^2 \eta(x - \bar{x})$ where $\eta_1 = \eta$, $b_1 = e^{2\eta\bar{x}}$. Note that for $N > 1$ the energy of the N-soliton state is simply the sum of the energies of the individual soliton components. This is not unexpected since the energy is conserved and, in the long time limit, the N-soliton state will asymptotically approach the linear sum of N separate solitons (see (3.108)). It is also worth remarking that, from (3.70a), the mass content of the solitons $\sum_1^N 4\eta_k$ is always greater than the actual mass contained by the solution $q(x)$ ($\int_{-\infty}^{\infty} q \, dx$) because $0 < |R|^2 < 1$. Therefore the contribution of the continuous spectrum to the mass content is always negative. This result has important ramifications when we consider perturbations on the KdV equation. It is now a good time to work out several examples in detail.

Exercises 3d.

1. Take (a) $v_x = u(x, t)v$ and (b) $v_t = (u^2 + u_x)v$. Note that the integrability condition is Burgers' equation (c) $u_t = 2uu_x + u_{xx}$. But, by differentiating (a), we see (b) is $v_t = v_{xx}$, the heat equation. Hence (c) can be solved directly by using the map (a).

2. Take $q(x) = Q_0 \delta(x)$ and show that $a(\zeta) = (Q_0 + 2i\zeta)/2i\zeta$, $R(\zeta) = -Q_0/(Q_0 + 2i\zeta)$. There is one bound state at $\zeta = iQ/2$. Also $R(0) = -1$. In fact, for all potentials except the reflectionless ones (where $R(\xi) \equiv 0$) $R(0) = -1$. Verify that (3.60) holds. Calculate $\phi(x, \zeta)$, $\psi(x, \zeta)$ and find b_k.

---

[5] H_{-1} gives rise to the flow, $q_t = 0$,

3. Take $q(x) = Q$, $0 < x < L$ and zero otherwise. Show

$$a(\zeta) = e^{i\zeta L}\left\{ \cos \sqrt{\zeta^2 + Q}\, L - \frac{i(2\zeta^2 + Q)}{2\zeta\sqrt{\zeta^2 + Q}} \sin \sqrt{\zeta^2 + Q}\, L \right\},$$

$$b(\zeta) = e^{-i\zeta L} \frac{iQ}{2\zeta\sqrt{\zeta^2 + Q}} \sin \sqrt{\zeta^2 + Q}\, L. \tag{3.71}$$

Why is $a(\zeta)$ analytic for $\mathrm{Im}\, \zeta > 0$? Note that as $\zeta \to 0$, $R(\zeta) \to -1$.

To find the zeros of $a(\zeta)$ set $\zeta = i\sqrt{Q} \cos \theta$ and find $a(\zeta) = \sin(2\theta + \alpha \sin \theta)/\sin 2\theta$ where $\alpha = \sqrt{Q}\, L$. Discuss the zeros of $a(\zeta)$ as a function of α by looking for the $0 < \theta < \pi/2$ intersections of the graphs $y = \alpha \sin \theta$ and $y = n\pi - 2\theta$, $n = 1, 2, \ldots$.

Remark. Note that in this case $a(\zeta)$, $b(\zeta)/a(\zeta)$ and all the eigenfunctions $\phi(x, \zeta)$, $\psi(x, \zeta)$ are analytic everywhere except at $\zeta = 0, \infty$. This property carries over for all $q(x)$ on compact support.

4. Take $q(x) = 2\eta^2 \, \mathrm{sech}^2\, \eta(x - \bar{x})$ and show that by the transformation $\tanh \eta(x - \bar{x}) = t$, (3.51) can be converted into an associated Legendre equation. Show $R(\xi) \equiv 0$. Find explicitly $a(\zeta)$, the eigenfunctions and eigenvalues (there is only one $\zeta_1 = i\eta$) and norming constant b_1 and show $q(x) = -4i(b_1/a_1')\eta_1\psi^2(x, \zeta_1)$.

5. Take $q(x) = A\, \mathrm{sech}^2\, x$ and show that (3.51) can be solved in terms of the hypergeometric function and that the reflection and transmission coefficients are given in terms of the Γ function with arguments depending on ζ and A. In particular show that when $A = n(n+1)$, n a positive integer, $R(\xi) \equiv 0$ and a pure n soliton solution arises. Note for A almost equal to $n(n+1)$, $R(0) = -1$ whereas for $A = n(n+1)$, $R(\xi) \equiv 0$. (For details, the reader may consult Lamb's book [69].)

3e.   The inverse transform. In this section, our main goal is to show how to reconstruct the potential $q(x)$ from the scattering data S. The end result will be the celebrated Gel'fand–Levitan [81] equation, although an equation we obtain along the way will be more useful for obtaining the formulae for the reflectionless (pure soliton) potentials. We will examine the two-soliton case in detail, and explain what is meant by the phase shift.

The first and longest part of the analysis is the reconstruction of the fundamental solution matrix

$$\Phi = \begin{pmatrix} \phi(x, \zeta) & \psi(x, \zeta) \\ \phi_x(x, \zeta) & \psi_x(x, \zeta) \end{pmatrix}. \tag{3.72}$$

Once Φ, and in particular its elements, are known, one can find out that it satisfies the Schrödinger equation and from this pick off the potential $q(x)$. The most convenient but not the only way of accomplishing the latter step is to isolate $q(x)$ by looking at the asymptotic expansion for $\psi(x)e^{-i\zeta x}$ given by (3.29a). From this

$$q(x) = \lim_{\delta \to \infty} -2i\zeta \frac{d}{dx}(\psi(x, \zeta)e^{-i\zeta x} - 1), \qquad \mathrm{Im}\, \zeta \geq 0. \tag{3.73}$$

A similar analysis to that which leads to (3.29) tells us that $\phi(x, \zeta)e^{i\zeta x}$, $\phi(x, -\zeta)e^{-i\zeta x}$, $\psi(x, -\zeta)e^{i\zeta x}$ each tends to unity in their half planes (Im $\zeta \lessgtr 0$) of analyticity.

What we want to find is $\phi(x, \zeta)e^{i\zeta x}$ (or $\psi(x, \zeta)e^{-i\zeta x}$) which we know is analytic in the upper half plane with asymptotic behavior $\phi e^{i\zeta x} \to 1$ as $\zeta \to \infty$, Im $\zeta > 0$. We also know $\psi(x, -\zeta)e^{i\zeta x}$ is analytic for Im $\zeta < 0$ and on the real ζ axis, $\zeta = \xi$, equation (3.59) specifies the jump between $(\phi/a)e^{+i\zeta x}$, which contains only a finite number of poles in Im $\zeta > 0$, and $\psi(x, -\zeta)e^{i\zeta x}$ to be the continuous function $R(\xi)\psi(x, \xi)e^{i\xi x}$. This is the classic Riemann–Hilbert problem whose solution is constructed as follows.

Consider, for Im $\zeta > 0$, the integral along the real ξ axis,

$$I = \frac{1}{2\pi i} \int_{-\infty}^{\infty} \frac{\phi(x, \xi)e^{i\xi x} \, d\zeta}{a(\xi)(\xi + \zeta)}, \qquad \text{Im } \zeta > 0. \tag{3.74}$$

We will evaluate I in two ways. First, we will take advantage of the fact that $\phi(x, \xi)e^{i\xi x}$ and $a(\xi)$ are analytic in the upper half plane Im $\xi > 0$ and move the integration contour to the circle $|\xi| = R$, $R \to \infty$, $0 < \text{Arg } \xi < \pi$. We find

$$I = \sum_{k=1}^{N} \frac{\gamma_k \psi_k e^{i\zeta_k x}}{\zeta + \zeta_k} - \frac{1}{2}. \tag{3.75}$$

The first term in (3.75) arises from the contributions from the zeros of $a(\zeta)$, which are simple, at which points ζ_k, $\phi_k = b_k \psi_k$ (ψ_k means $\psi(x, \zeta_k)$). We define γ_k by $b_k(a_k')^{-1}$ where $a_k' = da/d\zeta |_{\zeta_k}$. The second term comes from the integral along the semicircle at ∞ on which we know that the integrand tends to ξ^{-1}.

Next, we evaluate I by using (3.59) and writing,

$$I = \frac{1}{2\pi i} \int_{-\infty}^{\infty} \frac{\psi(x, -\xi)e^{i\xi x} \, d\zeta}{\xi + \zeta} + \frac{1}{2\pi i} \int_{-\infty}^{\infty} R(\xi) \frac{\psi(x, \xi)e^{i\xi x} \, d\zeta}{\xi + \zeta}. \tag{3.76}$$

From the analytic properties of the integrand in the first integral, we obtain

$$I = -\psi(x, \zeta)e^{-i\zeta x} + \frac{1}{2} + \frac{1}{2\pi i} \int_{-\infty}^{\infty} R(\xi) \frac{\psi(x, \xi)e^{i\xi x} \, d\zeta}{\xi + \zeta}. \tag{3.77}$$

Now we are finished because (3.75) and (3.77) give us a closed integral equation for $\psi(x, \zeta)$. It is

$$\psi(x, \zeta)e^{-i\zeta x} = 1 - \sum_{k=1}^{N} \frac{\gamma_k \psi_k e^{i\zeta_k x}}{\zeta + \zeta_k} + \frac{1}{2\pi i} \int_{-\infty}^{\infty} R(\xi) \frac{\psi(x, \xi)e^{i\xi x} \, d\xi}{\xi + \zeta}. \tag{3.78}$$

Note that the set of data we need to solve (3.78) is precisely $S = (R(\xi), \xi$ real, $(\zeta_k, \gamma_k)_1^N)$. In order to accomplish the task of solving (3.78) and of proving existence and uniqueness of the solution, it is sometimes convenient to make a Fourier transform of (3.74) and introduce the "time" variable we called τ in Section 3d. We make the ansatz

$$\psi(x, \zeta)e^{-i\zeta x} = 1 + \int_{x}^{\infty} K(x, s)e^{i\zeta(s-x)} \, ds, \qquad \text{Im } \zeta > 0. \tag{3.79}$$

It is not too difficult to show that such a $K(x, s)$, independent of ζ, exists [12]. The asymptotic expansion of (3.79) gives

$$\psi(x, \zeta)e^{-i\zeta x} = 1 - \frac{1}{i\zeta} K(x, x) + O\!\left(\frac{1}{\zeta^2}\right),$$

and a comparison with (3.73) gives

$$q(x) = 2\frac{d}{dx} K(x, x). \tag{3.80}$$

Introduce (3.79) into (3.78), multiply across the equations by $e^{i\zeta(x-\tau)}$ and integrate in ζ from $-\infty$ to ∞ along a line just above the real axis. Then using the facts that

$$\int_{-\infty}^{\infty} d\zeta e^{i\zeta(s-\tau)} = 2\pi\, \delta(s-\tau),$$

$$\int_{-\infty}^{\infty} d\zeta \frac{e^{i\zeta(x-\tau)}}{\zeta + \zeta_k} = -2\pi i e^{-i\zeta_k(x-\tau)}H(\tau - x),$$

$$\int_{-\infty}^{\infty} d\zeta \frac{e^{i\zeta(x-\tau)}}{\zeta + \xi} = -2\pi i e^{-i\xi(x-\tau)}H(\tau - x)$$

where $H(y)$ is the Heaviside function, we obtain

$$K(x, \tau) + B(x+\tau) + \int_{x}^{\infty} K(x, s)B(s+\tau)\, ds = 0, \qquad \tau > x, \tag{3.81}$$

where

$$B(z) = -i\sum_{k=1}^{N} \gamma_k e^{i\zeta_k z} + \frac{1}{2\pi}\int_{-\infty}^{\infty} R(\xi)e^{i\xi z}\, d\xi. \tag{3.82}$$

Equation (3.81) is the *Gel'fand–Levitan equation*. It is a Fredholm equation from which $K(x, \tau)$ is constructed as function of $\tau > x$ from a knowledge of $B(x+\tau)$. Then $q(x)$ is found from (3.80). It is sometimes worth writing (3.82) in the more compact form

$$B(z) = \frac{1}{2\pi}\int_{C} R(\xi)e^{i\xi z}\, d\xi \tag{3.83}$$

where C is a contour from $\xi = -\infty$ to $\xi = +\infty$ going over the poles of $R(\xi)$. This formula is only valid if $R(\xi)$ admits analytic extension to the upper half plane.

The class of reflectionless potentials. The class of reflectionless potentials $q(x)$ arise when $R(\zeta)$ is identically zero and in that case it is much easier to work with (3.78) directly;

$$\psi(x, \zeta)e^{-i\zeta x} = 1 - \sum_{1}^{N} \frac{\gamma_k \psi_k e^{i\zeta_k x}}{\zeta + \zeta_k}. \tag{3.84}$$

Note that from (3.71), if $\zeta_k = i\eta_k$,

$$q(x) = 2i\frac{d}{dx}\sum_{1}^{N} \gamma_k \psi_k e^{-\eta_k x}. \tag{3.85}$$

Let $\zeta = \zeta_j = i\eta_j$ in (3.84) and obtain a system of N linear equations

$$\sum_{k=1}^{N} \left(\delta_{jk} + \frac{\gamma_k e^{-(\eta_k + \eta_j)x}}{i(\eta_k + \eta_j)} \right) \psi_k = e^{-\eta_j x}, \qquad j = 1, \ldots, N$$

for ψ_k from which we can calculate $\psi(x, \zeta)e^{-i\zeta x}$ and $q(x)$ directly. From the properties of b_k and $a(\zeta)$, it easy to show that γ_k is pure imaginary and we will write it as

$$\gamma_k = 2i\eta_k e^{2\eta_k \bar{x}_k}. \tag{3.86}$$

The one-soliton solution $\zeta_1 = i\eta$ (each discrete eigenvalue gives rise to one soliton) is

$$\psi e^{-i\zeta x} = 1 - \frac{i\eta \operatorname{sech} \eta(x - \bar{x})e^{-\eta(x - \bar{x})}}{\zeta + i\eta},$$

$$\phi e^{i\xi x} = 1 - \frac{i\eta \operatorname{sech} \eta(x - \bar{x})e^{-\eta(x - \bar{x})}}{\xi + i\eta}$$

$$a'(\zeta_1) = \frac{1}{2i\eta}, \qquad b_1 = e^{2\eta \bar{x}}, \tag{3.87}$$

$$\psi(x, \zeta_1) = \frac{e^{-\eta \bar{x}}}{1 + e^{-2\eta(x - \bar{x})}},$$

$$\phi(x, \zeta_1) = b_1 \psi(x, \zeta_1) = e^{2\eta \bar{x}} \frac{e^{-\eta x}}{1 + e^{-2\eta(x - \bar{x})}},$$

$$q(x) = 2\eta^2 \operatorname{sech}^2 \eta(x - \bar{x}).$$

In particular, the parameter b_1, the ratio between ϕ and ψ at ζ_1, is $e^{2\eta \bar{x}}$ and thus specifies the soliton position. In the next section, we will find how it depends on time t.

In addition, one can show that for the N-soliton solution, $q(x)$ can be written as the second x derivative of the logarithm of a function we shall call $\tau(x)$ (not the variable τ in the Gel'fand–Levitan equation) which may be written

$$\tau = \sum_{\mu_j = 0,1} \exp \left\{ \sum_{j=1}^{N} \mu_j H_j + \sum_{1 \le i < j \le N} A_{ij} \mu_i \mu_j \right\}, \tag{3.88}$$

where

$$H_j = -2\eta_j(x - \bar{x}_j), \qquad e^{A_{ij}} = \left(\frac{\eta_i - \eta_j}{\eta_i + \eta_j} \right)^2 \tag{3.89}$$

and the first sum is taken over all $\mu_j = 0$ or 1. The reader is invited to check that, for $N = 2$,

$$\tau = 1 + e^{-2\eta_1(x - \bar{x}_1)} + e^{-2\eta_2(x - \bar{x}_2)} + e^{-2\eta_1(x - \bar{x}_1) - 2\eta_2(x - \bar{x}_2) + A_{12}}. \tag{3.90}$$

The two-soliton interaction. Let us for the moment anticipate that if $q(x, t)$ evolves according to the KdV equation, $\bar{x}_j = 4\eta_j^2 t + \bar{x}_j^{(0)}$ and let us examine the two-soliton interaction. Imagine that $\eta_1 > \eta_2$ and let us look near $x \approx \bar{x}_2$. Because $\bar{x}_2 - \bar{x}_1 = 4(\eta_2^2 - \eta_1^2)t$, the second and the fourth terms in (3.90) are exponentially small if $t \to -\infty$ and thus near $\bar{x}_2$,

$$\tau \approx 1 + e^{-2\eta_2(x - \bar{x}_2)}$$

and

$$q(x) = 2\eta_2 \operatorname{sech}^2 \eta_2(x - \bar{x}_2).$$

On the other hand, near $x = \bar{x}_1$, the third and fourth terms dominate and

$$\tau \approx e^{-2\eta_2(x - \bar{x}_2)}(1 + e^{-2\eta_1(x - \bar{x}_1) + A_{12}}).$$

The corresponding $q(x)$ is

$$q(x) = 2\eta_1 \operatorname{sech}^2 \eta_1\left(x - \bar{x}_1 - \frac{1}{2\eta_1} A_{12}\right).$$

Similarly as $t \to +\infty$, near $x = \bar{x}_1$,

$$q(x) = 2\eta_1 \operatorname{sech}^2 \eta_1(x - \bar{x}_1).$$

and near $x = \bar{x}_2$,

$$q(x) = 2\eta_2 \operatorname{sech}^2 \eta_2\left(x - \bar{x}_2 - \frac{1}{2\eta_2} A_{12}\right).$$

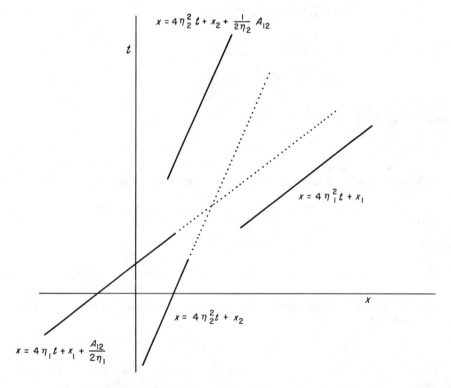

FIG. 3. *The two-soliton interaction.*

Therefore the smaller soliton η_2, which for large negative time lies to the right of soliton η_1, undergoes a phase shift by an amount $(1/2\eta_2)A_{12}$, which is negative, in its position. The larger soliton η_1 jumps forward by the positive amount $-A_{12}/2\eta_1$.

Finally, the reader should observe from (3.88) that if there are N solitons $\eta_1 > \eta_2 > \cdots > \eta_N$, then as t sweeps from $-\infty$ to $+\infty$, the total phase shift experienced by any soliton is the sum of the pairwise phase shifts.

It is also true, and I will tell you why in Chapter 4, that the phase shift function

$$A_{12} = 2 \ln \left| \frac{\eta_1 - \eta_2}{\eta_1 + \eta_2} \right|$$

is shared by every equation in the KdV family. Can you see now why this must be true?

3f. Time evolution of the scattering data and the effects of small variations in the potentials.

In this section, I first want to show you how to calculate how the scattering data change in time as the potentials $q(x, t_{2k+1})$ change according to any one of the equations (3.9). We will begin with the case when $q(x, t)$ satisfies (3.52). Next, I want to derive formulae by which infinitesimal changes in the scattering data resulting from an infinitesimal change δq in the potential can be calculated. The infinitesimal change in the potential is arbitrary and need not follow the direction of any of the special flows (3.9). These formulae will be important for

(i) establishing a framework for carrying out a perturbation theory in the case where $q(x, t)$ evolves like $q_t + 6qq_x + q_{xxx} = \varepsilon F(q, q_x, \ldots)$, $0 < \varepsilon \ll 1$;

(ii) proving (3.17) which established the relation between the fluxes $L^n q$ and the variational derivative of H_{2n+1}; and

(iii) writing q and δq in terms of a "squared eigenfunction" basis, the analogue of the Fourier transform.

Time dependence of scattering data. It is simple to calculate the time dependence of the scattering data using (3.53). The first task is to choose C so that the particular normalization is consistent with the definitions of the eigenfunctions $\phi(x, t, \zeta)$ and $\psi(x, t, \zeta)$. Recall that by definition (3.58), $\phi(x, t, \zeta) \sim e^{-i\zeta x}$ as $x \to -\infty$ for all t, and in order that this be so we must choose $C = 4i\zeta^3$. Similarly for $\psi(x, t, \zeta)$ we must choose $C = -4i\zeta^3$. Hence

$$\phi_t = (q_x + 4i\zeta^3)\phi + (4\zeta^2 - 2q)\phi_x, \tag{3.91}$$

$$\psi_t = (q_x - 4i\zeta^3)\psi + (4\zeta^2 - 2q)\psi_x. \tag{3.92}$$

To find the t derivatives of $a(\zeta, t)$, $b(\zeta, t)$ use (3.59) directly or its asymptotic form as $x \to +\infty$. Taking the t derivative of the left-hand side of (3.59) gives

$$(q_x + 4i\zeta^3)(a\psi(x, -\zeta) + b\psi(x, \zeta)) + (4\zeta^2 - 2q)(a\psi_x(x, -\zeta) + b\psi_x(x, \zeta)),$$

whereas on the right, we find

$$a_t\psi(x, -\zeta) + b_t\psi(x, \zeta) + a(q_x + 4i\zeta^3)\psi(x, -\zeta) + a(4\zeta^2 - 2q)\psi_x(x, -\zeta)$$
$$+ b(q_x - 4i\zeta^3)\psi(x, \zeta) + b(4\zeta^2 - 2q)\psi_x(x, \zeta).$$

Equating these two expressions gives

$$a_t = 0, \qquad b_t = 8i\zeta^3 b. \tag{3.93a}$$

A similar calculation on (3.66) gives

$$b_{kt} = 8i\zeta_k^3 b_k = 8\eta_k^3 b_k. \tag{3.93b}$$

Note that (3.93a) and (3.93b) are consistent in the sense that if the potential $q(x, t)$ were such that $b(\zeta, t)$ admitted analytic continuation to $\zeta = \zeta_k$, then (3.93a) would become (3.93b). Alternatively, substituting the asymptotic form of $\phi(x, t, \zeta)$, $a(\zeta, t)e^{-i\zeta x} + b(\zeta, t)e^{i\zeta x}$ into (3.91) also gives (3.93a). The key point to note is that even though the time evolutions of ϕ and ψ contain the unknown $q(x, t)$ and its x derivatives, the scattering data involve only the relative behavior of the eigenfunctions at $x = \pm\infty$, and at these points $q(x, t)$ and its derivatives are known for all time: they are zero. On the other hand, if one were solving the periodic problem over $0 \leq x \leq L$, there is no point x at which one knows $q(x, t)$ for all time, and this fact makes the time evolution of the new coordinates for that problem much more complicated.

The constancy of the function $a(\zeta)$ is central to the theory. First it ensures that the discrete eigenvalues $\zeta_k = i\eta_k$, $k = 1, \ldots, N$ are constants of the motion. Second, from (3.65) and (3.69), we see that the Hamiltonians H_{2n+1}, $n = 0, 1, \ldots$ and the mass $\int_{-\infty}^{\infty} q\, dx$ are constants of the motion. This is true for any of the flows in the KdV family. The only difference between $(\text{KdV})_3$ and $(\text{KdV})_{2n+1}$ as given by (3.9) (recall that the t in (3.9) for $n = 1$ is 4 times the t of (3.52), (3.53)) is that in the former case

$$b_{t_3} = 2i\zeta^3 b, \tag{3.94}$$

whereas in the latter

$$b_{t_{2n+1}} = 2i\zeta^{2n+1} b. \tag{3.95}$$

So, whereas in physical space it is difficult to see the connections between flows $n = 1, 2, 3$, etc. (compare (3.15), (3.16)), in scattering space their difference is trivial and involves only a different power of ζ and ζ_k in the phases of $b(\xi, t)$ and $b_k(t)$. For this reason we may easily take linear combinations of the flows; nothing changes except the phases of these two quantities. For example,

$$q_t = \alpha q_{t_3} + \beta q_{t_5} + \gamma q_{t_7} + \cdots$$

gives $a_t = 0$ and

$$b_t = 2i(\alpha\zeta^3 + \beta\zeta^5 + \gamma\zeta^7 + \cdots)b,$$
$$b_{kt} = 2i(\alpha\zeta_k^3 + \beta\zeta_k^5 + \gamma\zeta_k^7 + \cdots)b_k.$$

One consequence of this is we can write the general multisoliton solution (3.88) for all the flows in one formula. Simply use the appropriate phase speed $\bar{x}_k$. Recall that $b_k = e^{2\eta_k \bar{x}_n} = e^{-2i\zeta_k x_k}$. The multisoliton formula for $q(x, t_3, t_5, \ldots)$ as a function of the pure flows (3.9) is given by (3.88), (3.89) with

$$H_j = 2i(\zeta_j x + \zeta_j^3 t_3 + \zeta_j^5 t_5 + \cdots). \tag{3.96}$$

The effect of an infinitesimal change in the potential. Our next task is to determine formulae expressing infinitesimal changes in the scattering data as functionals of infinitesimal changes in the potential $q(x)$. This is an important calculation for a variety of reasons as we shall see. If q undergoes a change $q \to q + \delta q$, the corresponding infinitesimal change in $\psi(x, \zeta)$ for real ζ (fixed) satisfies the equation

$$(\delta\psi)_{xx} + (\zeta^2 + q)\delta\psi = -\delta q\psi. \tag{3.97}$$

We know two linearly independent solutions $\psi(x, \zeta)$, $\psi(x, -\zeta)$ for the homogeneous equation, and therefore (3.97) may be solved by variation of constants to give

$$\delta\psi = \frac{1}{2i\zeta} \psi(x, -\zeta) \int_{\infty}^{x} \delta q(y)\psi^2(y, \zeta)\, dy - \frac{1}{2i\zeta} \psi(x, \zeta) \int_{\infty}^{x} \delta q(y)\psi(y, \zeta)\psi(y, -\zeta)\, dy.$$

Letting $x \to -\infty$ gives

$$\delta a = \frac{1}{2i\zeta} \int_{-\infty}^{\infty} \delta q(x)\psi(x, \zeta)\phi(x, \zeta)\, dx, \tag{3.98a}$$

$$\delta b^* = \frac{1}{2i\zeta} \int_{-\infty}^{\infty} dq(x)\psi(x, \zeta)\phi(x, -\zeta)\, dx. \tag{3.98b}$$

From (3.98),

$$\delta\left(\frac{b^*}{a}\right) = \frac{1}{2i\zeta a^2} \int_{-\infty}^{\infty} \delta q\psi^2\, dx. \tag{3.99}$$

The easiest way to find $\delta\zeta_k$ and $\delta\gamma_k$ is to assume that the analytic extension to $\zeta = \zeta_k$ exists; even if it does not, the resulting formulae are correct as the change in q and ζ does not depend on whether the initial data q is on compact support or not. One small difficulty with this is that for ζ real, $b^*(\zeta)$ is $b(-\zeta)$. For ζ complex, the extension of $b(\zeta)$ to $\zeta = \zeta_k$ must satisfy $b(\zeta_k)b(-\zeta_k) = -1$ since $a(\zeta_k) = 0$. Hence the extension of b^* to $\zeta = \zeta_k$ is $-1/b_k$. I will leave it as an exercise to the reader to show

$$\delta\zeta_k = \frac{-\gamma_k}{2i\zeta_k} \int_{-\infty}^{\infty} \delta q\psi_k^2\, dx, \tag{3.100}$$

$$\delta\beta_k + \left(\frac{a_k''}{a_k'} + \frac{1}{\zeta_k}\right)\beta_k\,\delta\zeta_k = \frac{-\gamma_k}{2i\zeta_k a_k'^2} \int_{-\infty}^{\infty} \delta q\left(\frac{\partial\psi^2}{\partial\zeta}\right)_k dx, \tag{3.101}$$

where $\beta_k = -1/b_k a_k'$, a_k', a_k'' and $(\partial\psi^2/\partial\zeta)_k$ are the first and second derivatives of $a(\zeta)$ at ζ_k and the first derivative of ψ^2 with respect to ζ at ζ_k. Formula (3.100) is familiar to quantum mechanicists and expresses the change in energy level as a function of the change in potential.

We will now use (3.98a) to establish the relationship (3.17). It is not difficult to show directly that for L given by (3.12),

$$(L - \zeta^2)\left(\frac{\phi\psi}{a(\zeta)} - 1\right) = \frac{1}{2} q. \tag{3.102}$$

Solving iteratively, we obtain that

$$\frac{\phi\psi}{a(\zeta)} \sim 1 - \frac{1}{2} \sum_{0}^{\infty} \frac{1}{\zeta^{2n+2}} L^n q.$$

Therefore, using (3.98a),

$$\frac{\delta \ln a}{\delta q} = \frac{1}{2i\zeta} \frac{\phi\psi}{a(\zeta)} \sim \frac{1}{2i\zeta}\left(1 - \frac{1}{2} \sum_{0}^{\infty} \frac{1}{\zeta^{2n+2}} L^n q\right). \tag{3.103}$$

Now compare (3.103) with (3.65) and (3.69) and find

$$L^n q = \frac{\delta H_{2n+1}}{\delta q}. \tag{3.104}$$

This is (3.17) and it is very important for it enables us to write all the flows in Hamiltonian form.

In the next section, we are going to tackle the problem of what happens to solutions of exactly integrable equations when they are perturbed and will make heavy use of the material of this section. Before we go on, however, I want to mention another feature. As we have pointed out in Section 3d, in the small $q(x)$ limit, $(ik/2\pi)b(k/2)$ is $\hat{q}(k)$, its Fourier transform. This connection, together with the realization that the change in the scattering data is expressed in (3.99), (3.100), (3.101) as an inner product between δq and the squared eigenfunctions and their ζ derivatives suggests that an inverse relation is possible in which δq is expressed as a function of $\delta(b^*/a)$, $\delta\zeta_k$, $\delta\beta_k$. This is indeed possible; for details I refer the reader to reference [75]. We find

$$\delta q(x, t) = -\frac{1}{\pi} \int_{-\infty}^{\infty} \delta\left(\frac{b^*}{a}\right) \frac{\partial \phi^2(x, \xi)}{\partial x} d\xi + 2i \sum_{k=1}^{N} \delta\beta_k \frac{\partial \phi^2(x, \zeta_k)}{\partial x}$$

$$+ 2i \sum_{k=1}^{N} \beta_k \delta\zeta_k \frac{\partial}{\partial \zeta} \left(\frac{\partial \phi^2(x, \zeta)}{\partial x}\right)_{\zeta_k}. \tag{3.105}$$

The set of x derivatives of the squared eigenfunctions

$$\left\{\frac{\partial \phi^2}{\partial x}(x, \xi), \xi \text{ real}; \frac{\partial \phi^2(x, \zeta_k)}{\partial x}, \frac{\partial^2 \phi^2(x, \zeta_k)}{\partial \zeta \partial x}, k = 1, \ldots, N\right\} \tag{3.106}$$

form a basis for functions in the class (3.63). You should verify from (3.99)– (3.101) that if the change δq is brought about by any of the time evolutions of the KdV family, then

$$\delta\left(\frac{b^*}{a}\right) = -2i\zeta^{2n+1}\frac{b^*}{a}, \quad \delta\zeta_k = 0, \quad \delta\beta_k = -2i\zeta_k^{2n+1}\beta_k, \quad \delta = \frac{\partial}{\partial t}. \quad (3.107)$$

To see this when $n = 1$, for example, replace δq in (3.99) by $-\frac{1}{4}(q_{xx} + 3q^2)_x$, integrate by parts and use the equation that $(\psi^2)_x$ satisfies in order to calculate the integral.

Finally we point out that $q(x)$ itself can be written

$$q(x) = \frac{2}{i\pi}\int_{-\infty}^{\infty}\zeta R(\zeta)\psi^2(x, \zeta)\,d\zeta - 4\sum_{k=1}^{N}\gamma_k\zeta_k\psi^2(x, \zeta_k). \quad (3.108)$$

Equation (3.108) is easily found by considering

$$\frac{2}{i\pi}\int_{-\infty}^{\infty}\zeta R(\zeta)\psi^2(x, \zeta)\,d\zeta = \frac{2}{i\pi}P\int_{-\infty}^{\infty}\zeta\left(\frac{\psi(x, \zeta)\phi(x, \zeta)}{a(\zeta)} - 1\right)d\zeta, \quad (3.109)$$

where P is the Cauchy principal value, and evaluating the right-hand side by deforming the contour to the semicircle at $|\zeta| = \infty$, Im $\zeta > 0$. From (3.27) we know $\phi\psi a^{-1} - 1 \sim q/2\zeta^2 + \cdots$ as $|\zeta| \to \infty$.

Remark. From these results, we see that integrable flows lie on an infinite family of surfaces

$$|R(\xi)| = \text{constant}, \quad \zeta_k = \text{constant}. \quad (3.110)$$

Because of the obvious analogy with finite dimensional Hamiltonian systems, we can think of the intersection of the level surfaces (3.110) as an infinite dimensional torus. General perturbations which do not fall into the integrable class—like the effects of a slowly changing depth (see the following section)—can induce changes in the trajectories both within and orthogonal to this torus. From (3.105) we can see that changes which stay within the original torus form a vector space spanned by the vectors $\partial\phi^2/\partial x$, whereas the changes normal to the torus are spanned by the ζ derivatives of these quantities.

3g. Perturbation theory. Solitary waves in channels of slowly changing depth.

In Section 2b, we derived (2.16)

$$q_\tau + 6qq_\theta + q_{\theta\theta\theta} = -\frac{9}{4}\frac{D_\tau}{D}q \quad (3.111)$$

as a model for the right going propagation of small amplitude, long waves in a channel of varying depth. If the relative change in depth D_τ/D is small compared with the length of waves in question, the τ-dependent parameter $\Gamma = +\frac{9}{4}D_\tau/D$ is small, of order σ say, where $0 < \varepsilon \ll \sigma \ll 1$. One would imagine that such an innocent looking problem could be solved in a straightforward manner by standard perturbation techniques but such is not the case. It took almost ten years to clear up most of the difficulties associated with (3.111) and its relation to the full two-directional water wave problem. There are still some questions open. Therefore, it is an excellent vehicle for illustrating the various avenues for attacking perturbed soliton equations. The results I am about to describe were obtained in a series of papers with my colleague David Kaup [45] and later with my student Collen Knickerbocker [43]. Karpman and his colleagues [46] independently made many of the same discoveries about the same time.

The problem we consider is: imagine that a solitary wave

$$q(\theta, \tau) = 2\eta_0^2 \operatorname{sech}^2 \eta_0(\theta - 4\eta_0^2\tau - \theta_0) \qquad (3.112)$$

meets a change of depth at $\theta = \tau = 0$. (The reader should recall the connection of the new coordinates q, θ, τ to the old ones:

$$\theta = -t + \frac{1}{\varepsilon} \int^x \frac{dX}{D^{1/2}}, \qquad \tau = \frac{1}{6} \int^X D^{1/2}(X) \, dX, \, X = \varepsilon x;$$

the elevation N is proportional to D^2q where $D(X)$ is the nondimensional depth.) Our goal is to describe the subsequent evolution. Many approaches are possible; one is better than another in certain aspects but there is no uniquely correct method. In (i) and (ii) below I will describe the direct perturbation method and the inverse scattering method respectively and point out their merits and deficiencies. The method followed in (iii), which I call the judicious use of conservation laws, when used with some understanding of what to expect, an understanding originally gained from (ii), is the one I recommend. In (iv) I will describe briefly how to use (iii) to overcome the incompatibility between the constant mass flux $M = \int_{-\infty}^{\infty} D(x)U(x, t) \, dt$ for the full two-directional equations and that conserved by the unidirectional approximation $m \propto \int_{-\infty}^{\infty} D^{3/4}(x)U_+(x, t) \, dt$. In the exercises there will be several examples to test you.

(i) *The direct approach.* The most obvious thing to do is to look for solutions to (3.111)

$$q(\theta, \tau) = q^{(0)} + \sigma q^{(1)} + \cdots \qquad (3.113)$$

where one assumes $q^{(0)}$ to have a solitary wave form (3.112) except that η, the amplitude parameter, is a slowly varying function of $\tau(\eta(T), T = \sigma\tau)$ and the corresponding phase speed is

$$\bar{\theta}_\tau = 4\eta^2 + O(\sigma). \qquad (3.114)$$

It is natural, then, to use a coordinate system moving with the solitary wave $\xi = \theta - \bar{\theta}$, $s = \tau$ whence the equation for $q^{(1)}$ is

$$q_s^{(1)} - 4\eta^2 q_\xi^{(1)} + q_{\xi\xi\xi}^{(1)} + 6q_\xi^{(0)}q^{(1)} + 6q^{(0)}q_\xi^{(1)} = -\Gamma/\sigma q^{(0)} - q_T^{(0)}. \qquad (3.115)$$

Observe that for general $q^{(0)}(\theta, \tau)$ it is very difficult to solve (3.115) because it does not separate in any obvious way (method (ii) automatically will know how to achieve this). However in the present case, because $q^{(0)}$ is simply a function of ξ, this can be done. If we look for solutions $q^{(1)}$ which only depend on ξ and not on s, then we know the solvability of (3.115) demands that the RHS is orthogonal to solutions of the adjoint equation $L^A V = -V_{\xi\xi\xi} + 4\eta^2 V_\xi - 6q^{(0)} V_\xi = 0$ which decay as $\xi \to \pm\infty$. The only candidate is $q^{(0)}$ itself. Hence we require

$$\frac{\partial}{\partial T} \int_{-\infty}^{\infty} q^{(0)2} \, d\theta = -\frac{2\Gamma}{\sigma} \int_{-\infty}^{\infty} q^{(0)2} \, d\theta \qquad (3.116a)$$

which gives us that

$$\eta_\tau = -\tfrac{2}{3}\Gamma\eta. \qquad (3.116b)$$

If we do not demand (3.116a) and allow $q^{(1)}$ to depend on s, then it will grow linearly with s and destroy the uniformity of the asymptotic expansion (3.113) over long distances $s = \tau = O(1/\sigma)$.

Next we solve for $q^{(1)}$ and find

$$q^{(1)} = \frac{\Gamma}{6\eta} \{-1 + \tanh \eta\xi + \text{sech}^2 \eta\xi (3 - 3\eta\xi \tanh \eta\xi + 2\eta\xi - 2\eta^2\xi^2 \tanh \eta\xi)\}$$

$$+ \frac{\Gamma}{2}(1 - \tanh \eta\xi) \, \text{sech}^2 \eta\xi.$$

Note that as $\xi \to +\infty$, $q^{(1)} \to 0$ but that in the lee of the solitary wave $\xi \to -\infty$,

$$q^{(1)} \to -\frac{\Gamma}{3\eta}, \qquad (3.117)$$

a *nonzero* (almost) constant. Ouch! The series (3.113) becomes nonuniform in the lee of the solitary wave. This fact was originally discovered and verified numerically by Leibovich and Randall [82].

But there is worse to come. Let us check the exact conservation law,

$$\frac{\partial}{\partial \tau} \int_{-\infty}^{\infty} q(\theta) \, d\theta = -\Gamma \int_{-\infty}^{\infty} q(\theta) \, d\theta \qquad (3.118)$$

which is the equation describing mass flux conservation $m = \int D^{9/4} q \, d\theta$ for the right going flow. (I will use small m to distinguish it from the true mass flux M of the full two-directional problem.) If q can be approximated by $q^{(0)}$, then the LHS of (3.118) is $(\partial/\partial\tau)(4\eta) = 4\eta(-\tfrac{2}{3}\Gamma)$ whereas the RHS is $-\Gamma(4\eta)$. Therefore (imagine $\Gamma < 0$, D decreases), of the amount of extra water accumulated by the changing depth situation, only $\tfrac{2}{3}$ goes into the growing solitary wave. Where does the rest go?

Whereas the problem with (3.117) had been noted by the direct perturbationists, the last difficulty just described had not.

(ii) *The inverse scattering approach.* How are these two difficulties overcome? In 1976, David Kaup and I reattacked this problem from the point of view of inverse scattering theory. Our rationale was that the most natural way to examine the effects of perturbations was to convert the perturbed equation straight away into the action-angle variables or normal modes of the exactly solvable system. Kaup [73] had previously shown how to write down these

equations for the perturbed NLS equation. In this approach, we go directly to
(3.99), (3.100) and (3.101). Recall that if δq or q_t evolves according to any
member of the KdV family, say $(\text{KdV})_3$ as in (3.52), then all the integrals may
be calculated to give respectively, $8i\zeta^3 b^*/a$, 0 and $8i\zeta_k^3 b^*/a$. To calculate the
effect of the perturbation $-\Gamma q$, we simply use the leading approximation to q,
the one-soliton solution and calculate the corresponding squared eigenfunctions. We know for example from (3.110) that for a pure one-soliton solution
$(\zeta_1 = i\eta)$ $q = -4\gamma_1\zeta_1\psi^2(x, \zeta_1)$ and thus (3.100) becomes $\zeta_{1\tau} = i\eta_\tau = -\frac{2}{3}i\Gamma\eta$ which
is exactly (3.116b). When we used (3.99) to compute how much continuous
spectrum is excited (remember at $\tau = 0$, $b^*/a = 0$, we found that a singular
contribution arose from the neighborhood of $\zeta = 0$ and gave rise to a corresponding new flow field q_c which for *short* distances is

$$q_c(\theta, \tau) = \begin{cases} -\dfrac{\Gamma}{3\eta}, & 0 < \theta < \bar{\theta}, \\ 0 & \text{elsewhere.} \end{cases} \tag{3.119}$$

A shelf is created which stretches between $\theta = 0$ and the present soliton
position $\bar{\theta}$; in the original coordinates, it stretches between the solitary wave
and the point to which the longest linear wave would have travelled from the
position at which the depth change first occurred. It has exactly the amplitude
of that part of $q^{(1)}$, namely (3.117), which did not decay as $\xi \to -\infty$. This is the
shelf which Leibovich and Randall had observed, but they did not understand
in what sense it had a finite range. While the shelf is slowly varying in
amplitude, its range varies at an order one rate. This means that the response
of the solution to the slowly changing depth is not, repeat not, purely adiabatic!
It is this feature that allows us to satisfy the local mass flux law,

$$\frac{\partial}{\partial\tau}\int_{-\infty}^{\infty} q\, d\theta = -\frac{9}{4}\frac{D_\tau}{D}\int_{-\infty}^{\infty} q\, d\theta. \tag{3.120}$$

Since q_c is slowly varying and small, from equation (3.111) we expect,
$\partial q_c/\partial\tau = -\frac{9}{4}(D_\tau/D)q_c$ which is simply Green's law, that along right going characteristics $\theta = -t + \int dx/D^{1/2}$, $D^{9/4}q_c$ is constant. Hence the LHS of (3.120) is (call
$\Gamma = \frac{9}{4}D_\tau/D$)

$$\frac{\partial}{\partial\tau}\int_{-\infty}^{\infty} q_s\, d\theta + \frac{\partial}{\partial\tau}\int_0^{\bar{\theta}} q_c\, d\theta = 4\eta\left(-\frac{2\Gamma}{3}\right) + \frac{\partial\bar{\theta}}{\partial\tau}q_c(\bar{\theta}) + \int_0^{\bar{\theta}}\frac{\partial q_c}{\partial\tau}\, d\theta$$

$$= 4\eta\left(-\frac{2\Gamma}{3}\right) + 4\eta^2\left(-\frac{\Gamma}{3\eta}\right) - \Gamma\int_0^{\infty} q_c\, d\theta$$

$$= -\Gamma\int_{-\infty}^{\infty} q_s\, d\theta - \Gamma\int_0^{\bar{\theta}} q_c\, d\theta,$$

which is the RHS. Note the crucial importance of the second term $-\frac{4}{3}\eta\Gamma$ which
comes from the fact that the flow is not adiabatic (the range of q_c, $0 < \theta < \bar{\theta}$, is
not slowly varying). Indeed we could have calculated the initial amplitude

("initial" meaning its amplitude on creation immediately behind the solitary wave) of the shelf by using the local mass flux deficit; that is instead of verifying (3.120), we can use it to obtain $q_c(\bar{\theta})$.

The great strength of the inverse scattering method from the theoretical point of view is that it converts the perturbation equation directly into the "right" coordinates. In effect, it solves the most important problem facing the direct perturbation method, how to separate (3.115). The basis in which to expand $q^{(1)}$ is (3.105); *in that basis* (3.115) *automatically separates*, no matter how complicated $q^{(0)}(\theta, \tau)$ is!! Furthermore, the formulae one obtains for the time dependence of the coefficients of $q^{(1)}$ in that basis, namely the time evolution of the changes in b^*/a, ζ_k and β_k induced by the perturbation, are exactly the same as one obtains by expanding (3.99)–(3.101) in a perturbation series with $\delta = \partial/\partial\tau$, $q_\tau = (q_\tau)_I + (q_\tau)_p$ $((q_\tau)_I$ is the integrable part of (3.111), $(q_\tau)_p$ the perturbation part). The squared eigenfunctions are approximated by their one-soliton expressions.

The weakness in the method is a practical one. It is clear that over long distances new flow components (which are $O(\sigma)$ in magnitude but which carry an $O(1)$ mass flux) are created. Therefore, the approximation of using the one-soliton eigenfunctions in (3.99) is invalid over long times. The reason is that the correction to the squared eigenfunction has singular behavior near $\zeta = 0$. One can, however, use this approximation over long times in (3.100). Nevertheless, despite these weaknesses, the inverse scattering approach did provide the key into solving the vexing and crucial question concerning the mass flux balance and for that achievement alone deserves high marks. I now will describe an approach which uses to best advantage all the experience gained from methods (i) and (ii) and which can deal successfully with the problem of describing the evolution of the shelf over long $(O(\sigma^{-1}))$ distances.

(iii) *Judicious use of the conservation laws.* Knowing what we know, this turns out to be the best method. We know that the conservation laws are exact; using the one for energy

$$\frac{\partial}{\partial\tau}\int_{-\infty}^{\infty} q^2 \, d\theta = -2\Gamma \int_{-\infty}^{\infty} q^2 \, d\theta \tag{3.121}$$

and indeed any of the later ones we find (3.116b) by replacing q by the slowly varying single solitary wave. The reason this is valid is that the new flow components induced by the perturbation are $O(\sigma)$ in amplitude and at most $O(\sigma^{-1})$ in length. Therefore the only conserved density to which they contribute an order one amount is the mass flux. The correction to the energy and higher conserved densities is at most $O(\sigma)$. All the conservation laws from energy on give the same behavior for η, namely (3.116b). Now we know already that the changing solitary wave cannot absorb (think of $\Gamma = \frac{9}{4}D_\tau/D < 0$) sufficient water to satisfy the mass flux law

$$\frac{\partial}{\partial\tau}\int_{-\infty}^{\infty} q \, d\theta = -\Gamma \int_{-\infty}^{\infty} q \, d\theta \tag{3.122}$$

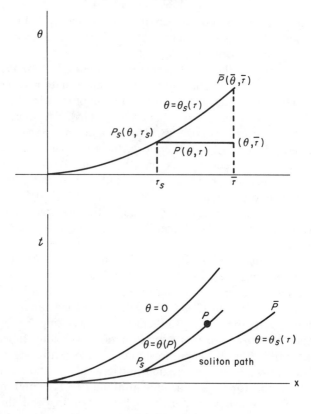

FIG. 4. *The right going shelf.*

and that, directly in the lee of the solitary wave, a shelf of amplitude $q_c = -\Gamma/3\eta$ is created. We can deduce this as follows. Assume that the field $q(\theta, \tau)$ consists of a solitary wave component $q_s(\theta, \tau)$ with slowly changing η given by (3.116b) and a shelf component $q_c(\theta, \tau)$ stretching between $\theta = 0$ and $\theta = \bar\theta$, the present location of the solitary wave. Since q_c is slowly varying in θ and small in amplitude, its evolution in τ is given by Green's law which, if $\Gamma = \frac{9}{4}D_\tau/D$, means that $\partial q_c/\partial \tau = -\Gamma q_c$ and that $D^{9/4}q_c$ is a function only of θ and therefore a constant along right-going characteristics. We now use (3.122): $(\partial/\partial\tau)4\eta + q_c(\bar\theta) \cdot \bar\theta_\tau = -\Gamma(4\eta)$ and therefore, $q_c(\bar\theta) = -\Gamma/3\eta$. In order to find $q_c(\theta, \tau)$ everywhere, consider Fig. 4. We know that $D^{9/4}q_c(\theta, \tau)$ is constant along the path $\theta = $ constant. Therefore, follow the right going characteristic through θ, τ back until it intersects the soliton path. At this point we know q_c. Hence

$$q_c(\theta, \tau) = \frac{D^{9/4}(\tau_s)}{D^{9/4}(\tau)} q_c(\tau_s),$$

where q_c is given as a function of θ by integrating and inverting the solitary wave path formula

$$\theta_\tau = 4\eta^2.$$

To make things concrete, let us consider an example. Let $-\frac{9}{4}D_\tau/D = \sigma$, a constant. Then

$$\eta = \eta_0 e^{2/3\sigma\tau}.$$

The solitary wave path is

$$\theta_s = \frac{3\eta_0^2}{\sigma}(e^{4/3\sigma\tau_s} - 1).$$

Now, we know that at $P_s(\theta_s, \tau_s)$

$$q_c(\tau_s) = \frac{\sigma}{3\eta(\tau_s)} = \frac{\sigma}{3\eta_0}e^{-2/3\sigma\tau_s}.$$

We also know that

$$q_c(\theta, \tau) = e^{\sigma(\tau-\tau_s)}q_c(\tau_s) = \frac{\sigma}{3\eta_0}e^{\sigma\tau}e^{-5/3\sigma\tau_s},$$

which from inverting the soliton path formula is

$$q_c(\theta, \tau) = \frac{\sigma}{3\eta_0}e^{\sigma\tau}\left(1 + \frac{\sigma\theta}{3\eta_0^2}\right)^{-5/4}.$$

Let us now calculate the mass flux associated with the shelf component,

$$m_c = D^{9/4}(\tau)\int_0^{\bar{\theta}} q(\theta, \tau)\,d\theta.$$

Using Green's law, namely the fact that along a $\theta = $ constant characteristic, $D^{9/4}(\tau)q(\theta, \tau) = D^{9/4}(\tau_s)q_c(\theta = \theta_s, \tau_s)$, we can write this as an integral over τ_s, the solitary wave path variable from $\tau_s = 0$, the point at which the depth change began and $\tau_s = \bar{\tau}$, the present position of the solitary wave.

$$m_c = \int_0^{\bar{\tau}} D^{9/4}(\tau_s)q_c(\theta_s, \tau_s)\frac{d\theta_s}{d\tau_s}\,d\tau_s$$

$$= \int_0^{\bar{\tau}} D^{9/4}(\tau_s)\left(-\frac{3D_\tau(\tau_s)}{4\eta(\tau_s)D(\tau_s)}\right)4\eta^2(\tau_s)\,d\tau_s$$

$$= -3\eta_0\int_0^{\bar{\tau}} D^{-1/4}(\tau_s)D_\tau(\tau_s)\,d\tau_s,$$

and since $\eta D^{3/2} = \eta_0 D_0^{3/2} = \eta_0$ (because $D_0 = 1$),

$$= 4\eta_0 - 4\eta_0 D^{3/4}(\bar{\tau}).$$

But the mass flux associated with the solitary wave is

$$m_s = D^{9/4}(\tau)\cdot 4\eta(\tau) = 4\eta_0 D^{3/4}(\bar{\tau}).$$

Therefore, the total mass flux associated with the right going component of the flow,

$$m = m_s + m_c = \int_{-\infty}^{\infty} D^{9/4}(\tau) q(\theta, \tau)\, d\theta = 4\eta_0,$$

is indeed a constant as equation (3.111) demands.

Before we tackle the problem of the discrepancy between the mass flux carried by the perturbed KdV equation which describes the right going component of the flow and that associated with the full two-directional flow, let us consider the nature of the shelf in terms of scattering data. Recall the first of the trace formulae (3.70) which expresses the mass content

$$\int_{-\infty}^{\infty} q\, d\theta = 4 \sum_{1}^{N} \eta_k + \frac{2}{\pi} \int_{0}^{\infty} \ln\left(1 - |R|^2\right) d\zeta$$

in terms of the scattering parameters. From (3.111), $\int_{-\infty}^{\infty} q\, d\theta$ is exactly $4\eta_0(1/D)^{9/4}$. The solitary wave component q_s of the field q with which we associate soliton η_1 has mass $4\eta_1 = 4\eta_0(1/D)^{3/2}$. If $1 > D$, the depth decreases; then since the contribution of the mass from the continuous spectrum $(2/\pi)$ $\int_{0}^{\infty} \ln(1 - |R|^2)\, d\zeta$ is always negative, q_c must be resolved into solitons in order that (3.70) can balance because $1/D^{9/4} > 1/D^{3/2}$. The shelf which in this case has a positive amplitude of order σ, and a width of order σ^{-1}, will decompose into a large number of solitary waves whose spectral representation is a set of $\zeta_k = i\eta_k$, $k = 2, \ldots, N$ which are densely packed along the imaginary axis between $\zeta = 0$ and $\zeta = 0(i\sigma)$. (Think of Exercise 3d(3) with $Q = \sigma$, $L = 1/\sigma$.) Eventually, in a time $(1/\sigma) \ln(1/\sigma)$, the solitary waves contained in the shelf will separate from each other.

It is quite surprising, I think, that what looked like a harmless perturbation is so very difficult to analyze. Although I do not fully understand the connection, part of the reason is that its influence on the "conserved quantity" $\int q\, dx$, the Casimir functional arising from the degeneracy of the Poisson bracket (3.25), is so very different from its influence on all the other constants $\int q^2\, dx$, $\int (\frac{1}{2} q_x^2 - q^3)\, dx, \ldots$.

(iv) *The reflected flow* [43]. We now seek to find the reflected flow field $\eta_-(x, t)$ and $u_-(x, t)$ which is generated by the interaction of the right going component $\eta_+(x, t)$ and $u_+(x, t)$ with the depth change. We use the same strategy as before. First, we calculate the value of u_- on the right going characteristic $\theta_+ = 0$ as follows. Consider Fig. 5.

Let us fix $x, 0 < x < \bar{x}$, where $\bar{x}$ is the present position of the solitary wave. Since the amplitude of the reflected flow turns out to be $O(\sigma \varepsilon)$, we can, for the purposes of this calculation, ignore the difference between the solitary wave path and $\theta_+ = 0$. Also $\eta_-(x, t)$ and $u_-(x, t)$ will satisfy the linear equations

$$\eta_{-t} + (Du_-)_x = 0, \tag{3.123a}$$

$$u_{-t} + \eta_{-x} = 0 \tag{3.123b}$$

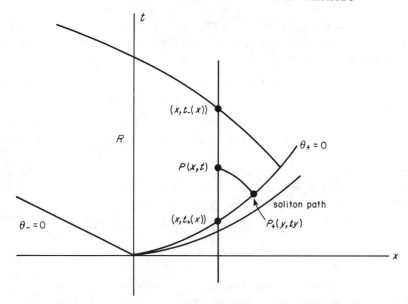

FIG. 5. *The reflected flow.*

(see (2.11), (2.12)) in the triangular region of the (x, t) plane shown in Fig. 5. Let $t_+(x)$ be the point at which x intersects $\theta_+ = 0$ and $t_-(x)$ the point at which x intersects the left going characteristic through the present position $(\bar{x}, \bar{t})$ of the solitary wave. Then since $\int_{-\infty}^{\infty} D(x)u(x, t) \, dt$ is independent of x to leading order in ε, we have

$$\frac{\partial}{\partial x} \int_{t_+(x)}^{t_-(x)} D(x)u_-(x, t) \, dt + \frac{\partial}{\partial x} D^{1/4}(x) \left(\int_{-\infty}^{\infty} D^{3/4}(x)u_+(x, t) \, dt \right) = 0.$$

(3.124)

The second term in this equation is

$$\tfrac{1}{4} D_x D^{-3/4} m$$

where $m = \tfrac{8}{3}\eta_0$, the mass flux associated with the right going flow, is independent of x. Equation (3.124) tells us that (recall $dt_+/dx = 1/\sqrt{D}$, $dt_-/dx = -1/\sqrt{D}$),

$$-D^{1/2}(x)u_-(x, t_-) - D^{1/2}(x)u_-(x, t_+) + \eta_-(x, t_-) + \eta_-(x, t_+)$$
$$= -\tfrac{2}{3}\eta_0 D_x D^{-3/4} - \eta_-(x, t_-) + \eta_-(x, t_+)$$
$$= -\tfrac{2}{3}\eta_0 D_x D^{-3/4},$$

where we have used (3.123a) to replace $(Du_-)_x$. Adding and subtracting $D^{1/2}(x)u_-(x, t_+)$, we have

$$D(x)u_-(x, t_+) = \tfrac{1}{3}\eta_0 D^{-1/4}D_x - \frac{\sqrt{D}}{2}(\eta_- + \sqrt{D}\,u_-)_{t_-} + \frac{\sqrt{D}}{2}(\eta_- + \sqrt{D}\,u_-)_{t_+}.$$

(3.125)

Now we can take x as close as we like to $\bar{x}$, in which case $t_+ \to t_- \to \bar{t}$ and (3.125) furnishes two pieces of information. First, immediately behind the right going solution component

$$D(\bar{x})u_-(\bar{x}, \bar{t}) = \tfrac{1}{3}\eta_0 D^{-1/4} D_x, \tag{3.126a}$$

and second

$$\frac{\partial}{\partial t}(\eta_- + \sqrt{D}u_-) = 0. \tag{3.126b}$$

Since the same argument would apply no matter where the right going component is along $\theta_+ = 0$, we have that $u_-(x, t)$ on $\theta_+ = 0$ is given by (3.126a) and that (3.126b) holds throughout the triangular region in Fig. 5. A little calculation with (3.123) and (3.126b) shows that

$$\left(\frac{\partial}{\partial t} - \sqrt{D}\frac{\partial}{\partial x}\right)\binom{\eta_-}{Du_-} = 0, \tag{3.127}$$

which means that both η_- and Du_- are constant along θ_- characteristics.

These facts were checked numerically by solving (3.123) as a Goursat problem; given u_-, $\eta_- = 0$ along $\theta_- = 0$ and given u_- from (3.126a) on $\theta_+ = 0$, find η_-, u_-.

Therefore, Green's law does not hold for the reflected wave. The reason is that Green's law follows from a geometrical optics argument in which the background changes slowly compared to the horizontal gradients of the wave. Here both $(1/u_-)\,\partial u_-/\partial x$ and D_x/D are of the same order of magnitude; both are $O(\sigma\varepsilon)$. From (3.123a), it is clear also that the amplitude of the reflected wave is $O(\sigma\varepsilon)$. However its length is $O(\sigma\varepsilon)^{-1}$ and therefore it carries an order one mass flux, which we now compute.

Let $x = x_f$ be the point at which the depth again becomes a constant D_f. Then the total mass flux through any station x should be equal to the total mass flux of the right going component through x_f because after this point there is no further reflection. We have already computed the flux of the right going flow component to be

$$\tfrac{8}{3}\eta_0 D^{1/4}(x).$$

We seek to show that the mass flux of the reflected wave is

$$\tfrac{8}{3}\eta_0(D_f^{1/4} - D^{1/4}(x)).$$

Consider

$$\int_{t_+(x)}^{t_f} D(x)u_-(x, t)\,dt$$

and write this as an integral along the path $\theta_+ = 0$ from $x = x$ to $x = x_f$ by using the fact that Du_- is a constant along a θ_- characteristic. Note that the t-coordinate at $P(x, t)$ in Fig. 5 is related to the x-component of $P_+(y, t_y)$, the point at which the left going characteristic $t + \int^x D^{-1/2}(s)\,ds = t_y + \int^y D^{-1/2}(s)\,ds$

through P meets the curve $\theta_+ = 0$, $t_y = \int^y D^{-1/2}(s)\, ds$, by

$$t + \int^x D^{-1/2}(s)\, ds = 2 \int^y D^{-1/2}(s)\, ds.$$

Therefore,

$$\int_x^{x_f} \left(\frac{\eta_0}{3} D^{-1/4}(y) D_y(y) \right) \frac{2}{D^{1/2}(y)}\, dy = \int_x^{x_f} \frac{2\eta_0}{3} D^{3/4}(y) D_y\, dy = \frac{8\eta_0}{3} (D_f^{1/4} - D^{1/4})$$

as required.

The reader should note that the total mass flux through a station x is constant and equal to

$$\tfrac{8}{3}\eta_0 D_f^{1/4}.$$

This means that as D_f gets small, most of the water in the incoming wave is reflected and very little propagates through to the beach.

Exercises 3g.

1. Find how the solitary wave parameter η changes and the shape of the shelf for the following examples given that $q(x, 0) = 2\eta_0^2 \operatorname{sech}^2 \eta_0 x$:

(a) $q_t + 6qq_x + q_{xxx} = \sigma q_{xx}$, $0 < \sigma \ll 1$;
(b) $q_t + 6q^2 q_x + q_{xxx} = \sigma q$, $0 < \sigma \ll 1$.

Answer.

(a)
$$\eta = \eta_0 \left(1 + \frac{16\eta_0^2 \sigma t}{15} \right)^{-1/2},$$

$$q_c = \frac{8\eta_0 \sigma}{15} \exp\left(\frac{-2\sigma x}{15} \right), \qquad 0 < x < \bar{x},$$

$$\bar{x}_t = 4\eta^2.$$

(b)
$$\eta = \eta_0 e^{2\sigma t},$$

$$q_c = \frac{\pi \sigma e^{2\sigma t}}{\eta_0^2 + 4\sigma x}, \qquad 0 < x < \bar{x},$$

$$\bar{x}_t = \eta^2.$$

2. Use the conservation laws

$$\frac{\partial}{\partial t} \int qq^*\, dx, \quad \frac{\partial}{\partial t} \int (qq_x^* - q^* q_x)\, dx$$

to find how the parameters η, ξ of an NLS solitary wave

$$q(x, t) = 2\eta \operatorname{sech} 2\eta(x - \bar{x}) \exp(-2i\xi x - 2i\bar{\sigma})$$

satisfying

$$q_t = iq_{xx} + 2iq^2 q^* - \Gamma q - E e^{i\omega_0 t}, \qquad \Gamma, E \ll 1$$

change. Show that $(\xi\eta)_t = -\Gamma(\xi\eta)$. Assume that $\xi = 0$; then show that

$$\eta_t = -2\Gamma\eta + \tfrac{1}{2}\pi E \sin(\omega_0 t + 2\bar{\sigma}), \qquad \bar{\sigma}_t = -2\eta^2.$$

Analyze these equations and show how the solitary wave phase locks to the forcing frequency ω_0. For details, see reference [45].

3h. Multisoliton, rational and finite gap solutions [25]–[29], [83]–[85].
What we are going to do. The first goal of this section is to introduce multisoliton solutions for the KdV family in a new and instructive way. The approach is instructive because it focuses attention on the unique structure of the eigenfunctions $\psi(x, t_{2k+1}; \zeta)$ which are associated with the multisoliton solution. Under proper normalization, they are the product of a polynomial in ζ^{-1} with a simple exponential. Indeed the derivation of the solution formula makes explicit use of this structure. The rational solutions are then introduced in the same way and their connection with the multisoliton solution is noted. They are a special kind of limit. In both cases, the formula for $q(x, t_{2k+1})$ is

$$q(x, t_{2k+1}) = 2\frac{d^2}{dx^2}\ln\tau$$

where τ is a $(N\times N)$ determinant which when expanded assumes the form (3.88). In the rational solution limit, τ is a polynomial.

The second and by far the longest part of this section is devoted to the derivation of the finite gap or the multiphase, quasiperiodic solutions. This means that $q(x, t_{2k+1})$ is a periodic function of N phases θ_i, $i = 1, \ldots, N$, each linear in x and t_{2k+1}, $\theta_i = \sum_{j \text{ odd}} c_{ij} t_j$, $t_1 = x$. Because the c_{ij} are not necessarily commensurate, q is only quasiperiodic in x and t_{2k+1}. The one-gap solution for the KdV equation $q_t + 6qq_x + q_{xxx} = 0$ is

$$q(x, t) = \beta + (\alpha - \beta)\operatorname{cn}^2\left\{\sqrt{\frac{\alpha - \gamma}{2}}(x - 2(\alpha + \beta + \gamma)t - x_0); \frac{\alpha - \beta}{\alpha - \gamma}\right\}.$$

I have broken the calculation of these solutions into three segments. In the first, I show you how the N-gap solutions are connected with a Riemann surface

$$R : y^2 = -\prod_{j=0}^{2N}(\lambda - \lambda_j),$$

which is independent of t_{2k+1}, $k = 0, \ldots, N$. In the second, I introduce the new coordinates μ_j, $j = 1, \ldots, N$ which lie in the fixed intervals $(\lambda_{2j-1}, \lambda_{2j})$ $j = 1, \ldots, N$. The calculation of their dependence on t_{2k+1}, $k = 0, \ldots, N$, given by (3.167), is neat and swift. At first sight, however, these equations do not seem to be any more simple than the equations from which they were derived. In segment three, however, I show you that if we construct a map from the Riemann surface R, on which the μ's live, to a new manifold called the Jacobi variety, then the coordinates on the Jacobi variety which correspond to the μ's move linearly with the times. The solution for $q(x, t_{2k+1})$ is

$$c + 2\frac{d^2}{dx^2}\ln\Theta(\theta_1, \theta_2, \ldots, \theta_N),$$

where c is a constant, Θ is the Riemann Θ function and $\theta_i = \sum_{j\ odd} c_{ij}t_j$, with c, c_{ij} determined. We note that, yet again, the solution has the form $2(d^2/dx^2)\ln\tau$. This time $\tau = \Theta e^{(c/4)x^2}$.

The multisoliton and rational solutions. If you look again at (3.84), you will notice that the eigenfunction $\psi(x, t, \zeta)$ as a function of ζ has the form of a meromorphic function with poles at $\zeta = -\zeta_k = -i\eta_k$, $k = 1, \ldots, N$, the eigenvalues. One could also renormalize $\psi(x, t; \zeta)$ by multiplying (3.84) by $\zeta^{-N}\prod_1^N(\zeta + i\eta_k)$, whence it would look like $e^{i\zeta x}$ times a polynomial of degree N in ζ^{-1}. Since the renormalization does not involve x or t, ψ will still satisfy (3.1) and (3.92).

Motivated by these comments, let us seek multisoliton solutions to the full KdV family (3.9) (the first three are listed in (3.14)–(3.16)) by looking for solutions of (3.1) and the companion family (3.3)

$$v_{t_k} = \tfrac{1}{2}B_x^{(k)}v - B^{(k)}v_x, \tag{3.128}$$

where

$$B^{(k)} = -\lambda^k + B_1\lambda^{k-1} + \cdots + B_k, \tag{3.129}$$

in the form

$$v(x, t_3, t_5, \ldots) = e^{H(\zeta)}\left(1 + \frac{C_1}{i\zeta} + \frac{C_2}{(i\zeta)^2} + \cdots + \frac{C_N}{(i\zeta)^N}\right), \tag{3.130}$$

with

$$H = i\zeta x + i\zeta^3 t_3 + \cdots + i\zeta^{2n+1}t_{2n+1} + \cdots. \tag{3.131}$$

The translation flow $q_{t_1} = q_x$ is not included here; it can be reintroduced by substituting $x + t_1$ for x. The compatibility of (3.1) and (3.128) ensures that $q(x, t_3, t_5, \ldots)$ as function of $t_3, t_5, \ldots$ is a solution of the KdV family. Substituting (3.130) into (3.1) and comparing various powers of ζ^{-1} gives us relations between $C_1, C_2, \ldots, C_N$ and q and its x derivatives; indeed $C_1, C_2, \ldots, C_N$ are the first N terms in the asymptotic expansion derived from (3.27) of $v(x; \zeta)$. In particular, the first two are:

$$q = -2C_{1x}, \qquad C_2 = C_1^2 + \frac{q}{4}. \tag{3.132}$$

The fact that $C_{N+r} = 0$, $r \geq 1$, means that the solutions $q(x, t_3, \ldots)$, which emerge from this procedure, satisfy a set of nonlinear ordinary differential equations. We will return to this point later in the section.

Since $v(\mathbf{x}, \zeta)$ ($\mathbf{x} = (x, t_3, \ldots)$) satisfies (3.1), so does the linearly independent $v(\mathbf{x}, -\zeta)$. If we think of $v(\mathbf{x}, \zeta)$ as being proportional to $\psi(x, \zeta)$ as defined in Section d, then $v(x, -\zeta)$ has the form of the asymptotic expansion for $\phi(x, \zeta)$.

We know that at the eigenvalues $\zeta = i\eta_k$, $k = 1, \ldots, N$, ϕ and ψ are proportional. Motivated by this observation, we determine the functions $C_1, \ldots, C_N$ by demanding that at the N distinct locations $\zeta = i\eta_k$, $\eta_k > 0$, $k = 1, \ldots, N$,

$$v(\mathbf{x}, i\eta_k) = e^{-2\eta_k \bar{x}_k} v(\mathbf{x}, -i\eta_k), \qquad (3.133)$$

$e^{-2\eta_k \bar{x}_k}$ being the constant of proportionality. (3.133) is then a set of N equations in N unknowns which can be readily solved to give

$$q = -2C_{1x} = 2\frac{d^2}{dx^2} \ln W(\theta_1, \theta_2, \ldots, \theta_N). \qquad (3.134)$$

In this formula, for N odd,

$$W = \det \begin{pmatrix} \cosh \theta_1 & -\eta_1 \sinh \theta_1 & \eta_1^2 \cosh \theta_1 \cdots \\ \vdots & \vdots & \\ \cosh \theta_N & -\eta_N \sinh \theta_N & \cdots \end{pmatrix}. \qquad (3.135)$$

For N even, the first column is $\{\sinh \theta_j\}$ and the other columns change alternately between $\sinh \theta_j$ and $\cosh \theta_j$. The phase θ_j is linear in all the independent variables and is equal to (the H was defined in (3.131))

$$\theta_j = H(i\eta_j) + \eta_j \bar{x}_j = \eta_j(x - \bar{x}_j) + \eta_j^3 t_3 - \eta_j^5 t_5 \cdots.$$

The reader can calculate directly the first few: $N = 1$: then (3.133) is

$$e^{\theta_1}\left(1 - \frac{C_1}{\eta_1}\right) = e^{-\theta_1}\left(1 + \frac{C_1}{\eta_1}\right),$$

whence $C_1 = \eta_1 \tanh \theta_1$, $q = -2C_{1x} = 2\eta_1^2 \operatorname{sech}^2 \theta_1$, the one-soliton solution.

Now let us turn things around. Consider $v(\mathbf{x}, \zeta)$, $v(\mathbf{x}, -\zeta)$ given by (3.130) and demand that (3.133) holds. Then from our previous analysis, $v(\mathbf{x}, \zeta)$ is unique (the $C_1, \ldots, C_N$'s are uniquely determined); *there is one and only one function $v(\mathbf{x}, \zeta)$ which satisfies (3.130), (3.133)*. I now claim that the function $v(\mathbf{x}, \zeta)$ defined in this way satisfies (3.1) and (3.128). Let us check by direct calculation;

$$v_{xx} + \zeta^2 v = 2i\zeta e^H\left(\frac{C_{1x}}{i\zeta} + \frac{C_{2x}}{(i\zeta)^2} + \cdots + \frac{C_{Nx}}{(i\zeta)^N}\right) + e^H\left(\frac{C_{1xx}}{i\zeta} + \cdots + \frac{C_{Nxx}}{(i\zeta)^N}\right).$$

Therefore the function $w(x)$, which is defined to be

$$w(\mathbf{x}, \zeta) = v_{xx} + (\zeta^2 - 2C_{1x})v = e^H\left(\frac{d_1}{i\zeta} + \cdots + \frac{d_N}{(i\zeta)^N}\right),$$

where $d_j = C_{jxx} + 2C_{j+1x}$, $j = 1, \ldots, N-1$, $d_N = C_{Nxx}$, has the form of a polynomial of degree N in ζ^{-1}. But all the d_j, $j = 1, \ldots, N$ must be zero, for otherwise we could add $w(\mathbf{x}, \zeta)$ to the previous $v(\mathbf{x}, \zeta)$ and the sum $v(\mathbf{x}, \zeta) + w(\mathbf{x}, \zeta)$ would also satisfy (3.130), (3.133). But $v(\mathbf{x}, \zeta)$ is unique and hence $w(\mathbf{x}, \zeta) \equiv 0$. Thus $v(\mathbf{x}, \zeta)$ satisfies (3.1) with $q = -2C_{1x}$. As an exercise, show by a similar argument that

$$v_{t_3} - \frac{q_x}{4} v - \left(\zeta^2 - \frac{q}{2}\right)v_x = 0.$$

It is worth going through this argument carefully. Such kinds of arguments using the uniqueness of functions appear again and again in the beautiful theory of Krichever for finding the finite gap solutions of the KdV family.

The rational solutions are obtained as a special limit of the multisoliton solutions. One allows all the ζ_k to tend to zero in a proportional way and the constant of proportionality in (3.133) becomes $(-1)^N$. The reason for this becomes obvious as we calculate. Take $N = 1$ and apply (3.133), whence

$$e^{H(\zeta_1)}\left(1 + \frac{C_1}{i\zeta_1}\right) = e^{+2i\zeta_1\bar{x}_1}e^{H(-\zeta_1)}\left(1 - \frac{C_1}{i\zeta_1}\right).$$

Now, expand about $\zeta_1 = 0$; in order for the ζ_1^{-1} terms to balance, we must have $e^{2i\zeta_1\bar{x}_1} \to -1$. It is as if we made the phase shift $\bar{x}_1 = \pi/2\zeta_1$. We obtain, after taking the limit $\zeta_1 \to 0$, $C_1 x + 1 = 0$ or $C_1 = -1/x = -(d/dx)\ln x$. Hence

$$q = 2\frac{d^2}{dx^2}\ln x = -\frac{2}{x^2}. \tag{3.136a}$$

The reader should verify that for $N = 2$

$$q = 2\frac{d^2}{dx^2}\ln (x^3 + 3t). \tag{3.136b}$$

The limiting procedure is tedious but straightforward. The N-phase rational solution is given by

$$q = 2\frac{d^2}{dx^2}\tau_N, \tag{3.137}$$

where τ_N can most easily be obtained by successive applications of the Bäcklund transformation (4.107).

The finite gap solutions and their connection with a fixed Riemann surface. We now turn to the *finite gap solutions* of which the multisoliton solutions are a special limiting case. The name arises from a study of the periodic problem for (3.1). Given $q(x)$, periodic in x on an interval $[0, P]$, then it is known that the spectrum (the set of values $\zeta^2 = \lambda$ for which at least one of the eigenfunctions of (3.1) is periodic or antiperiodic) consists of a discrete set $\lambda_0 < \lambda_1 \leq \lambda_2 < \lambda_3 \leq \lambda_4 \cdots < \lambda_{2n-1} \leq \lambda_{2n} \cdots$. $(\lambda_0, \lambda_3, \lambda_4, \lambda_7, \lambda_8, \ldots$ correspond to the periodic eigenfunctions; $\lambda_1, \lambda_2, \lambda_5, \lambda_6, \ldots$ to the antiperiodic ones.) The bands $(\lambda_{2n-1}, \lambda_{2n})$, which may have zero length, are called the *unstable* bands as, in these regions, the corresponding *Bloch* eigenfunctions, defined by the conditions

$$\psi_{\pm}(x, \zeta) = 1, \qquad x = x_0, \quad 0 \leq x_0 \leq P, \quad x_0 \text{ fixed},$$

$$\psi_{\pm}(x + P, \zeta) = \rho\psi_{\pm}(x, \zeta),$$

grow exponentially with x (i.e. ρ, which depends on ζ, is greater than 1 in absolute value). If $q(x)$ is such that only a finite number of unstable bands, say N, are open, then it is called an N-gap potential. Since under any flow in the KdV family the spectrum is invariant, $q(x, t_3, t_5, \ldots)$ remains an N-gap potential for all values of $t_3, t_5, \ldots$ and, as we shall see, represents a solution which is quasiperiodic in the time variables. The general solution of the periodic problem is a limit of the N-gap solutions as $N \to \infty$. The reader is referred to reference [29].

The class of solutions which we investigate in this section arises by relaxing the condition that $q(x, t_3, \ldots)$ has a fixed period P in x. The resulting N-gap solution will be a quasiperiodic function of x as well as of $t_3, \ldots, t_{2N+1}$. We begin by writing equations (3.1) and (3.128) in system form

$$V_x = \begin{pmatrix} -i\zeta & q \\ -1 & i\zeta \end{pmatrix} V = Q^{(1)} V, \tag{3.138}$$

$$V_{t_3} = \begin{pmatrix} -i\zeta^3 + \dfrac{iq\zeta}{2} - \dfrac{q_x}{4} & \zeta^2 q + \dfrac{i\zeta q_x}{2} - \dfrac{q_{xx} + 2q^2}{4} \\ -\zeta^2 + \dfrac{q}{2} & i\zeta^3 - \dfrac{iq\zeta}{2} + \dfrac{q_x}{4} \end{pmatrix} V = Q^{(3)} V, \tag{3.139}$$

and, in general,

$$V_{t_{2N+1}} = Q^{(2N+1)} V. \tag{3.140}$$

Suppose we were to seek a solution to

$$q_{t_3} = -\tfrac{1}{4}(q_{xxx} + 6qq_x), \tag{3.141}$$

the solvability condition of (3.138) (3.139) in the form $q(X = x - ct_3)$. Let $X = x - ct_3$, $T = t_3$. whence (3.138), (3.139) become

$$V_X = Q^{(1)} V, \qquad V_T = (Q^{(3)} + cQ^{(1)}) V. \tag{3.142}$$

But the coefficient matrices only depend on X and therefore one can solve the T equation by separation of variables, $V = U e^{yT}$, whence (3.142) is

$$U_X = Q^{(1)} U, \tag{3.143}$$

$$yU = (Q^{(3)} + cQ^{(1)}) U = QU. \tag{3.144}$$

The compatibility of (3.143), (3.144) has Lax form

$$Q_X = [Q^{(1)}, Q] \tag{3.145}$$

which, if you work it out, is

$$q_{xxx} + 6qq_x - 4cq_x = 0 \tag{3.146}$$

and admits solutions

$$Q(X, \zeta) = U(X, \zeta) Q(X_0, \zeta) U^{-1}(X, \zeta), \tag{3.147}$$

where $Q^{(1)}$ and U are related by (3.143). Hence the characteristic polynomial of Q is independent of X and

$$R(y, \zeta) = \det(Q - yI) = 0 \qquad (3.148)$$

is an algebraic curve with coefficients constant in X. In the present case, (3.148) has the form

$$y^2 = h^2 + ef, \qquad Q = \begin{pmatrix} h & e \\ f & -h \end{pmatrix},$$

$$= -\lambda^3 - 2c\lambda^2 + \lambda\left(\frac{q_{xx} + 3q^2}{4} - cq - c^2\right) + \left(\frac{q_x^2}{16} + \left(\frac{q}{2} - c\right)\left(-\frac{q_{xx} + 2q^2}{4} + cq\right)\right)$$

where $\lambda = \zeta^2$. But from (3.146)

$$-\frac{q_{xx} + 3q^2}{4} + cq = E_1$$

and

$$-\frac{q_x^2}{16} - \frac{q^3}{8} + \frac{cq^2}{4} - \frac{qE_1}{2} = E_2$$

whence

$$y^2 = -\lambda^3 - 2c\lambda^2 - (E_1 + c^2)\lambda - (E_2 + cE_1). \qquad (3.149)$$

Equation (3.149) defines a Riemann surface of genus one (topologically equivalent to a torus or doughnut) which is independent of X.

Conversely, suppose we add the constraint $yV = (Q^{(3)} + cQ^{(1)})V$ to (3.138) (3.139), then q depends on x, t_3 only through $X = x - ct_3$ and (3.146) holds. To see this more generally, let us add to the list (3.138)–(3.140) the constraint

$$yV = QV, \qquad Q = \begin{pmatrix} h & e \\ f & -h \end{pmatrix}, \qquad (3.150)$$

with

$$Q = \sum_0^N u_{2r+1} Q^{(2r+1)}, \qquad Q^{(2r-1)} = \begin{pmatrix} h_r & e_r \\ f_r & -h_r \end{pmatrix}, \qquad (3.151)$$

where the u_{2r+1} are constants. Differentiate (3.150) and use (3.138) and the fact that

$$Q^{(1)}_{t_{2j+1}} - Q^{(2j+1)}_x + [Q^{(1)}, Q^{(2j+1)}] = 0 \qquad (3.152)$$

to show

$$\sum_0^N u_{2r+1} Q^{(1)}_{t_{2r+1}} = 0$$

or

$$\sum_0^N u_{2r+1} q_{t_{2r+1}} = 0. \qquad (3.153)$$

We can look at (3.153) two ways. First, as a first order partial differential equation in $x_1, t_3, t_5, \ldots, t_{2N+1}$, it tells us that q is a function of the N phases constructed from

$$\frac{dx}{u_1} = \frac{dt_3}{u_3} = \cdots = \frac{dt_{2N+1}}{u_{2N+1}} \tag{3.154}$$

which depend linearly on $x, t_3, \ldots, t_{2N+1}$. But we can also interpret (3.153) as an autonomous nonlinear ordinary differential equation of order $(2N+1)$ in x; replace $q_{t_{2r+1}}$ by

$$\frac{\partial}{\partial x} \frac{\delta H_{2r+1}}{\delta q} = \frac{\partial}{\partial x} L^r q \tag{3.155}$$

where the operator L is given by (3.12). The N-gap solution $q(x, t_3, \ldots, t_{2N+1})$ with the independent variables constrained by (3.154) is, therefore, when integrated once, a nonlinear autonomous ordinary differential equation for q as function of x,

$$\sum_0^N u_{2r+1} L^r q = \text{constant.} \tag{3.156}$$

Equation (3.156) is known as the *Lax–Novikov* equation. Since all the flows commute and are compatible with (3.156), equation (3.156) describes the shape of the N-gap solution for all times $t_1, t_3, t_5, \ldots, t_{2N-1}$. In fact, we shall see shortly how these times parametrize its solutions. Furthermore, the time flow with respect to t_{2m+1} for $m \geq N$,

$$q_{t_{2m+1}} = \frac{\partial}{\partial x} L^m q \tag{3.157}$$

may be written as a linear combination of the time flows $q_{t_{2r+1}}, r = 0, \ldots, N-1$ ($t_1 = x$) by using (3.156). Therefore the N-gap solutions are solutions not only of the first N members of the KdV family but of the infinite KdV family.

   The new coordinates and their time dependence. Now how do we find these solutions? In inverse scattering theory we started from the x equation (3.138) and derived from it the scattering data whose time evolution was found from (3.139), (3.140). This is the way we would also proceed if we were doing the periodic-in-x problem, although as I have mentioned previously, the time evolution is difficult to obtain because we do not have a point at ∞ ($x = \pm\infty$) where q is known for all time. However, for studying the N-gap solutions with the restriction of a fixed period in x relaxed, the convenient starting point is not (3.138) but rather the algebraic system of equations (3.150). We immediately find that in order for nontrivial solutions V to exist,

$$y^2 = \det Q = h^2 + ef = -\prod_0^{2N}(\lambda - \lambda_j). \tag{3.158}$$

Equation (3.158), the characteristic polynomial of Q, is an algebraic curve in (y, λ) and defines a hyperelliptic Riemann surface R of genus N. The determinant of Q is a polynomial of order $(2N+1)$ in λ; it is easy to check that its leading coefficient is -1, and we assume that the roots λ_j, $j = 0, \ldots, 2N$ are real. The Riemann surface R plays the same role for the finite gap solutions as the spectrum does for the initial value problem. First and foremost, it is a constant of the motion, independent of x and $t_1, t_3, \ldots$. To see this, cross-differentiate (3.150) with any of the time flows and find

$$Q_{t_{2r+1}} = [Q^{(2r+1)}, Q] \qquad (3.159)$$

with solution

$$Q = VQ_0V^{-1}, \qquad (3.160)$$

where Q_0 is independent of x and $t_1, t_3, t_5, \ldots$. Hence the characteristic polynomial of Q is indeed a constant of the motion. As a consequence, the roots $\lambda_0, \lambda_1, \ldots, \lambda_{2N}$ of $\det Q$ are also constants of the motion and for q periodic in x are the simple spectrum of (3.1) corresponding to the periodic and antiperiodic eigenfunctions.

Next, we introduce new variables μ_j, $j = 1, \ldots, N$ which are the (real) roots of $f(\lambda)$, the (2,1) element in Q (see (3.150)). (In our example (3.144), $N = 1$, $f = -\lambda + q/2 - c$, and there is only one μ which is equal to $q/2 - c$.) In order to discover some of their properties, we have to do some calculations. If we translate (3.159) into three equations for h_k, e_k, f_k (recall

$$Q^{(2k-1)} = \begin{pmatrix} h_k & e_k \\ f_k & -h_k \end{pmatrix})$$

and the polynomials h, e, f, we obtain

$$h_x = qf + e,$$
$$e_x + 2i\zeta e = -2hq, \qquad (3.161)$$
$$f_x - 2i\zeta f = -2h,$$

and three equations for h_{t_k}, e_{t_k}, f_{t_k}, the first of which is

$$h_{t_{2k-1}} = e_k f - f_k e, \qquad (3.162)$$

the only one we need. Now a little calculation on (3.161) shows that

$$y^2 = h^2 + ef = -\tfrac{1}{2}ff_{xx} + \tfrac{1}{4}f_x^2 - (\lambda + q)f^2 \qquad (3.163)$$

and, since f is always real, $y^2(\mu_j) = \tfrac{1}{2}f_x^2(\lambda = \mu_j) > 0$. Hence the roots of f lie between λ_{2k-1}, λ_{2k}, $k = 1, \ldots, N$ and we list them so that $\lambda_{2k-1} \leq \mu_k \leq \lambda_{2k}$, $k = 1, \ldots, N$.

$$f(\lambda) = -\prod_1^N (\lambda - \mu_j). \qquad (3.164)$$

Now from (3.162), (3.163), (3.164), we find, on comparing the coefficient of λ^{2N}, that

$$q = -\sum_0^{2N} \lambda_j + 2\sum_1^N \mu_j. \qquad (3.165)$$

Check this for $N = 1$: $\mu = q/2 - c$, and we have $q(x) = -\lambda_0 - \lambda_1 - \lambda_2 + q(x) - 2c$; but we have seen that the sum of the roots is $-2c$, and therefore (3.165) holds.

The μ_j are confined to the intervals $(\lambda_{2j-1}, \lambda_{2j})$ and they will move with x, $t_3, \ldots, t_{2N+1}$. We now seek this dependence. Since $h^2 + ef$ is constant,

$$2hh_{t_{2k-1}} + ef_{t_{2k-1}} + e_{t_{2k-1}}f = 0. \tag{3.166}$$

At $\lambda = \mu_j$, using (3.162), (3.166) becomes

$$2h(\mu_j)(-ef_k(\mu_j)) + ef_{t_{2k-1}}(\mu_j) = 0.$$

But, recalling (3.164),

$$f_{t_{2k-1}}(\mu_j) = -\mu_{j,t_{2k-1}}\prod_{l \neq j}(\mu_j - \mu_l)$$

and, from (3.158),

$$h(\mu_j) = \left(\prod_{l=0}^{2N}(\lambda_l - \mu_j)\right)^{1/2}.$$

Therefore

$$\mu_{j,t_{2k-1}} = \mp 2 \frac{(\prod_0^{2N}(\lambda_l - \mu_j))^{1/2}}{\prod_{l \neq j}(\mu_j - \mu_l)} f_k(\mu_j), \qquad k = 1, \ldots, N, \tag{3.167}$$

and we have the $t_1 = x, \ldots, t_{2N-1}$ dependence of μ's. In particular, for $k = 1$ ($t_1 = x$),

$$\mu_{jx} = \mp \frac{(\prod_0^{2N}(\lambda_l - \mu_j))^{1/2}(-1)}{\prod_{l \neq j}(\mu_j - \mu_l)}. \tag{3.168}$$

For $k = 2$, the KdV flow,

$$\mu_{jt_3} = \mp \frac{(\prod_0^{2N}(\lambda_l - \mu_j))^{1/2}}{\prod_{l \neq j}(\mu_j - \mu_l)}\left(\frac{q}{2} - \mu_j\right) \tag{3.169}$$

and q should be expressed in terms of λ's and μ's by (3.165).

The Abel map from the Riemann surface to the Jacobi variety. At first sight, these equations (3.167) look awful. Nevertheless, we are about to find that, after some manipulations, a little order and structure begins to appear. To begin, let me remind you that with u_{2N+1} set equal to one,

$$f = u_1 f_1 + u_3 f_2 + \cdots + u_{2N-1}f_N + f_{N+1}, \tag{3.170}$$

where

$$f_k = \lambda^{k-1}L^0(-1) + \lambda^{k-2}L^1(-1) + \cdots + \lambda^0 L^{k-1}(-1). \tag{3.171}$$

In (3.171), L is the operator (3.12), and $L^0(-1) = -1$ $L(-1) = q/2$. Note that $f_2 = q/2 - \lambda$. How do we see this? Note that if we write the "t" equations (3.3) in system form with $v_2 = v$,

$$\begin{pmatrix} v_1 \\ v_2 \end{pmatrix}_{t_{2k-1}} = \begin{pmatrix} h_k & e_k \\ f_k & -h_k \end{pmatrix}\begin{pmatrix} v_1 \\ v_2 \end{pmatrix}$$

then $f_k = B^{(k)} = B_0 \lambda^{k-1} + \cdots + B_{k-1}$, where the B_r are defined in (3.13). Also, since

$$f(\lambda) = -\prod_{j=1}^{N} (\lambda - \mu_j), \tag{3.172}$$

we find, on comparing powers of λ,

$$S_1 = L(-1) + u_{2N-1} L(-1),$$
$$-S_2 = L^2(-1) + u_{2N-1} L^1(-1) + u_{2N-3} L^0(-1), \tag{3.173}$$
$$\vdots$$
$$(-1)^{N-1} S_N = L^N(-1) + u_{2N-1} L^{N-1}(-1) + \cdots + u_1 L^0(-1),$$

where the $\{S_r\}_{r=1}^{N}$ are the sums of symmetric products of the roots

$$S_1 = \sum \mu_k, \quad S_2 = \sum_{k=l} \mu_k \mu_l, \quad \ldots, \quad S_N = \mu_1 \cdots \mu_N. \tag{3.174}$$

Finally, it is convenient to define the sequence $\{A_r\}_1^{\infty}$ by the relation

$$\left(1 - \frac{u_{2N-1}}{\lambda} + \frac{u_{2N-3}}{\lambda^2} + \cdots + (-1)^N \frac{u_1}{\lambda^N}\right)^{-1} = \sum_0 \frac{A_r}{\lambda^r}. \tag{3.175}$$

The first few are

$$A_0 = 1, \quad A_1 = u_{2N-1}, \quad A_2 = -u_{2N-3} + u_{2N-1}^2,$$
$$A_3 = u_{2N-5} - 2u_{2N-1} u_{2N-3} + u_{2N-1}^3. \tag{3.176}$$

With these definitions, we can invert (3.173) to give

$$L(-1) = S_1 + A_1,$$
$$-L^2(-1) = S_2 + A_1 S_1 + A_2,$$
$$L^3(-1) = S_3 + A_1 S_2 + A_2 S_1 + A_3,$$
$$\vdots$$
$$(-1)^{N-1} L^N(-1) = S_N + A_1 S_{N-1} + \cdots + A_N. \tag{3.177}$$

Note that the first equation in this set is (3.165): $\frac{1}{2} A_1 = \frac{1}{2} u_{2N-1} = -\sum_0^{2N} \lambda_j$ since the λ_j are the roots of $h^2 + ef$.

Now let us examine (3.167). Write the equation as

$$\frac{d\mu_j}{y(\mu_j)} = \frac{2f_k(\mu_j)}{\prod_{l \neq j} (\mu_j - \mu_l)} \, dt_{2k-1}. \tag{3.178}$$

We coordinatize a point on the hyperelliptic Riemann surface R

$$y^2(\lambda) = \prod_0^{2N} (\lambda_l - \lambda)$$

by the ordered pair (y, λ). Next, form the N linearly independent holomorphic differentials

$$\omega_s(\lambda) = \frac{\lambda^s \, d\lambda}{y(\lambda)}, \quad s = 0, \ldots, N-1, \tag{3.179}$$

over R. From (3.178) we have

$$d\phi_s = \sum_{j=1}^{N} \omega_s(\mu_j) = \sum_{j=1}^{N} \frac{\mu_j^s \, d\mu_j}{y(\mu_j)} = \sum_{j=1}^{N} \frac{\mu_j^s \sum_1^N \mu_{j,t_{2k-1}} \, dt_{2k-1}}{y(\mu_j)}$$

$$= 2 \sum_{k=1}^{N} dt_{2k-1} \sum_{j=1}^{N} \frac{\mu_j^s f_k(\mu_j)}{\prod_{l \ne j} (\mu_j - \mu_l)}$$

$$= 2 \sum_{k=1}^{N} dt_{2k-1} \sum_{j=1}^{N} \frac{\mu_j^s(\mu_j^{k-1} L^0(-1) + \mu_j^{k-2} L(-1) + \cdots + L^{k-1}(-1))}{\prod_{l \ne j} (\mu_j - \mu_l)},$$

$$s = 0, \ldots, N-1. \quad (3.180)$$

We have a remarkable result: the quantities

$$\sum_{j=1}^{N} \frac{\mu_j^s f_k(\mu_j)}{\prod_{l \ne j} (\mu_j - \mu_l)} \quad (3.181)$$

are independent of x, $t_3, \ldots, t_{2r-1}, \ldots$. Therefore (3.180) can be simply integrated; i.e., $\int d\phi_s = \phi_s$, $\int dt_{2k-1} = t_{2k-1}$.

In order to prove this, we need the result that

$$I_s = \sum_{j=1}^{N} \frac{\mu_j^s}{\prod_{l \ne j} (\mu_j - \mu_l)} = \delta_{s,N-1} \quad \text{for} \quad s \le N-1 \quad (3.182)$$

and the remaining members of the sequence $I_N, I_{N+1}, \ldots, I_{2N-1}$ satisfy the recurrence relations,

$$I_N = S_1 I_{N-1},$$
$$I_{N+1} = S_1 I_N - S_2 I_{N-1},$$
$$I_{N+2} = S_1 I_{N+1} - S_2 I_N + S_3 I_{N-1},$$
$$\vdots$$
$$I_{2N-1} = S_1 I_{2N-2} - S_2 I_{2N-3} + \cdots + (-1)^{N-1} S_N I_{N-1}. \quad (3.183)$$

I leave the proof of this as an exercise to the reader. (3.182) is proved by considering

$$\frac{1}{2\pi i} \int_C \frac{z^s \, dz}{\prod_{l=1}^{N} (z - \mu_l)},$$

where C is the circle at infinity. (3.183) is proved by subtracting appropriate multiples of $z^p \prod_1^N (z - \mu_k)$, $p = 0, \ldots, r$ from the numerator z^{N+r} so as to make the integral converge as $|z| \to \infty$.

It is easiest to show (3.181) by calculating the first few. For $k = 1$,

$$\sum_{j=1}^{N} \frac{\mu_j^s(-1)}{\prod_{l \ne j} (\mu_j - \mu_l)} = -\delta_{s,N-1} \quad \text{from (3.128)};$$

for $k = 2$,

$$\sum_{j=1}^{N} \frac{\mu_j^{s+1}(-1) + L(-1)\mu_j^s}{\prod_{l \ne j} (\mu_j - \mu_l)} = \begin{cases} 0, & s \le N-3, \\ -1, & s = N-2, \\ A_1 & s = N-1, \end{cases} \quad (3.184)$$

by using (3.183), (3.182) and the first equation in (3.177). For $k = 3$, replacing $L(-1)$, $L^2(-1)$ from (3.177),

$$\sum_{j=1}^{N} \frac{\mu_j^{s+2}(-1) + L(-1)\mu_j^{s+1} + L^2(-1)\mu_j^s}{\prod_{l \neq j}(\mu_j - \mu_l)} = \begin{cases} 0, & s \leq N-4, \\ -1, & s = N-3, \\ A_1, & s = N-2, \\ -A_2, & s = N-1, \end{cases} \quad (3.185)$$

because $-I_{N+1} + S_1 I_N - S_2 I_{N-1}$ and $I_N - S_1$ are both zero from (3.183). The pattern is now clear, and by induction it is readily shown that the matrix (rows $s = 0, \ldots, N-1$, columns $k = 1, \ldots, N$)

$$M_{sk} = \sum_{j=1}^{N} \frac{\mu_j^s f_k(\mu_j)}{\prod_{l \neq j}(\mu_j - \mu_l)} = \begin{pmatrix} 0 & 0 & 0 & \cdots & -1 \\ 0 & 0 & & & A_1 \\ \vdots & & & & -A_2 \\ \vdots & & -1 & & \vdots \\ -1 & A_1 & & & A_{N-1} \end{pmatrix}.$$

$$(3.186)$$

Return to (3.180) and integrate, for it is now a separable equation. The right-hand sides integrate to

$$2t_{2k-1}M_{sk}, \quad s = 0, \ldots, N-1, \quad k = 1, \ldots, N, \quad t_1 = x.$$

The left-hand sides, namely $\sum_{j=1}^{N} \omega_s(\mu_j)$, we integrate from a fixed point on the Riemann surface $p_0(y(\mu_0), \mu_0)$ to $p_j(y(\mu_j), \mu_j)$:

$$\phi_s(p_1, \ldots, p_N) = \sum_{j=1}^{N} \int_{p_0}^{p_j} \omega_s(\mu_j) = 2 \sum_{k=1}^{N} t_{2k-1}M_{sk}. \quad (3.187)$$

The phases $\phi_s(p_1, \ldots, p_N)$ are simply linear combinations of $x, t_3, \ldots, t_{2N-1}$.

But wait! The integrals on the left-hand side of (3.187) are not uniquely defined because the integration paths are not specified. Consider Fig. 6, with the contours $\{a_j\}_1^N$, $\{b_j\}_1^N$. The a_r contour surrounds the branch cut $(\lambda_{2r-1}, \lambda_{2r})$ whereas the b_r path comes from $-\infty$ to the branch cut $(\lambda_{2r-1}, \lambda_{2r})$ on one sheet and returns on the other. Therefore the left-hand sides are only defined up to

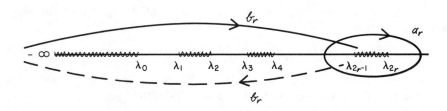

FIG. 6. The $\{a_j\}$, $\{b_j\}$ contours in the λ-plane.

the sum of

$$\sum_{1}^{N} n_k \int_{a_k} \omega_s(\mu_j) + \sum_{1}^{N} m_k \int_{b_k} \omega_s(\mu_j).$$

It turns out to be convenient to normalize the closed integrals about the a_k cycles as follows: Let

$$U_r = \sum_{0}^{N-1} C_{rs}\omega_s$$

and choose C_{ms} so that

$$\int_{a_n} U_n = \delta_{mn}.$$

Now, let us define the phases θ_r by

$$\theta_r(p_1, \ldots, p_N) = \sum_{j=1}^{N} \int_{p_0}^{p_j} U_r(\mu_j)$$

and, from (3.187) we find

$$\theta_r = \sum_{0}^{N-1} C_{rs}\phi_s, \qquad C = (C_{rs}),$$

$$= 2 \sum_{k=1}^{N} t_{2k-1}N_{rk} \tag{3.188}$$

where N is the product CM. (3.188) can be written $\boldsymbol{\theta} = 2N\mathbf{t}$ where $\boldsymbol{\theta}$, $\mathbf{t}$ are $\theta_1, \ldots, \theta_N$ and $t_1, \ldots, t_{2N-1}$ respectively.

Now that we have integrated the equations, it remains for us to determine the sums of symmetric products of the μ's which are the quantities q, $Lq, \ldots, L^N q$ of interest. The question then is, given

$$\theta_1, \theta_2, \ldots, \theta_N$$

can we determine $p_1, p_2, \ldots, p_N$ and in particular

$$\sum_{1}^{N} \mu_j = \tfrac{1}{2}q + \tfrac{1}{2}\sum_{0}^{2N} \lambda_j.$$

In order to answer this question we have to look at the nature of the map, called the *Abel map*,

$$(p_1, \ldots, p_N) \rightarrow (\theta_1, \theta_2, \ldots, \theta_N) \tag{3.189}$$

and its inverse. Since any permutation of $1, \cdots, N$ on the left of (3.189) gives the same set of θ's, the map is from $R \times R \times \cdots \times R/P_N$, namely N copies of the Riemann surface, modulo the permutation group on N symbols P_N, into C^N, N-dimensional complex space. Since the right-hand side depends on the paths of integration we can add to θ_r any linear combination $\sum_{i=1}^{n} (n_i \int_{a_i} U_r + m_i \int_{b_i} U_r)$ when m_i, n_i are integers. We will call

$$\int_{b_i} U_r = B_{ri}, \tag{3.190}$$

and state without proof that B_{ri} is symmetric and that its imaginary part is positive definite [85]. Recall that we have normalized $\int_{a_i} U_r = \delta_{ri}$. Therefore the resulting point in C^N is determined only up to an integral linear combination of the $2N$ vectors

$$\begin{pmatrix} 1 \\ 0 \\ \vdots \\ 0 \end{pmatrix}, \quad \dots, \quad \begin{pmatrix} 0 \\ \vdots \\ \vdots \\ 1 \end{pmatrix}, \quad \begin{pmatrix} B_{11} \\ B_{12} \\ \vdots \\ B_{N1} \end{pmatrix}, \quad \begin{pmatrix} B_{1N} \\ \vdots \\ \vdots \\ B_{NN} \end{pmatrix}.$$

Those $2N$ vectors define a lattice Λ in $C^N \simeq R^{2N}$. For example, if $N = 1$, the complex plane is covered by the period parallelogram familiar from elliptic functions. Thus $p_1, \dots, p_N$ is determined by where in the N-dimensional period parallelogram the point $\theta_1, \dots, \theta_N$ sits and is unchanged if the point $\theta_1, \dots, \theta_N$ is moved to a congruent point in another period parallelogram. Hence the symmetric products of the p's are periodic functions of the θ's and it is convenient to think of identifying opposite edges of the period parallelogram. Now one sees that the point $\theta_1, \dots, \theta_N$ really lives on an N-torus denoted C^N/Λ called the *Jacobi variety* of the curve (3.148).

The finite-gap solution for the KdV family is therefore equivalent to a linear flow on the Jacobi variety. The solution $q(x, t_3, \dots)$ can be expressed in terms of the Riemann Θ function

$$q(x, t_3, \dots) = 2 \frac{\partial^2}{\partial x^2} \ln \Theta(\theta_1, \theta_2, \dots, \theta_N) + c \tag{3.191}$$

where

$$\Theta(\theta_1, \dots, \theta_N) = \sum_{\nu_1, \nu_k \in Z}^{\infty} \exp\left(\sum_{k=1}^{N} 2\pi i \nu_k \theta_k + i\pi \sum_{k,j=1}^{N} B_{kj} \nu_k \nu_j \right), \tag{3.192}$$

where Z is the set of all integers and c is a complicated constant [85]. Notice the close correspondence between the finite-gap solution and the N-soliton solution. Note also that the τ-function, defined by $q = 2(\partial^2/\partial x^2) \ln \tau$, which I have briefly talked about and which will be a most important function for the rest of these lectures is the Θ-function times $e^{c/4x^2}$ when q is a finite gap potential.

CHAPTER 4

The τ-Function, the Hirota Method, the Painlevé Property and Bäcklund Transformations for the Korteweg–deVries Family of Soliton Equations

4a. Introduction. Up to this point, it has been our point of view that the class of problems we are attempting to solve are initial-boundary value problems. Namely, we think of x in all the equations as a space coordinate and associate with it boundary conditions such as periodicity or decay at infinity. This point of view emphasizes, and indeed makes necessary, considerations of the analytic nature of the functions we are dealing with. For instance, we found that if $q(x, 0)$ decays sufficiently rapidly as $x \to \pm\infty$ and obeys a certain integral condition, the scattering data have certain holomorphic properties. But in reality, the equations we are studying are magic because of their local propeties; for example, the fact that the KdV equation has multisoliton and multirational solutions has nothing to do with boundary conditions but is simply a consequence of the very special balance which occurs between the various terms in the equation. Disturbing this balance by adding a q or qq_{xx} term destroys the magic properties. Disturbing the boundary condition at infinity may make the equation more difficult to solve as an initial-boundary value problem but does not destroy its locally integrable character.

Accordingly, in this chapter, we seek to focus on those methods which depend on local rather than global properties of the equations. The most important character in the list of dramatis personae is the τ-function who is now ready to claim his rightful place in the main floodlight, center stage. He is ubiquitous and pops up in just about every scene, often independent of our original intentions. Somehow, he seems to know just how important he is.

4b. The τ-function. This multifaceted function was first discovered by Hirota as a means of generating soliton solutions and we will discuss that method in the next section. However, his true significance and central role in soliton theory was not appreciated until the work of the group consisting of M. Sato, Miwa, Jimbo, Kashiwara, Date, Y. Sato [39] at Kyoto University, and I think it is fair to say that even at this time, the import of this function has yet to be fully understood. Like the devil himself, he appears at many times in many disguises. Sometimes he is just a simple polynomial (the rational solutions of the KdV equation), sometimes he is a finite sum of exponentials (the multisoliton solution). Other times he gets more complicated ; he can be a Riemann Θ function (times a harmless factor) and a correlation function. Sounds intriguing, doesn't he?

We will get to know him first as a *potential function* from whose second logarithmic derivatives all conserved densities and fluxes can be calculated. Consider the family of KdV flows

$$q_{t_{2n+1}} = \frac{\partial}{\partial x} L^n q = 2 \frac{\partial}{\partial x} B_{n+1}. \tag{4.1}$$

For convenience, we introduce

$$w = \int^x q \, dx, \tag{4.2}$$

whence (4.1) is

$$w_{t_{2n+1}} = L^n q = 2 B_{n+1}. \tag{4.3}$$

Hence the function $w(x, t_3, \ldots)$ may be considered as a potential function for the infinite sequence $\{L^n q = 2 B_{n+1}\}_0^\infty$. Now, it is a known fact that the derivatives of all these functions with respect to the various times namely, $(\partial/\partial t_{2m+1}) L^n q$ can be written as the x or t_1 derivative of local quantities. Therefore, it is natural to use the potential function $\tau(t_1, t_3, \ldots)$ defined by

$$w = 2 \frac{\partial}{\partial t_1} \ln \tau(t_1, t_3, t_5, \ldots) \tag{4.4}$$

instead of w itself. In this way, you see,

$$\frac{\partial}{\partial t_{2m+1}} L^n q = \frac{\partial}{\partial t_1} 2 \frac{\partial^2 \ln \tau}{\partial t_{2m+1} \partial t_{2n+1}}$$

and so all the time derivatives of the sequence $\{L^n q\}$ are given by an equation in conservation form with respect to the special variable x. Notice, however, that the expression for the flux corresponding to the rate of change of the conserved density $L^n q$ in the $(KdV)_3$ flow is most conveniently given as a derivative with respect to the time t_{2n+1} corresponding to the $(n+1)$st flow in the KdV hierarchy. This emphasizes the fact that, when one seeks solutions for an integrable equation, it is good to understand that one is really seeking common solutions for the whole hierarchy of flows in its family.

For the first time in the literature, I will give you a formula for the *flux tensor*

$$F_{2m+1,2n+1} = 2 \frac{\partial^2 \ln \tau}{\partial t_{2m+1} \partial t_{2n+1}}. \tag{4.5}$$

Note, in particular, that

$$L^n q = 2 \frac{\partial^2 \ln \tau}{\partial t_1 \partial t_{2n+1}}. \tag{4.6}$$

In order to derive this formula, we start by writing a general equation for the B_n's. Since

$$v_{t_{2n+1}} = \tfrac{1}{2} B_x^{(n)} v - B^{(n)} v_x, \tag{4.7a}$$

$$v_{t_{2m+1}} = \tfrac{1}{2} B_x^{(m)} v - B^{(m)} v_x, \tag{4.7b}$$

where

$$B^{(n)} = \lambda^n \sum_0^n \frac{B_r}{\lambda^r},$$

it is not hard to show that

$$B_{t_{2m+1}}^{(n)} - B_{t_{2n+1}}^{(m)} + B^{(m)}B_x^{(n)} - B_x^{(m)}B^{(n)} = 0. \tag{4.8}$$

Now define

$$B = \lim_{n \to \infty} \frac{B^{(n)}}{\lambda^n}$$

and divide (4.8) by λ^n and take the limit $\lambda \to \infty$, thinking of $|\lambda| > 1$. Then

$$B_{t_{2m+1}} = B_x^{(m)}B - B^{(m)}B_x. \tag{4.9}$$

By writing (4.7) in system form for the vector $V = (v_1, v_2)^T$, $v_2 = v$, $v_1 = -v_{2x} + i\zeta v_2$, we have $(\lambda = \zeta^2)$

$$V_{t_{2n+1}} = Q^{(2n+1)}V, \tag{4.10a}$$

where

$$Q^{(2n+1)} = \left(i\zeta B^{(n)} - \frac{B_x^{(n)}}{2}\right)H + \left(i\zeta B_x^{(n)} - \frac{B_{xx}^{(n)}}{2} - qB^{(n)}\right)E + B^{(n)}F \tag{4.10b}$$

and H, E, F are the basis

$$\begin{pmatrix} 1 & 0 \\ 0 & -1 \end{pmatrix}, \quad \begin{pmatrix} 0 & 1 \\ 0 & 0 \end{pmatrix}, \quad \begin{pmatrix} 0 & 0 \\ 1 & 0 \end{pmatrix}$$

for $sl(2, C)$. It is easy to show that, with

$$Q = \lim_{n \to \infty} \frac{Q^{(2n+1)}}{\zeta^{2n+1}} = \sum_0^\infty \frac{Q_r}{\zeta^r}, \tag{4.11}$$

equation (4.9) takes the Lax form with the usual matrix commutator

$$Q_{t_{2n+1}} = [Q^{(2n+1)}, Q]. \tag{4.12}$$

By equating the λ^{-m-1} components of (4.9) we find

$$\frac{\partial B_{m+1}}{\partial t_{2n+1}} = B_{0x}B_{m+n+1} + \cdots + B_{nx}B_{m+1} - B_0 B_{m+n+1x} - \cdots - B_n B_{m+1x}. \tag{4.13}$$

I leave it as an exercise, admittedly a difficult one, to show that the right-hand side can be written as half the x derivative of

$$F_{2m+1,2n+1} = \frac{1}{2}\text{Tr}\left\{\sum_{s=0}^{2m+1} sQ_{2m+1-s}Q_{2n+1+s} + \sum_{s=0}^{2n+1} sQ_{2m+1+s}Q_{2n+1-s}\right\}, \tag{4.14}$$

where $\text{Tr}(Q_iQ_j)$ is the trace of the matrix product and the Q_j are defined by (4.11). It is clear that the expression (4.14) is symmetric under interchange of m and n. The simple rearrangement of indices gives us the conservation law for every flux tensor,

$$\frac{\partial}{\partial t_{2k+1}} F_{2m+1,2n+1} = \frac{\partial}{\partial t_{2m+1}} F_{2n+1,2k+1} = \frac{\partial}{\partial t_{2n+1}} F_{2k+1,2m+1}.$$

Knowing τ as a function of the times $x = t_1, t_3, t_5, t_{2n+1}, \ldots$ means we know everything about solutions of every member of the KdV family. In a sense, then, the τ function acts as a potential from which all components and all gradients with respect to all the times of the infinite dimensional vector B can be derived. It also has an important second interpretation which we will discuss when we get to the topic of Bäcklund transformations. In order to prepare the way, however, the following result which relates the eigenfunctions $v(x, t_3, \ldots)$ to τ is very useful.

The result I now give is formal because it uses (3.29a), the asymptotic expansion for $v(x, t_3, \ldots, \zeta)$ near $\zeta = \infty$. (Depending on the nature of the essential singularity at ∞, the formal expansion (3.29a) may not be uniformly valid in all neighborhoods of $\zeta = \infty$; however, for special classes of solutions, including the multisoliton solutions, the formal expansion (3.29a) is the Laurent expansion at $\zeta = \infty$.) The dependency of the asymptotic expansion on the times $t_3, t_5, \ldots$ is taken care of by changing the exponent $-i\zeta x$ in (3.29) to $-i \sum_0^\infty \zeta^{2k+1} t_{2k+1}$. We find

$$v(x, t_3, \ldots) \sim \exp\left(i \sum \zeta^{2k+1} t_{2n+1}\right) e^\Phi \tag{4.15a}$$

where

$$\Phi_x \sim \sum_1^\infty \frac{R_n}{(2i\zeta)^n}. \tag{4.15b}$$

(For multisoliton solutions, (4.15b) holds in all sectors of $\zeta = \infty$ and therefore is a Laurent expansion. One can then replace the asymptotic symbol in (4.15) with an equal sign.) Let us write the integrals of the first three terms of (4.15b) in terms of the τ function:

$$\Phi \sim \Phi_0 - \frac{\int q\, dx}{2i\zeta} - \frac{1}{4\zeta^2} q + \frac{1}{8i\zeta^3} \int (q_{xx} + q^2)\, dx + \cdots$$

$$= \Phi_0 - \frac{1}{i\zeta} \frac{\partial}{\partial t_1} \ln \tau - \frac{1}{2\zeta^2} \frac{\partial^2 \ln \tau}{\partial t_1^2} + \frac{1}{24 i\zeta^3} \int (q_{xx} + 3q^2)\, dx + \frac{1}{12 i\zeta^3} q_x + \cdots$$

$$= \Phi_0 - \frac{1}{i\zeta} \frac{\partial}{\partial t_1} \ln \tau - \frac{1}{2\zeta^2} \frac{\partial^2 \ln \tau}{\partial t_1^2} + \frac{1}{6i\zeta^3} \frac{\partial^3 \ln \tau}{\partial t_1^3} - \frac{1}{3i\zeta^3} \frac{\partial \ln \tau}{\partial t_3} \cdots \tag{4.16}$$

where we have used the fact that

$$q_{xx} + 3q^2 = -4 \int^x q_{t_3}\, dx = -8 \frac{\partial^2 \ln \tau}{\partial t_1 \partial t_3}.$$

Continuing this process (see Flaschka [86] for a proof) gives

$$\Phi - \Phi_0 \sim \ln \tau \left(t_{2k+1} - \frac{1}{i(2k+1)\zeta^{2k+1}} \right) - \ln \tau(t_{2k+1}).$$

Hence

$$v(x, t_3, \ldots ; \zeta) \sim \exp \left(i \sum_0 \zeta^{2k+1} t_{2k+1} \right) \cdot \frac{\tau(t_{2k+1} - 1/i(2k+1)\zeta^{2k+1})}{\tau(t_{2k+1})}.$$

$$(4.17)$$

We introduce the operator

$$X(\zeta) = \exp \left(i \sum \zeta^{2k+1} t_{2k+1} \right) \exp \left(\sum \frac{-1}{i(2k+1)\zeta^{2k+1}} \frac{\partial}{\partial t_{2k+1}} \right). \qquad (4.18a)$$

Operators of this type are closely related to the objects called vertex operators in the literature [101], [102]. We now see that

$$v(x, t_3, \ldots ; \zeta) \sim \frac{1}{\tau} X(\zeta) \cdot \tau \qquad (4.18b)$$

gives us a (formal) relation between the function which generates solutions of the KdV family and the eigenfunctions $v(x, t_3, \ldots ; \zeta)$. This particular formula will be used when we introduce Bäcklund transformations in Section 4f.

4c. Symmetries, conservation laws and constants of the motion. Consider a particle of unit mass with coordinate vector (q_1, q_2) and momentum vector (p_1, p_2) constrained to move in a plane under the influence of a central conservative force field. The Hamiltonian for this system is

$$H(q_1, q_2, p_1, p_2) = \tfrac{1}{2}(p_1^2 + p_2^2) + V(\sqrt{q_1^2 + q_2^2}) \qquad (4.19)$$

and the motion is given by

$$z^{\cdot} = J \nabla H \qquad (4.20)$$

where $z = (q_1, q_2, p_1, p_2)$, ∇ is the gradient with respect to these four variables and $J = \left(\begin{smallmatrix} 0 & I \\ -I & 0 \end{smallmatrix} \right)$ where I is the identity matrix $\left(\begin{smallmatrix} 1 & 0 \\ 0 & 1 \end{smallmatrix} \right)$.

Now it is intuitively obvious that we could have chosen to describe the motion from any reference frame which is a pure planar rotation through an angle θ of the coordinates z,

$$z' = Rz$$

where

$$R = \begin{pmatrix} M & 0 \\ 0 & M \end{pmatrix}, \qquad M = \begin{pmatrix} \cos \theta & \sin \theta \\ -\sin \theta & \cos \theta \end{pmatrix}$$

and θ, which coordinatizes the amount of rotation, is arbitrary. Since both the Hamiltonian H and the equations of motion are invariant under the action of the group of rotations (this means that $H'(z'(z)) = H(z)$ and $z' = J \nabla' H'$), the infinitesimal change $\partial z'/\partial \theta$, evaluated at the identity $\theta = 0$, satisfies the linearized version of equation (4.20). The reader might check this for himself. For θ small, $q_1' = q_1 + \theta q_2$, $\partial/\partial p_1' = \partial/\partial p_1 + \theta(\partial/\partial p_2)$ and then $\dot{q}_1' = \partial H'/\partial p_1'$ becomes $(\dot{q}_1 + \theta \dot{q}_2) = ((\partial/\partial p_1) + \theta(\partial/\partial p_2))H$ (remember $H'(z') = H(z)$), which is indeed true.

One can see, therefore, that the property that the Hamiltonian and equations of motion are invariant under the action of the rotation group can be expressed by the fact that the partial derivative of the solution $\partial z/\partial \theta \,|_{\theta=0}$, evaluated at the identity, solves the linearized equations of the motion. It is emphasized that the linearization can take place about any solution of the original equations (4.20).

We call the group action under which the Hamiltonian and the equations of motion are unchanged a *symmetry* of the system. As we have remarked, a necessary and sufficient condition for a continuous group action to be a symmetry is that its infinitesimal action, measured here by $\sigma(z) = \partial z/\partial \theta \,|_{\theta=0}$, is a solution of the linearized equations of motion. We shall also refer to the function $\sigma(z)$ itself as a symmetry.

Symmetries are very useful. In Hamiltonian systems, each symmetry is associated with a constant of the motion (Noether's theorem) through which the dimension of the system can be reduced by two. In the example quoted, the constant of the motion associated with the rotation group is the angular momentum. By a suitable choice of coordinates, the angle variable corresponding to the angular momentum (one of the two action variables for (4.19), the other is H itself), can also be removed (it becomes a cyclic or ignorable coordinate in the language of Hamilton). Therefore, by fixing the angular momentum h, one obtains a reduced equation of dimension two. Indeed, in this case, using the polar coordinate $r = (q_1^2 + q_2^2)^{1/2}$, the equation is

$$ \ddot{r} - \frac{h^3}{r^3} + \frac{\partial V}{\partial r} = 0 $$

from which the orbital motion of the particle is readily inferred by a phase plane analysis and, for certain potentials V, the motion $r(t)$ can be explicitly computed in terms of known functions.

The idea that a symmetry can be used to reduce the dimension of the mechanical system has been known for a long time [87]. In most of the classical examples, however, the symmetries are fairly obvious and have simple geometric interpretations (the motion is invariant under translation, rotation). So, too, are the corresponding conservation laws which have a correspondingly simple physical interpretation like conservation of linear and angular momentum. In soliton equations, however, things are not that simple. I have already pointed out to you that after the first couple of conservation laws for the KdV equation (which correspond to the conservation of mass (or momentum) and energy), the infinite number which follow have no physical interpretation. Neither do the symmetries. For this reason, they are called hidden (connotation: nonobvious) symmetries. We will see at the beginning of Chapter 5 that they are associated with the action of certain infinite dimensional Lie groups. In the case

of the KdV family, the symmetry group is the Kac–Moody group correspond-
ing to the graded Lie algebra $\widetilde{sl}(2, C)$, the loop algebra of $sl(2, C)$. We will also
talk later in Chapter 5 about the method of reduction in these cases and will
show how, using the Marsden, Weinstein [88] generalization of the classical
reduction method, the equation families given by (3.9) and (3.49) are reductions of much simpler flows on a higher dimensional manifold.

Let me now define and identify the symmetries and corresponding conservation laws for the KdV family. Motivated by the earlier example, we say the
function $v(\mathbf{u})$, meaning $v(u, u_x, u_{xx}, \ldots)$ is a symmetry of the scalar equation

$$u_t = Q(\mathbf{u}) \tag{4.21}$$

if $u + \varepsilon v(\mathbf{u})$ also satisfies (4.21) for all solutions of u of (4.21) for an arbitrarily
small ε. This means that $v(\mathbf{u})$ has to satisfy the linearized version of (4.21),
which is

$$v_t = Q'(\mathbf{u})[\mathbf{v}]. \tag{4.22}$$

The right-hand side of (4.22) means the directional (Fréchet) derivative of Q at
the point $\mathbf{u}$ in the direction $\mathbf{v}$ i.e. $Q_u v + Q_{u_x} v_x + \cdots$ and is defined as

$$\lim_{\varepsilon \to 0} \frac{1}{\varepsilon}(Q(u + \varepsilon v(\mathbf{u})) - Q(u)).$$

Note that the left-hand side of (4.22) can also be written $v'(\mathbf{u})[\mathbf{Q}]$ since

$$v_t = v_u u_t + v_{u_x} u_{xt} + \cdots = v_u Q + v_{u_x} Q_x + \cdots.$$

We have many candidates for symmetries for the typical member of the KdV
family (3.9)

$$q_{t_{2k+1}} = 2NB_{k+1} = \frac{\partial}{\partial x} L^k q \tag{4.23}$$

because we know all the flows (4.23) commute and therefore q can be
considered a function of the infinite number of independent variables $x = t_1$,
$t_3, \ldots, t_{2k+1}, \ldots$. Thus we can differentiate (4.23) with respect to t_{2j+1} and find
that $\partial q/\partial t_{2j+1}$ satisfies the linearized equation. Hence the symmetries of each
and every member of the KdV family are $\sigma_{2j+1} = \partial q/\partial t_{2j+1}$, $j = 0, 1, 2, \ldots$.
Equation (4.23) can also be integrated once to give

$$\frac{\partial w}{\partial t_{2k+1}} = L^k q.$$

Associated with each symmetry σ_{2j+1} is the local conservation law

$$\frac{\partial}{\partial t_{2k+1}} \cdot \frac{\partial w}{\partial t_{2j+1}} = \frac{\partial}{\partial t_1} \cdot \frac{\partial^2 \ln \tau}{\partial t_{2k+1} \partial t_{2j+1}}$$

and the constants of the motion (when x is considered as the special variable)
are

$$\int_{-\infty}^{\infty} \frac{\partial w}{\partial t_{2j+1}} \, dx = \int_{-\infty}^{\infty} L^j q \, dx.$$

The reader can check that these are the constants of the motion; $j = 0$ is total mass (or momentum), $j = 1$ is energy and $j = 2$ is proportional to the first hidden symmetry H_3, the Hamiltonian which generates the KdV equation. (It should be pointed out, however, that H_3 is a very useful functional in proving the stability of solitary waves (see [126]). It was called the moment of stability by Boussinesq.)

There are other symmetries as well. These are connected with Bäcklund transformations which I will discuss in Section 4f. For the moment, however, let me introduce the idea as follows. Let $q(x, t_1, t_3, \ldots ; \eta, x_0)$ be a one-soliton solution of the KdV family (3.9), $n = 0, 1, 2, 3, \ldots,$

$$q = 2\eta^2 \operatorname{sech}^2 \eta \left(x - x_0 + t_1 + \sum_1^\infty (-1)^k \eta^{2k} t_{2k+1} \right).$$

Since it is a solution for all values of the amplitude η and position x_0 parameters, $\partial q / \partial \eta$ and $\partial q / \partial x_0$ are solutions of the linearized KdV family equations and therefore also are symmetries. In particular, they are solutions of the equations linearized about the identity (either $\eta = 0$ or $x_0 = \infty$) state. A Bäcklund transformation is a transformation which builds new and richer (in the sense that the transformation can add components to the scattering data which were not there before) solutions from old solutions of the KdV family. They can also be built in a continuous way, starting with the identity. This means, for example, that we can add a solution with arbitrarily small values of the amplitude parameter η or at arbitrarily large distances so that the parameter $b = \exp(2\eta x_0)$ is as small as we wish.

Therefore, in addition to the symmetries associated with the flows (the translation of the time coordinates), there are continuous symmetries associated with transformations which take one solution type to another in a continuous way. In the last section of this chapter, Section 4g, I will indicate how both sets of symmetries combine to form the Kac–Moody algebra associated with the loop algebra of $sl(2, C)$.

4d. The Hirota story [34], [89]. You will recall that, in Chapter 1, I briefly mentioned the ingenious method of Hirota for obtaining multisoliton for the KdV family. Motivated both by the form which the N-soliton solution takes and by similar transformations for Burgers' equation, Hirota associated with the solution $q(x, t_3)$ of the KdV equation a function $\tau(x, t_3)$ defined as follows:

$$q(x, t_3) = 2 \frac{\partial^2}{\partial x^2} \ln \tau. \tag{4.24}$$

As we have seen in Section 4a, this choice is a very natural one in the sense that the entire flux tensor for the KdV family can be written in terms of one scalar function.

We will now develop the Hirota formalism. It is convenient to define

$$q = w_x, \tag{4.25}$$

whereupon the KdV equation

$$q_t + 6qq_x + q_{xxx} = 0 \tag{4.26}$$

after one integration becomes

$$w_t + 3w_x^2 + w_{xxx} = 0; \tag{4.27}$$

the constant of integration is taken to be zero because this is a property of the class of solutions we are interested in. (In order to avoid fractional coefficients in the calculations that follow, I have rescaled the times of (3.9) $t_{2n+1} \to 2^{2n} t_{2n+1}$.) Now calculate the following quantities:

$$\tfrac{1}{2}w = \frac{\tau_x}{\tau},$$

$$\tfrac{1}{2}q = \frac{\tau\tau_{xx} - \tau_x^2}{\tau^2} = \frac{\tau_{xx}}{\tau} - \frac{\tau_x^2}{\tau^2},$$

$$\tfrac{1}{2}q_x = \frac{\tau_{xxx}}{\tau} - \frac{3\tau_x\tau_{xx}}{\tau^2} + 2\frac{\tau_x^3}{\tau^3},$$

$$\tfrac{1}{2}q_{xx} = \frac{\tau_{xxxx}}{\tau} - \frac{4\tau_x\tau_{xxx}}{\tau^2} - \frac{3\tau_{xx}^2}{\tau^2} + \frac{12\tau_x^2\tau_{xx}}{\tau^3} - 6\frac{\tau_x^4}{\tau^4},$$

$$\tfrac{3}{2}q^2 = 6\frac{\tau_{xx}^2}{\tau^2} - 12\frac{\tau_x^2\tau_{xx}}{\tau^3} + 6\frac{\tau_x^4}{\tau^4}.$$

Observe that if one adds the last two quantities, which is exactly the combination of linear dispersion term and quadratic nonlinear term as appears in (4.27), all the ratios cubic and higher in τ and its derivatives vanish and we find,

$$\tfrac{1}{2}(w_{xxx} + 3w_x^2) = \tfrac{1}{2}q_{xx} + \tfrac{3}{2}q^2 = \frac{\tau_{xxxx}}{\tau} - 4\frac{\tau_x\tau_{xxx}}{\tau^2} + 3\frac{\tau_{xx}^2}{\tau^2}.$$

Hence the KdV equation becomes in the new variable τ,

$$\tau\tau_{xt} - \tau_x\tau_t + \tau\tau_{xxxx} - 4\tau_x\tau_{xxx} + 3\tau_{xx}^2 = 0. \tag{4.28}$$

The first interesting feature this equation has is that it is quadratic in τ. Observe that $\tau \equiv 1$ is a solution which corresponds to a zero q field. Next, let

$$\tau = 1 + e^{\theta(x,t)} \tag{4.29}$$

with $\theta(x, t) = kx + \omega t + \theta_0$, linear in x and t. The form (4.29) is an exact solution provided that $\omega = -k^3$; the coefficient of the second harmonic terms $e^{2\theta}$ automatically vanishes. Let us be bolder and try

$$\tau = 1 + e^{\theta_1} + e^{\theta_2}$$

where $\theta_j = k_j x - k_j^3 t + \theta_{0j}$. This is not a solution because even though the coefficients $e^{2\theta_1}$ and $e^{2\theta_2}$ vanish, the coefficient of $e^{\theta_1+\theta_2}$ does not. Compensate for this by adding a constant times this term to the ansatz with the constant to be chosen so as to eliminate the coefficient of $e^{\theta_1+\theta_2}$ which now comes from two sources, the quadratic interactions of e^{θ_1} and e^{θ_2} and 1 and $e^{\theta_1+\theta_2}$. Miraculously, the coefficients of $e^{2\theta_1+\theta_2}$, $e^{\theta_1+2\theta_2}$ and $e^{2\theta_1+2\theta_2}$ vanish. We will see why shortly. Therefore, the form

$$\tau(x,t)=1+e^{\theta_1}+e^{\theta_2}+e^{\theta_1+\theta_2+A_{12}} \tag{4.30}$$

provides an exact solution for (4.26) provided

$$e^{A_{12}}=\left(\frac{k_1-k_2}{k_1+k_2}\right)^2. \tag{4.31}$$

In Section 3d, we already discussed the nature of the solutions (4.29) and (4.30) and interpreted $A_{12}(k_1,k_2)$ as the phase shift function. There $k_j=-2\eta_j$. In particular, the solution (4.30) can be thought of as a nonlinear superposition of two solitons, with amplitudes $\frac{1}{2}k_1^2$ and $\frac{1}{2}k_2^2$. If $k_1^2>k_2^2$, then as t traverses from $-\infty$ to $+\infty$, pulse one overtakes, interacts with and passes pulse two in a manner discussed already in Chapter 3. After the interaction, the larger pulse is ahead by an amount $-A_{12}/|k_1|$ of where it would have been had it travelled unimpeded, and the smaller one an amount $-A_{12}/|k_2|$ behind. Remember $A_{12}<0$.

Once an equation has been put in quadratic form, there is always a two-soliton solution. However, that is not the case for a three-soliton solution for which

$$\tau=1+\sum_{j=1}^{3} e^{\theta_j}+\sum_{1\le j<k\le 3} e^{\theta_j+\theta_k+A_{jk}}+e^{\theta_1+\theta_2+\theta_3+A_{23}+A_{31}+A_{12}}. \tag{4.32}$$

Here in order that the coefficient of $e^{\theta_1+\theta_2+\theta_3}$ vanish, the coefficients in the original quadratic equation (4.28) have to be just right. We also observe that the coefficient of $e^{\theta_1+\theta_2+\theta_3}$ is the exponential of the sum of the two-soliton interaction phase shift function. This property holds in general and the N-phase soliton solution for KdV is

$$\tau=\sum_{\mu_j=0,1} \exp\left(\sum_{j=1}^{N} \mu_j\theta_j+\sum_{1\le i<j\le n} A_{ij}\mu_i\mu_j\right). \tag{4.33}$$

Consider $k_1^2>k_2^2>\cdots>k_n^2$. As the time t traverses from $-\infty$ to $+\infty$, the largest soliton will undergo a total phase shift which consists of the sum of the phase shifts it experiences with each of the other solitons it passes.

I will show you how to construct these solutions in a moment. But first let me tell you that the N-soliton solution for each of the members in the KdV family has exactly the same form. The only change is

$$\theta_j(x,t_3,t_5,\ldots)-\theta_{j0}=\sum_{n=0}^{\infty} (-1)^n k_j^{2n+1} t_{2n+1}. \tag{4.34}$$

(I want to emphasize again that the t_{2n+1} in (4.34) are equal to 2^{-2n} times the t_{2n+1} defined in §3b. This change is made so as to remove the factors 1, 2^{-2}, $2^{-4},\ldots,2^{-2n}$ in (3.14), (3.15), (3.16), (3.22) which would result in integral factors in the analogous equation to (4.28) for the $(KdV)_{2n+1}$ flow.)

Now the remarkable thing is that not only is (4.33), with θ_j defined by (4.34), the N-soliton solution for all the flows in the KdV family, but the phase-shift function $A_{ij}(k_i, k_j)$ and the resulting phase shifts themselves,

$$\delta_i = \sum_{N \geq j > i} -\frac{1}{|k_i|} A_{ij} + \sum_{i > j \geq 1} \frac{1}{|k_i|} A_{ij}, \qquad i = 1, \ldots, N, \tag{4.35}$$

are the same for every equation in the family. This is less remarkable when one understands that it is a direct consequence of the commutativity of the flows and the fact that $q(x, t_3, t_5, t_7, \ldots)$ is a common solution. To see this imagine that we begin with a two-soliton shape $q(x, 0, 0, \ldots)$ of the KdV family. Now run the evolution two ways. First, insert the given shape as an initial condition for $(KdV)_3$ and allow it to run for a time t_3 sufficient for the interaction to have taken place. Then take the solution $q(x, t_3, 0, \ldots)$ as initial condition and run it for a sufficiently long time t_7 in the $(KdV)_7$ flow. Since the velocity for $(KdV)_7$ is also positive, no further interaction takes place. Now reverse the sequence. The resulting shape $q(x, t_3, 0, t_7, 0, \ldots)$ must be the same and therefore the phase shifts associated with the t_3 and t_7 flows must also be.

We will use this property frequently in what follows, but first I want to introduce to you a new calculus invented by Hirota and show you how to construct N-soliton solutions. Hirota noted that the terms in (4.28) were very like the Leibnitz formulae for derivatives of products. Except for signs, (4.28) looks somewhat like

$$\frac{\partial^2}{\partial x \, \partial t} \tau^2 + \frac{\partial^4}{\partial x^4} \tau^2.$$

He invented a new operator D_x defined on ordered pairs of functions $\sigma(x)$, $\tau(x)$ as follows:

$$D_x \sigma \cdot \tau = \lim_{\varepsilon \to 0} \frac{\partial}{\partial \varepsilon} \sigma(x + \varepsilon)\tau(x - \varepsilon) = \sigma_x \tau - \sigma \tau_x. \tag{4.36}$$

This definition can be extended to functions $\sigma(x_1, x_2, \ldots)$, $\tau(x_1, x_2, \ldots)$ of infinitely many variables and to higher powers of the operator:

$$D_{x_1}^{\alpha_1} D_{x_2}^{\alpha_2} \cdots D_{x_n}^{\alpha_n} \sigma \cdot \tau$$

$$= \prod_{r=1}^{n} \lim_{\varepsilon_r \to 0} \frac{\partial^{\alpha_r}}{\partial \varepsilon_r^{\alpha_r}} \sigma(x_r + \varepsilon_r)\tau(x_r - \varepsilon_r). \tag{4.37}$$

For example,

$$D_x D_t \tau \cdot \tau = 2(\tau\tau_{xt} - \tau_x\tau_t),$$

$$D_x^4 \tau \cdot \tau = 2(\tau\tau_{xxxx} - 4\tau_x\tau_{xxx} + 3\tau_{xx}^2).$$

In this notation, the KdV equation (4.28) takes on a very compact form,

$$(D_x D_t + D_x^4)\tau \cdot \tau = 0. \tag{4.38}$$

The calculations of the multisoliton solutions are also made easier. To see this, we look at how the operators D_x, D_t act on exponentials. It is easy to show that

$$D_x^m e^{k_1 x} \cdot e^{k_2 x} = (k_1 - k_2)^m e^{(k_1 + k_2)x}. \tag{4.39}$$

Indeed, for a general polynomial

$$P(D_x, D_t)e^{k_1 x + \omega_1 t} \cdot e^{k_2 x + \omega_2 t} = P(k_1 - k_2, \omega_1 - \omega_2)e^{(k_1 + k_2)x + (\omega_1 + \omega_2)t}. \tag{4.40}$$

In general, if we take

$$\theta_i = \sum_0^\infty (-1)^n k_i^{2n+1} t_{2n+1}, \tag{4.41}$$

then

$$P(D_{t_1}, \ldots, D_{t_{2r+1}})e^{\theta_i} \cdot e^{\theta_j} = P(k_i - k_j, \ldots, (-1)^r(k_i^{2r+1} - k_j^{2r+1}), \ldots)e^{\theta_i + \theta_j}. \tag{4.42}$$

From these formulae, we can see why the coefficients of $e^{2\theta_1}$, $e^{2\theta_1 + \theta_2}$, $e^{\theta_1 + 2\theta_2}$, obtained when looking for two-soliton solutions, automatically vanish. First, we note that for the class of equations we shall be dealing with,

$$P(-D_{t_1}, -D_{t_3}, \ldots) = P(D_{t_1}, D_{t_3}, \ldots), \tag{4.43}$$
$$P(0, 0, \ldots) = 0, \tag{4.44}$$

and

$$P(\mathbf{k}) \equiv P(k, -k^3, k^5, \ldots) = 0. \tag{4.45}$$

The last equation (4.45) expresses the fact that the dispersion relation for the soliton solution

$$\tau = 1 + e^\theta, \qquad \theta = kx + \omega_3 t_3 + \omega_5 t_5 + \cdots$$

is satisfied by $\omega_{2r+1} = (-1)^r k^{2r+1}$. Furthermore, it gives a one-parameter family of surfaces on which the algebraic functions

$$x_1 x_3 + x_1^4,$$

or in general,

$$P(x_1, x_3, \ldots),$$

which are associated in an obvious way with the Hirota equations (4.38), vanish.

Let us calculate the phase shift function $A_{12}(k_1, k_2)$ for the general equation

$$P(D_{t_1}, D_{t_3}, \ldots)\tau \cdot \tau = 0 \tag{4.46}$$

which has Hirota form. P is polynomial in its arguments. Take

$$\tau(t_1, t_3, t_5, \ldots) = 1 + e^{\theta_1} + e^{\theta_2} + e^{\theta_1 + \theta_2 + A_{12}} \tag{4.47}$$

with $\theta_j = \sum (-1)^r k_j^{2r+1} t_{2r+1}$. The coefficients of e^0, $e^{2\theta_1}$, $e^{2\theta_2}$ and $e^{2\theta_1 + 2\theta_2}$ are $P(\mathbf{0})$ and therefore zero. The coefficient of a term such as $e^{2\theta_1 + \theta_2}$, arising from the product $e^{\theta_1 + \theta_2}$ with e^{θ_1} is

$$P(k_1 + k_2 - k_2, -k_1^3 + k_2^3 - k_2^3, \ldots) = P(k_1, -k_1^3, \ldots)$$

which is also zero. The only surviving term is $e^{\theta_1+\theta_2}$ which has the coefficient

$$P(\mathbf{k}_1-\mathbf{k}_2)+e^{A_{12}}P(\mathbf{k}_1+\mathbf{k}_2)$$

where we define

$$P(\mathbf{k}_1-\mathbf{k}_2)=P(k_1-k_2,\,-k_1^3+k_2^3,\,\ldots,\,(-1)^r(k_1^{2r+1}-k_2^{2r+1}),\,\ldots).\qquad(4.48)$$

Hence, in general

$$e^{A_{12}}=-\frac{P(\mathbf{k}_1-\mathbf{k}_2)}{P(\mathbf{k}_1+\mathbf{k}_2)}.\qquad(4.49)$$

Since for the $(KdV)_3$ equation (4.38), $P(x_1, x_3, x_5, \ldots)$ is $x_1x_3+x_1^4$, we have, for that case,

$$e^{A_{12}}=\left(\frac{k_1-k_2}{k_1+k_2}\right)^2.\qquad(4.50)$$

Further, we have the powerful result that, since the phase shift of each member of the KdV family is the same, the members of that family are characterized by all polynomials $P(D_{t_1}, D_{t_3}, \ldots)$ which have the property

$$(k_1-k_2)^2P(\mathbf{k}_1+\mathbf{k}_2)+(k_1+k_2)^2P(\mathbf{k}_1-\mathbf{k}_2)=0,\qquad(4.51)$$

in addition to satisfying (4.43)–(4.45).

In order to put this statement in perspective, let me review what we have done. We found that, by introducing the transformation (4.24), the KdV equation could be written in Hirota form

$$P(D_{t_1}, D_{t_3}, \ldots)\tau\cdot\tau=0\qquad(4.52)$$

with $P(x_1, x_3, \ldots)=x_1x_3+x_1^4$. Further, I stated that it possesses an N-soliton solution, for arbitrary N. Questions which naturally arise are:

(i) Does every even polynomial $P(x_1, x_3, \ldots)$ give rise to N-soliton solutions for $N>2$? We know from the last calculation that for any even P, one can always obtain a two-soliton solution. But how about $N>2$? The answer is "no". The polynomial P will have to satisfy severe constraints.

(ii) Can one conveniently characterize these constraints and thereby write down all polynomials $P(x_1, x_3, \ldots)$ which admit N-soliton solutions? We will call these *Hirota polynomials*. The answer is a qualified "yes". We can ask a more restricted question.

(iii) Given polynomial P, say $x_1x_3+x_1^4$, can we (a) determine if it has N-soliton solutions for arbitrary $N>2$, (b) find, in Hirota form, all the other members of its family and (c) find all the other Hirota polynomials compatible with the given one and determine how many of them there are? The answers seem to be YES, YES and YES. I will show you how to go about the proof but I have not yet made it rigorous or complete. I should explain further what question (c) means. If τ is such that it satisfies

$$(D_{t_1}D_{t_3}+D_{t_1}^4)\tau\cdot\tau=0,$$

then does it satisfy another equation which is of weight 6?. Notice that the Hirota polynomial is homogeneous in the sense that if we assign the weight $2k+1$ to $D_{t_{2k+1}}$ and add the weights in a product, then each term in the Hirota equation has the same weight. For example, the weight associated with (4.38) is 4. The thrust of the question is then, given that τ satisfies (4.38), how does one "apply" the operator D_x^2 to it? It is not by direct multiplication.

A fourth question, which brings us to the question of how all the different approaches are related, is

(iv) Is there an algebraic way of explaining the particular form of (that is, the coefficients in) the Hirota polynomials? I hope so. What I would like to have is an answer to question (iv) which relates these constraints to the algebras associated with $sl(2)$. The reason for this is that I believe (and the reader will see the reasons for this in Chapter 5) that this property is the common denominator of all the "methods" for analyzing soliton equations introduced in this chapter.

We will now return to questions (i), (ii) and (iii). Assume P satisfies (4.43)–(4.45) and look for a three-soliton solution for (4.51),

$$\tau = 1 + e^{\theta_1} + e^{\theta_2} + e^{\theta_3} + e^{\theta_2 + \theta_3 + A_{23}}$$
$$+ e^{\theta_3 + \theta_1 + A_{31}} + e^{\theta_1 + \theta_2 + A_{12}} + e^{\theta_1 + \theta_2 + \theta_3 + A_{12} + A_{31} + A_{23}}.$$

Using properties (4.43)–(4.45) and (4.50), all terms have identically zero coefficients except $e^{\theta_1 + \theta_2 + \theta_3}$ whose coefficient is built from four quadratic interactions and is

$$p_{123} e^{A_{23}} P(\mathbf{k}_1 - \mathbf{k}_2 - \mathbf{k}_3) + e^{A_{12} + A_{13} + A_{23}} P(\mathbf{k}_1 + \mathbf{k}_2 + \mathbf{k}_3),$$

where p_{123} is the cyclic permutation over 1, 2, 3 and

$$P(\mathbf{k}_1 + \mathbf{k}_2 + \mathbf{k}_3) = P(k_1 + k_2 + k_3, -k_1^3 - k_2^3 - k_3^3, k_1^5 + k_2^5 + k_3^5, \ldots).$$

This can also be written (using (4.50)) as

$$p_{123} P(\mathbf{k}_2 + \mathbf{k}_3) P(\mathbf{k}_1 - \mathbf{k}_2) P(\mathbf{k}_3 - \mathbf{k}_1) P(\mathbf{k}_1 - \mathbf{k}_2 - \mathbf{k}_3)$$
$$+ P(\mathbf{k}_2 - \mathbf{k}_3) P(\mathbf{k}_3 - \mathbf{k}_1) P(\mathbf{k}_1 - \mathbf{k}_2) P(\mathbf{k}_1 + \mathbf{k}_2 + \mathbf{k}_3). \qquad (4.53)$$

Therefore, the condition that (4.52) has a three-soliton solution is that the expression (4.53) is zero. Further, by similar considerations it can be shown that the condition that (4.52) has a N-soliton solution is

$$\sum_{\mu_i = -1, 1} P\left(\sum_1^N \mu_l \mathbf{k}_l\right) \prod_{j > i} P(\mu_j \mathbf{k}_j - \mu_i \mathbf{k}_i) \mu_i \mu_j = 0. \qquad (4.54)$$

We will call this the Hirota condition. We will say that an equation which can be put in Hirota form with a P satisfying (4.54) (and certain auxiliary properties like (4.23)–(4.25)) has the H-property. In particular cases, such as $(KdV)_3$ where $P = x_1 x_3 + x_1^4$, (4.54) can be shown to hold. The proof is usually given by induction. I think it is clear, however, that the condition is rather awkward and clumsy and difficult to prove in general so it is useful to have an alternative approach.

I claim, but have not yet completely proved, the following. Given a polynomial $P_L(x_1, x_3, \ldots)$ of a given weight, compute its phase shift function

$$e^{A_{12}} = -\frac{P_L(\mathbf{k}_1 - \mathbf{k}_2)}{P_L(\mathbf{k}_1 + \mathbf{k}_2)}. \tag{4.55}$$

Then calculate all the polynomials P_M which share the same phase shift. Often there will be more than one at each weight level. If there is at least one such polynomial at an infinite sequence of weight levels, then the following three statements are true:

(i) P_L is a Hirota polynomial, i.e. has an N-soliton solution (satisfies (4.54)) for arbitrary N.

(ii) Each P_M gives a Hirota equation $P_M \tau \cdot \tau = 0$ in the P_L family.

(iii) Each equation $P_M \tau \cdot \tau = 0$ in the list is a Hirota polynomial (i.e. satisfies (4.54) and therefore has an N-soliton solution for arbitrary N).

Let me illustrate these statements with some concrete examples. Let P_4 be $x_1 x_3 + x_1^4$. Then, assume a form for P_6 (from (4.43), only even levels are allowed),

$$P_6 = x_1 x_5 + a x_3^2 + b x_1^3 x_3 + c x_1^6.$$

Clearly (4.43) and (4.44) are satisfied. So is (4.45) if

$$1 + a - b + c = 0. \tag{4.56}$$

Now demand that

$$(k_1 + k_2)^2 P_6(\mathbf{k}_1 - \mathbf{k}_2) + (k_1 - k_2)^2 P_6(\mathbf{k}_1 + \mathbf{k}_2) = 0$$

holds for all k_1, k_2. The left-hand side can be written

$$2(k_1 + k_2)^2 (k_1 - k_2)^2 \{ k_1^4 + k_1^2 k_2^2 + k_2^4 + a(k_1^4 + 3k_1^2 k_2^2 + k_2^4)$$
$$- b(k_1^4 + k_2^4) + c(k_1^4 + 6k_1^2 k_2^2 + k_2^4) \}.$$

Therefore, in addition to (4.56) we must choose,

$$a = -2c - \tfrac{1}{3}. \tag{4.57}$$

Therefore, since c is arbitrary,

$$P_6 = x_1 x_5 - x_1^6 + \tfrac{5}{3}(x_1^3 x_3 + x_3^2) + (c + 1)(x_1^6 - 2x_3^2 - x_1^3 x_3), \tag{4.58}$$

is a one-parameter family of Hirota polynomials at weight level 6 spanned by

$$P_6^{(1)}(x_1, x_3, x_5, \ldots) = x_1 x_5 - x_1^6 + \tfrac{3}{5}(x_1^3 x_3 + x_3^2), \tag{4.59a}$$

$$P_6^{(1)}(x_1, x_3, x_5) = x_1^6 - 2x_3^2 - x_1^3 x_3. \tag{4.59b}$$

The Hirota form for $(KdV)_5$ is

$$P_6^{(1)}(D_{t_1}, D_{t_3}, D_{t_5}) \tau \cdot \tau = 0 \tag{4.60}$$

and (4.59b) is the equation one obtains by "applying" $D_{t_1}^2$ to

$$P_4(D_1, D_{t_3})\tau \cdot \tau = 0.$$

Notice how important it is to have three times in the problem. $(KdV)_5$ cannot be expressed in Hirota form in terms of t_1 and t_5 alone!

I will leave it to the reader to show

$$P_8^{(1)} = x_1 x_7 - \tfrac{1}{3} x_3 x_5 + \tfrac{2}{3} x_1^3 x_5, \tag{4.61a}$$

$$P_8^{(2)} = x_1^8 + 4 x_1^3 x_5 + 5 x_3 x_5, \tag{4.61b}$$

$$P_8^{(3)} = -x_3 x_5 + x_3 x_1^5, \tag{4.61c}$$

$$P_8^{(4)} = -x_1^3 x_5 + x_3^2 x_1^2, \tag{4.61d}$$

where $P_8^{(r)}\tau \cdot \tau = 0$, $r = 1, 2, 3, 4$. $P_8^{(1)}\tau \cdot \tau = 0$ is the Hirota form for $(KdV)_7$. $P_8^{(r)}$, $r = 2, 3, 4$ are what one gets by "applying" $D_{t_1}^2$, $D_{t_1}^4$ and $D_{t_3}^2$ to appropriate combinations of (4.59a,b) and $(D_{t_1}D_{t_3} + D_{t_1}^4)\tau \cdot \tau = 0$.

Repeating this process, it is not too hard to see that at each weight P_{2M}, $M > 2$, there exists many Hirota polynomials. I will show you a better way of interpreting this than simply "applying $D_{t_1}^2$ to (4.37)" when we discuss this matter again in Chapter 5. We will also get an idea of how to count the number at each weight level.

As an exercise, I invite the reader to calculate the sequences given the polynomials

$$P = x_1 x_5 - x_1^6 \tag{4.62a}$$

and

$$P = x_1 x_7 + x_1^8. \tag{4.62b}$$

The P of (4.62a) generates the Kotera–Sawada sequence [104] and has nontrivial polynomials at an infinite sequence of weight levels which do not include all of the even numbers. Can you find what they are? On the other hand, (4.62b) does not have an infinite string of polynomials with the same phase shift function. In fact, it appears to have none. Therefore, it only has a two-soliton solution.

The following questions also remain open.

(i) Can one characterize the properties that an equation must have in order to be expressible in Hirota form? Once it is in this form, one knows that it has a two-soliton solution. Whether it has an N-soliton solution for arbitrary N depends on whether one can find an infinite sequence P_M satisfying

$$P_L(\mathbf{k}_1 - \mathbf{k}_2)P_M(\mathbf{k}_1 + \mathbf{k}_2) - P_L(\mathbf{k}_1 + \mathbf{k}_2)P_M(\mathbf{k}_1 - \mathbf{k}_2) = 0$$

where P_L is the given polynomial. This expression is rather suggestive as it expresses what appears to be a condition for the commuting of polynomials and leads naturally to a definition of a Poisson bracket on the polynomial manifold.

(ii) Suppose one could only find a finite number of such P_M. Is this possible or, if one finds one P_M, must there be an infinite number? If not the latter, then, if there are M P_M's, does that mean there are N-soliton solutions up to $N \leqq N(M)$?

4e. The Painlevé property.[6]

(i) *Classical work.* Consider the system of Fuchsian differential equations

$$\frac{dV}{dz} = \sum_{j=1}^{n} \frac{A_j}{z - a_j} V, \tag{4.63}$$

where V is an m-vector and the A_j are constant $(m \times m)$ matrices. In general, the fundamental solution Φ to (4.63) is a multivalued function of complex z. Indeed, if we circle the regular singular point at a_j, then $\Phi(a_j + (z - a_j)e^{2\pi i})$, while a fundamental solution matrix of (4.63), is not equal to $\Phi(z)$; rather, its columns are linear combinations of the columns of $\Phi(z)$. The matrix M_j relating the two

$$\Phi(a_j + (z - a_j)e^{2\pi i}) = \Phi(z)M_j \tag{4.64}$$

is called the monodromy matrix. One can ask the following question. How can one write A_j as function of the location of the poles a_r such that the group (it is easy to see they form a group) of monodromy matrices remains fixed? The general answer to this question was given by Schlesinger [90]

$$\frac{\partial A_j}{\partial a_r} = \frac{[A_r, A_j]}{a_j - a_r}, \qquad \sum_{1}^{n} \frac{\partial A_j}{\partial a_j} = 0. \tag{4.65}$$

For $m = 2$, the linear equation is a 2×2 system and its regular singular points can be located at the fixed points $z = 0$, 1, ∞ and one moving point $z = s$. A priori, there are twelve adjustable elements in the coefficient matrices A_j, $j = 1$, 2, 3 but they can all be expressed in terms of one function $y(s)$ which satisfies the equation

$$y'' + \left(\frac{1}{s} + \frac{1}{s-1} + \frac{1}{y-s}\right)y' - \frac{1}{2}\left(\frac{1}{y} + \frac{1}{y-1} + \frac{1}{y-s}\right)y'^2$$

$$-2\frac{y(y-1)(y-s)}{s^2(s-1)^2}\left(\alpha - \beta\frac{s}{y^2} + \gamma\frac{s-1}{(y-1)^2} - \delta\frac{y(y-1)}{(y-s)^2}\right) = 0. \tag{4.66}$$

Equation (4.66) is the most general second order equation

$$y'' = R(y, y', s) \tag{4.67}$$

with R rational in y, y' and analytic in s which has the following property:

The Painlevé property. The location of any algebraic, logarithmic or essential singularity of its solutions is independent of the initial conditions. This means that only the location of its poles can depend on the arbitrary constants of integration.

Second order equations of the form (4.67) with these properties (the conditions on R and the Painlevé properties) were studied in exhaustive detail by

---

[6] Please see p. 144 for a historical note on the pioneering work of S. Kowalevski.

Painlevé and Gambier [91]. There are fifty canonical types which include equations such as $y'' = y$, equations solved by elliptic functions

$$y'' = 2y^3 + cy - \nu \tag{4.68}$$

and six equation types whose solutions could not be expressed (except in special limiting cases) in terms of known special functions. These six equations are called Painlevé equations and their solutions Painlevé transcendents. The reader can find a list of these equations in Ince [92]. Two which arise in these lectures are the second

$$q_{xx} = xq + 2q^3 - \nu \tag{4.69}$$

and third (after a transformation $z = e^u$)

$$(xu_x)_x = -\sinh u \tag{4.70}$$

Painlevé equations.

Now, what has all this to do with completely integrable partial differential equations or more generally with completely solvable models in physics? The amazing fact is this: it is an observation that the nonlinear ordinary differential equations which arise in a very natural way in these solvable models have the Painlevé property. This fact is not believed to be coincidental. Rather, it is believed that there are deep and intimate connections between exactly solvable models and the Painlevé property. I will comment further on this idea later in the section.

(ii) *The ARS conjecture.* In 1977, Ablowitz and Segur [93], [35] noted that since equations such as

$$q_t + 6qq_x + q_{xxx} = 0 \tag{4.71}$$

and

$$v_t - 6v^2 v_x + v_{xxx} = 0 \tag{4.72}$$

were exactly solvable (under certain boundary conditions), one could also solve the nonlinear ordinary differential equations obtained by imposing the various symmetry properties of the equations. For example, Galilean invariance means that (4.71) has solutions of the form $q(x, t) = f(X = x - ct)$ satisfying

$$-cf_X + 6ff_X + f_{XXX} = 0.$$

The scaling invariance means that if $q(x, t)$ satisfies (4.71), then so does $\beta^2 q(\beta x, \beta^3 t)$ and if $v(x, t)$ satisfies (4.72), then so does $\beta v(\beta x, \beta^3 t)$. Setting

$$v(x, t) = \frac{1}{(3t)^{1/3}} f\left(X = \frac{x}{(3t)^{1/3}}\right),$$

we obtain after one integration

$$f_{XX} = Xf + 2f^3 - \nu, \tag{4.73}$$

the second Painlevé equation (4.69). As an exercise, I will ask the reader to show that by a suitable transformation (hint: look at the Miura transformation), solutions of (4.71) of the form $q(x, t) = (1/(3t)^{2/3}) g(x/(3t)^{1/3})$ obey (4.73) with $\nu = 0$.

Ablowitz and Segur argued that the one-parameter family of solutions of (4.73) with $\nu = 0$ which decay exponentially as $x \to +\infty$ and algebraically at $x = -\infty$ could be found by the direct application of inverse scattering theory. The reason for this restriction is that inverse scattering theory requires the solutions $q(x, t)$ to decay at both infinities. I will tell you how to find the general solution to the initial value problem for (4.73) in Section 5f (iii).

Further, knowing of the particular properties of solutions to the Painlevé equation and observing that all the ordinary differential equations derivable from known completely integrable partial differential equations had this property, Ablowitz and Segur, who by this stage had been joined by Ramani [35], made the conjecture that this is always true: namely, all ordinary differential equations derived from completely integrable partial differential equations have the Painlevé property. Sometimes, it may be necessary to be rather clever in choosing the dependent variable (see (4.70)). The great advantage of the idea is that it gives a simple and readily applicable test for integrability.

Let me illustrate using (4.73) with $\nu = 0$. Suppose X_0 is a pole singularity of $f(X)$. Then we should be able to construct a Laurent expansion for f in the neighborhood of X_0,

$$f(X) = \sum_{n=-N}^{\infty} a_n (X - X_0)^n. \qquad (4.74)$$

It is easy to show that in order for (4.74) to satisfy (4.73), N must be one. Substitution of (4.74) into (4.73) gives the set of nonlinear algebraic equations

$$(n+1)(n+2)a_{n+2} = a_{n-1} + X_0 a_n + 2 \sum_{j,k,l=-1} a_j a_k a_l, \qquad j + k + l = n,$$
$$(4.75)$$

for $n \geq -3$ ($a_n = 0$, $n < -1$) which we proceed to solve iteratively for a_{n+2} in terms of the lower s's. This can be done only if there is compatibility at the n values $n = -3$, $n = 1$. This is necessary because in each case we find the coefficient of a_{n+2} to be zero unless

$$a_{-1}^2 = 1, \qquad (4.76)$$
$$a_0 + X_0 a_1 + 6 a_{-1} a_1^2 = 0. \qquad (4.77)$$

We find that for $n = -2, -1, 0$

$$a_0 = 0, \quad a_1 = -\frac{X_0 a_{-1}}{6}, \quad a_2 = -\frac{a_{-1}}{4}, \qquad (4.78)$$

and so (4.76) and (4.77) are satisfied with the choice $a_{-1} = \pm 1$. After this point, all the a_n are uniquely determined and it is not too difficult to show that the resulting series converges for sufficiently small $X - X_0 \neq 0$. This family of solutions has two free parameters X_0 and a_3 which is to be expected from a second order equation. If the compatibility condition (4.77) had not held, it would have been necessary to include $\ln (X - X_0)$ terms in the local expansion for $f(X)$ which would mean that the solution would see X_0 as a branch point. If this were the case, the location of a branch point would depend on the initial condition and the equation would not have the Painlevé property.

The great advantage of the ARS conjecture is that it is simple to apply. Its drawback is that in order to test the integrability of a partial differential equation, one has to ask what one means by testing *all* the ordinary differential equations associated with its symmetries. It would be better if one could directly attack and test the partial differential equation itself. Let us do this for (4.71) using the expansion

$$q(x, t) = \frac{a_{-2}}{(x - x_0)^2} + \frac{a_{-1}}{(x - x_0)} + a_0 + a_1(x - x_0) + \cdots, \qquad (4.79)$$

where x_0 and all the coefficients are allowed to be functions of t. Upon substitution in (4.71), we obtain

$$n(n-1)(n-2)a_n + 6 \sum_{\substack{-2 \\ r+s=n-2}}^{\infty} r a_r a_s$$
$$+ a_{n-3,t} - (n-2)a_{n-2}x_{0t} = 0, \qquad n \geq -2, \qquad (4.80)$$

which we solve for a_n in terms of the a_{n-r}. We find for $n = -2, -1, 0, 1$, $a_{-2} = -2$, $a_{-1} = 0$, $a_0 = \frac{1}{6}x_{0t}$, $a_1 = 0$. At $n = 2$, the coefficient of a_2 is zero but so is the sum of the other terms in the equation. Hence, $a_2(t)$ is arbitrary. At $n = 3$, we find $a_3 = \frac{1}{36}x_{0tt}$. At $n = 4$, again the coefficient of a_4 is zero but the equation reads

$$24a_4 + 6(2a_0a_2 + 2a_{-2}a_4) + a_{1t} - 2a_2x_{0t} = 0,$$

which is satisfied exactly. All the later a_n's are uniquely determined. Hence $q(x, t)$ has a local solution which can be written in a Laurent expansion with three arbitrary functions of t, $x_0(t)$, $a_2(t)$, $a_4(t)$. This approach was followed by Weiss, Tabor and Carnivale [96] with modifications by Kruskal and I refer the reader to their paper. In essence, the ARS conjecture has been modified to read that $q(x, t)$ has a local Laurent expansion in the neighborhood of those surfaces in (x, t) space where it has pole-like behavior.

What I want to do here is draw your attention to the connections between these results and those of the last section because I believe there is a direct correspondence between the Hirota property and the Painlevé property of a given equation. In both cases something magic has to happen in order for the property to hold. In the former, one has to be able to (i) write the given equations in Hirota form and (ii) show that the resulting polynomial P is one of the class which admits N-soliton solutions for arbitrary N. This means its coefficients, which it inherits from the underlying equation from which it was derived, have to bear special relations with each other. This is also what must happen when one applies the Painlevé test. The coefficients have to be exactly right so that the series (4.74) (or (4.79)) is a Laurent one.

But the connection goes deeper than this simple observation. Notice that

$$q(x, t) = -2\frac{\tau_x^2}{\tau^2} + 2\frac{\tau_{xx}}{\tau}, \tag{4.81}$$

and that therefore a double pole of $q(x, t)$ is a simple zero of the ubiquitous $\tau(x, t_3, t_5, \dots)$. Further, we know that (4.71) has an infinite sequence of rational solutions (see Section 3h) corresponding to the infinite set of multisoliton solutions and the requirements on P for the existence of the former and the latter are the same, namely, the Hirota condition (4.54). The rational solutions are obtained by expressing τ as a finite polynomial function of $x = t_1, t_3, t_5, \dots$ of a given weight (e.g. x, $x^3 + 12t_3, \dots$). The conditions that one can find a sequence of finite polynomial solutions are exactly the Hirota conditions. But now look at this from the point of view of the Painlevé property. If τ is expressible as a finite polynomial in x, $t_3, \dots$, then it is clear that it admits a Taylor series expansion in the neighborhood of points which lie in the $\tau = 0$ surfaces. But if τ has Taylor series expansions near surfaces where it is zero, then the corresponding $q(x, t_3, \dots)$ has a local Laurent series near its poles.

The conjecture that the τ function is analytic in each of its arguments has yet to be proved. One of the difficulties is that it is only true for certain solution classes. One has to find a convenient way of first removing all those points where τ and hence q has algebraic, logarithmic or essential singularities. This set of points, of course, is fixed and independent of the initial conditions.

I believe, however, that the Painlevé test contains more information than a simple yes or no as to whether the equations are integrable. Although the work is still preliminary, there is every indication that, just as the Hirota polynomials appear to have an underlying algebraic structure, so too does the Painlevé property. In other words, it is my guess that some version of the Painlevé test gives rise to the very same algebraic structure (in the case of KdV and the AKNS hierarchy, it will be the infinite dimensional loop algebra associated with $sl(2)$) which would arise by the application of the Wahlquist–Estabrook method which is discussed in the first part of Chapter 5.

However, there are still many nagging questions. Kruskal has constantly questioned the need to eliminate logarithmic and other singularities. After all, who would deny that $dy/dx = (y - \alpha)(y - \beta)(y - \gamma) \cdots$ is integrable (is it?) and yet x written as function of y has logarithmic singularities? Moreover, the number of counterexamples to the conjecture that completely integrable ordinary differential equations have the Painlevé property (as it has been described in the last couple of pages) is growing and the most recent word out of Paris (where the Ramani group works) is that the Painlevé test is dead! Whereas this surely is an exaggeration, it is clear that some changes are required. Kruskal has suggested that the Painlevé test is too strong and argues that a more subtle test in which one looks at the behavior of the equation near a confluence of poles (i.e., the equation's worst singular behavior) is needed. His ideas are new and I will not attempt to describe them here as he has yet to publish them. They have at least two attractive features. First, they contain the original

Painlevé test when it applies. More importantly, however, they go to the heart of the nature of integrability (what is a completely integrable system?) and make direct contact with the subtleties that distinguish between ergodic and integrable flows on compact manifolds. (For example, $x' = \alpha$, $y' = \beta$ is an integrable flow on the torus $0 < x$, $y < 1$ (in which opposite boundaries are identified) only if α and β are rationally related. Why? The reason is that on the torus the motion constant $C = \beta x - \alpha y$ behaves very erratically and is, in fact, not measurable if α/β is irrational.)

4f. Bäcklund transformations. A central and recurring theme of the inverse scattering method is that interesting nonlinear equations arise as integrability conditions of overdetermined linear systems. We have shown how the KdV equation and all the members in its family are integrability conditions of the linear equations

$$v_{xx} + (\zeta^2 + q(x, t_3, \ldots))v = 0 \tag{4.82}$$

and

$$v_{t_{2k+1}} = (\tfrac{1}{2}B_x^{(k)} + c)v - B^{(k)}v_x. \tag{4.83}$$

Let us write the equation pair for the KdV equation $(4t = t_3)$

$$q_t + 6qq_x + q_{xxx} = 0 \tag{4.84}$$

in system form as

$$V_x = \begin{pmatrix} -i\zeta & q \\ -1 & i\zeta \end{pmatrix} V, \tag{4.85}$$

and

$$V_t = \begin{pmatrix} -4i\zeta^3 + 2iq\zeta - q_x & 4\zeta^2 q + 2iq_x\zeta - q_{xx} - 2q^2 \\ -4\zeta^2 + 2q & 4i\zeta^3 - 2iq\zeta + q_x \end{pmatrix} V \tag{4.86}$$

with $V = (v_1 = -v_x + i\zeta v, \ v_2 = v)^T$. Bear with me while I make the following little calculation. Define

$$\gamma = \frac{v_1}{v_2} \tag{4.87}$$

and find

$$\gamma_x = -2i\zeta\gamma + q + \gamma^2, \tag{4.88a}$$

$$\gamma_t = (-8i\zeta^3 + 4iq\zeta - 2q_x)\gamma$$
$$+ (4\zeta^2 q + 2iq_x\zeta - q_{xx} - 2q^2) - (2q - 4\zeta^2)\gamma^2. \tag{4.88b}$$

The condition for the integrability of these Riccati equations is also (4.84). Next set

$$q = -u_x, \qquad \tilde{q} = -\tilde{u}_x,$$

whence $u(x, t)$ satisfies

$$u_t - 3u_x^2 + u_{xxx} = 0 \tag{4.89}$$

(we have put the integration constant zero), or, depending on which t we take in the KdV hierarchy, the corresponding member of that family integrated once with respect to x. Now define $\tilde{u}$ by

$$\gamma - i\zeta = \frac{\tilde{u} - u}{2}. \tag{4.90}$$

The claim is that if $u(x, t)$ satisfies (4.89) (and, with respect to the different t's, the other members of the KdV family), then $\tilde{u}(x, t)$ also satisfies (4.89) (and, with respect to the different t's, the other members of the family). We will prove this in a moment. First, however, let us examine the consequences of this assertion further. Substitute (4.90) into (4.88) and find

$$\frac{\tilde{u}_x + u_x}{2} = \left(\frac{\tilde{u} - u}{2}\right)^2 + \zeta^2 \tag{4.91a}$$

and

$$\frac{\tilde{u}_t + u_t}{2} = 2q_x\left(\frac{u - \tilde{u}}{2}\right) - 2q\left(\frac{\tilde{u} - u}{2}\right)^2 + q^2 - 2\zeta^2\tilde{q}. \tag{4.91b}$$

Now, equations (4.91) are a one-parameter (ζ^2) set of relations

$$R_j(u, u_{xx}, u_t; \tilde{u}, \tilde{u}_x, \tilde{u}_{xx}, \tilde{u}_t; \zeta^2) = 0 \tag{4.92}$$

involving $\tilde{u}$, u and their partial derivatives. There are fewer relations than variables. Further, we know $u(x, t)$ satisfies (4.89). I invite the reader to show by direct computation that $\tilde{u}(x, t)$ also does. There is also another proof which lends more insight into the close relation between equations (4.91) and (4.85), (4.86), (4.88). Using (4.91a), we can show that (4.91) can also be written

$$\frac{\tilde{u}_t + u_t}{2} = 2\tilde{q}_x\left(\frac{\tilde{u} - u}{2}\right) - 2\tilde{q}\left(\frac{\tilde{u} - u}{2}\right)^2 + \tilde{q}^2 - 2\zeta^2 q, \tag{4.91c}$$

which is simply (4.91b) with u and $\tilde{u}$ interchanged. Therefore, we can write equations (4.91a,c) as (4.88a,b) with $\gamma \to -\gamma$ and $\zeta \to -\zeta$ and where $\tilde{q}$ replaces q. But the change of sign of γ and ζ does not change the solvability conditions and therefore $\tilde{q}(x, t)$ satisfies (4.84) and $\tilde{u}(x, t)$ satisfies (4.89).

We call a set of relations such as (4.91) a Bäcklund transformation. It allows us to build more complicated solutions from simpler ones. For example, if we take $u = 0$ and solve the resulting pair of first order equations (4.90) in x and t, we find $\tilde{u}(x, t) = -2\eta \tanh \eta(x - 4\eta^2 t - x_0)$, $\zeta = i\eta$. For a more general $u(x, t)$, the equations for $\tilde{u}$ are more difficult to solve and there are much better ways of building new solutions from old ones than solving (4.91) directly. I will show you how shortly.

For the moment, however, I want to return to the notion of a Bäcklund transformation. The term has existed a long time in the literature. It is very hard to find a clear definition. The one I am giving you was given by Hanno Rund [95] and it is now the generally accepted one. Let $u(x, t)$ and $\tilde{u}(x, t)$

satisfy the partial differential equations

$$E(u) = 0 \qquad (4.93a)$$

and

$$D(\tilde{u}) = 0 \qquad (4.93b)$$

respectively. Then the set of relations

$$R_j((u), (\tilde{u}), (\zeta)) = 0, \qquad j = 1, \ldots, n, \qquad (4.94)$$

where (u) and $(\tilde{u})$ denote strings, not necessarily of equal length, consisting of u, $\tilde{u}$ and their various partial derivatives, is called a Bäcklund transformation if these relations ensure that $\tilde{u}$ satisfies (4.93b) whenever u satisfies (4.93a) and vice versa. If u and $\tilde{u}$ satisfy the same equation, the adjective "auto" is inserted in front of Bäcklund. The set of relations (4.91) is an auto Bäcklund transformation relating solutions of corresponding members of the KdV family. The first half of these relations (4.91a) relates solutions of all members of each family. The Miura transformation

$$q(x, t) = v^2(x, t) - iv_x(x, t) \qquad (4.95)$$

connects solutions of every member of the KdV family with every member of the modified KdV family, the first nontrivial equation of which is

$$v_t + 6v^2 v_x + v_{xxx} = 0. \qquad (4.96)$$

We know from (1.12) in Chapter 1 (an equation which holds between each corresponding member of the respective families, see [96]) that if $v(x, t)$ satisfies (4.96), then $q(x, t)$, given by (4.95), satisfies (4.84). On the other hand, a $q(x, t)$ which satisfies (4.84) can give rise to a $v(x, t)$ which does not necessarily satisfy (4.96). Indeed solving (1.12) gives

$$v_t + 6v^2 v_x + v_{xxx} = A \exp\left(-2i \int^x q \, dy\right),$$

and we therefore cannot call (4.95) by itself a Bäcklund transformation unless we complete the set of relations (here by adding one relation) with another relation (such as the equation (4.96) itself) which ensures that $v(x, t)$ will satisfy the modified KdV equation.

It should be clear that if we are to retain the central notion that q is a function of an infinite number of independent variables, the Bäcklund transformation which interconnects solutions of the whole family, must necessarily be an infinite set of relations involving all the time derivatives. One can formally derive these in exactly the same way as we found (4.91b). A much faster and elegant way is presented in Section 5g, where we define a Bäcklund transformation in terms of operations on the τ-function. Recall that

$$\gamma - i\zeta = \frac{v_1}{v_2} - i\zeta = -\frac{v_x}{v}, \qquad (4.97)$$

and since both $u(x, t)$ and $\tilde{u}(x, t)$ satisfy (4.89), we can write

$$u(x, t) = -2\frac{\partial}{\partial x}\ln\tau, \qquad \tilde{u}(x, t) = -2\frac{\partial}{\partial x}\ln\tilde{\tau}. \tag{4.98}$$

Therefore the equation (4.90) which defines $\tilde{u}$ is simply

$$\tilde{\tau} = \tau v. \tag{4.99}$$

One should be able to invert the process; namely,

$$\tau = \tilde{\tau}\tilde{v}(x, \zeta), \tag{4.100}$$

where $\tilde{v}$ is a solution of (4.82) with q replaced by $\tilde{q}$. But from (4.91a) we can write

$$\tilde{q} = -q - 2\frac{v_x^2}{v^2} - 2\zeta^2 \tag{4.101}$$

and therefore $1/v$ satisfies

$$\left(\frac{1}{v}\right)_{xx} + (\tilde{q} + \zeta^2)\frac{1}{v} = 0.$$

It is important to emphasize that the ζ in (4.97)–(4.101) is a specific ζ, a parameter which describes the soliton by which the solution $\tilde{q}$ is richer than the solution q. Let us therefore now call it ζ_1. I will leave it as a simple exercise for the reader to show that if $v(x, \zeta_1) = v_1(x)$ satisfies (4.82) with $\zeta = \zeta_1$ and $q = q$, $v(x, \zeta)$ satisfies (4.82), then

$$\tilde{v}(x, \zeta) = v_x(x, \zeta) - \frac{v_{1x}}{v_1}v(x, \zeta) \tag{4.102}$$

satisfies (4.82) with q replaced by $\tilde{q}$ (given by (4.90) and (4.92) or (4.101)). This is a result of Faddeev [96]. Let us now combine these results with those of Section 3d and determine what the Bäcklund transformation (4.91) does to the scattering data. I will follow closely the work of Flaschka and McLaughlin [96].

Let the potential $q(x)$ have the scattering data

$$S[q] = \{R(\zeta), \zeta \text{ real; } (\zeta_j = i\eta_j, \gamma_j)_{j=2}^N\}.$$

In order that $\tilde{v}(x, \zeta)$ does not have a pole at a zero of $v_1(x)$, we must demand that ζ_1^2 lie to the left of the spectrum associated with $q(x)$, in which case $v_1(x)$ does not have a zero in $(-\infty, \infty)$. Let us take (assume $A \neq 0$)

$$v_1(x) = A\psi(x, \zeta_1) + B\phi(x, \zeta_1).$$

Recall that since ζ_1 is not in the spectrum associated with q, $\phi(x, \zeta_1)$ and $\psi(x, \zeta_1)$ are not proportional. Next let us choose $v(x, \zeta)$ in such a way that $\tilde{v}(x, \zeta) = \tilde{\psi}(x, \zeta)$, i.e., $\tilde{v}(x, \zeta) \sim e^{i\zeta x}$ as $x \to +\infty$ for real ζ. It is a simple exercise to show that $v(x, \zeta) = (i\zeta - \eta_1)^{-1}\psi(x, \zeta)$ if $B \neq 0$, and is equal to $(i\zeta + \eta_1)^{-1}\psi(x, \zeta)$ if $B = 0$. If $B \neq 0$, then, as $x \to -\infty$,

$$\psi(x, \zeta) \to \frac{\zeta - i\eta_1}{\zeta + i\eta_1}a(\zeta)e^{i\zeta x} + b(-\zeta)e^{-i\zeta x}$$

which, by comparison with (3.62), means that (remember $R(\zeta) = b(\zeta)/a(\zeta)$)

$$\bar{a}(\zeta) = \frac{\zeta - i\eta_1}{\zeta + i\eta_1} a(\zeta), \quad \bar{b}(-\zeta) = -b(-\zeta), \quad R(\zeta) = \frac{\zeta + i\eta_1}{\zeta - i\eta_1} R(\zeta) \quad (4.103)$$

and the bound states are $\zeta_j = i\eta_j$ for $j = 1, \ldots, N$ with $\tilde{\gamma}_j = (\eta_1 + \eta_j)/(\eta_1 - \eta_j)\gamma_j$ (simply take the residue of $R(\zeta)$ at $\zeta = i\eta_j$), $j = 2, \ldots, N$ and γ_1 depends on A and B. Similarly, if $B = 0$,

$$\tilde{\psi}(x, \zeta) \to a(\zeta)e^{-i\zeta x} + \frac{\zeta + i\eta_i}{\zeta - i\eta_i} b(-\zeta)e^{-i\zeta x},$$

which means that

$$\bar{a}(\zeta) = a(\zeta), \quad \tilde{R}(\zeta) = \frac{\zeta - i\eta_1}{\zeta + i\eta_1} R(\zeta), \quad \tilde{\gamma}_j = \frac{\eta_1 - \eta_j}{\eta_1 + \eta_j} \gamma_j \quad (4.104)$$

and no new bound states are added.

Note, therefore, that whereas (4.103) adds a bound state $\zeta_1 = i\eta_1$ to the spectrum, it also changes the reflection coefficient by a phase factor. In order that $\tilde{q}$ has the same scattering data as q except for the addition of one bound state, one must first apply the Bäcklund transformation (4.104) (with $B = 0$) and then the Bäcklund transformation (4.103). The composite transformation gives a $\tilde{q}$ with scattering data

$$S[\tilde{q}] = \{R(\zeta), \zeta \text{ real}; (\zeta_j = i\eta_j, \gamma_j)_{j=1}^N\}$$

where γ_1 depends on the choice of A and B in (4.103). Note, however, that if one begins from a reflectionless potential $R(\zeta)$, no unwinding of the phase shift induced by the soliton adding Bäcklund transformation is required.

Next, we will look at how a Bäcklund transformation affects the τ-function. Since we make heavy use of (4.15) and (4.18), a formal result which is really only applicable for soliton and rational solutions (in which case the asymptotic expansion at $\zeta = \infty$ is a Laurent expansion), we restrict our results to reflectionless potentials. I will leave it to the reader to attempt to extend the domains of validity of the formulae. We begin by using (4.18) to write (4.99) as

$$\tilde{\tau} = \tau v(x, \zeta) = \tau(A\psi(x, \zeta) + B\psi(x, -\zeta)), \quad (4.105)$$

where $v(x, \zeta)$ is written as a linear combination of the two linearly independent solutions $\psi(x, \zeta)$, $\psi(x, -\zeta)$ whose (formal or asymptotic) connection with the τ-function is given by (4.18), namely,

$$\psi(x, \zeta) = \frac{X(\zeta)\tau}{\tau}, \quad \psi(x, -\zeta) = \frac{X(-\zeta)\tau}{\tau}, \quad (4.106)$$

where, from (4.18),

$$X(\zeta) = \exp\left(i \sum \zeta^{2k+1} t_{2k+1}\right) \exp\left(\sum \frac{-1}{i(2k+1)\zeta^{2k+1}} \frac{\partial}{\partial t_{2k+1}}\right).$$

Since we have chosen in most of this chapter to work with $2^{2k}t_{2k+1}$ rather than t_{2k+1} (remember this removes the fractions $1/2^{2k}$ from in front of the RHS of the t_{2k+1} flow (3.14)), we will write

$$X(\zeta) = \exp\left(i\sum \zeta(2\zeta)^{2k}t_{2k+1}\right)\exp\left(\sum \frac{-1}{i\zeta(2k+1)(2\zeta)^{2k}}\frac{\partial}{\partial t_{2k+1}}\right).$$

Using (4.106), (4.105) may be written

$$\tau_{new} = (AX(\zeta) + BX(-\zeta))\tau_{old}. \tag{4.107}$$

Let us do a few examples to show how all this works in practice. First take $\tau_0 = 1$. Then

$$\tau_1 = Ae^\theta + Be^{-\theta} \tag{4.108}$$

where

$$\theta = \sum_0^\infty i\zeta(2\zeta)^{2k}t_{2k+1}, \qquad t_1 = x. \tag{4.109}$$

Choose $A = \alpha e^{-i\zeta x_0}$, $B = \alpha e^{i\zeta x_0}$ and call $\zeta = i\eta$ and $-i\zeta x_0 = \theta_0$

$$\tau_1 = 2\alpha\cosh(\theta - \theta_0)$$

the one-soliton solution. Next

$$\tau_2 = (A_2X(\zeta_2) + B_2X(-\zeta_2))(A_1X(\zeta_1) + B_1X(-\zeta_1)).$$

It is a simple exercise to show that

$$X(\zeta)X(\zeta') = \left|\frac{\zeta - \zeta'}{\zeta + \zeta'}\right|^{1/2}e^{\theta + \theta'}. \tag{4.110}$$

Thus, using the obvious notation,

$$\tau_2 = A_1A_2\left|\frac{\zeta_1 - \zeta_2}{\zeta_1 + \zeta_2}\right|^{1/2}e^{\theta_1 + \theta_2} + A_2B_1\left|\frac{\zeta_1 + \zeta_2}{\zeta_1 - \zeta_2}\right|^{1/2}e^{\theta_2 - \theta_1}$$

$$+ A_1B_2\left|\frac{\zeta_1 + \zeta_2}{\zeta_1 - \zeta_2}\right|^{1/2}e^{\theta_1 - \theta_2} + B_1B_2\left|\frac{\zeta_1 - \zeta_2}{\zeta_1 + \zeta_2}\right|^{1/2}e^{-\theta_1 - \theta_2},$$

which, after choosing the coefficients of the first three terms to be unity and dividing out by $e^{\theta_1 + \theta_2}$, becomes ($\zeta_j = i\eta_j$, $j = 1, 2$)

$$\tau_2 = e^{\theta_1 + \theta_2}\left(1 + e^{-2\theta_1} + e^{-2\theta_2} + \left|\frac{\eta_1 - \eta_2}{\eta_1 + \eta_2}\right|^2 e^{-2\theta_1 - 2\theta_2}\right)$$

the two-soliton solution. Recall that the exponential factor out in front whose exponent is linear in t_{2k+1} $k = 0, 1, \ldots$ does not contribute to the field $q(x, t_3, t_5, \ldots)$. This process can be repeated.

Before we end this section on Bäcklund transformations, I want to add one more calculation which derives the formula for Bäcklund transformations between numbers of the AKNS hierarchy when $r = -q$. Recall from (3.36b) that the appropriate eigenvalue problem in that case is

$$v_{1x} + i\zeta v_1 = qv_2,$$
$$v_{2x} - i\zeta v_2 = -qv_1. \tag{4.111}$$

Then γ, defined to be v_2/v_1, satisfies

$$\gamma_x = 2i\zeta\gamma - q(1+\gamma^2). \tag{4.112}$$

If we set

$$\gamma = \tan\frac{u+\tilde{u}}{4}, \tag{4.113}$$

where $q = -\frac{1}{2}u_x$, $\tilde{q} = -\frac{1}{2}\tilde{u}_x$, then (4.112) is

$$\tilde{u}_x - u_x = 4i\zeta \sin\frac{u+\tilde{u}}{2}. \tag{4.114}$$

The corresponding relations between $\tilde{u}_{t_{2k+1}}$ and $u_{t_{2k+1}}$ can be found by simply converting the equations for the time dependence of v_1, v_2 into Riccati form. It is easy to show that if $q(x, t_{2k+1}) = -\frac{1}{2}u_x$ (x, t_{2k+1}) (k can also be negative, e.g. $k = -1$ gives the sine-Gordon equation, see [97]) satisfies the t_{2k+1} flow in the AKNS hierarchy with $r = -q$ (to say in this solution manifold, only the odd flows are allowed), then so does $\tilde{q}(x, t_{2k+1}) = -\frac{1}{2}\tilde{u}_x$ (x, t_{2k+1}). Hence (4.114) and the corresponding time relations are a Bäcklund transformation.

I give this calculation here because of the obvious connections in its method of derivation with the Bäcklund transformation for the KdV hierarchy. We will, however, tackle this same question again in Chapter 5 from a different and more general point of view which has the advantage that the complete set of relations corresponding to (4.114) and its companions relating $u_{t_{2k+1}}$ and $\tilde{u}_{t_{2k+1}}$ is given by one formula. Furthermore, the choice of transformation (4.113) becomes an obvious consequence. The trouble with (4.113) as it stands is that while it is natural to make a tan substitution to solve the Riccati equation (because of the ratio $\gamma_x : 2\gamma : 1+\gamma^2$), it is a priori mysterious as to how a $\tilde{u}$ introduced in this way will also satisfy the same equation that u does. It is not until a posteriori, when one sees the symmetry of (4.114) and its time companions, that this fact becomes clear.

4g. The appearance of a Kac–Moody algebra. In this chapter, we have seen that the τ-function, $\tau(t_1, t_3, t_5, \ldots, t_{2k+1}, \ldots)$, carries all the information we ever need to know about the space of solutions for the KdV family. It changes in either one of two ways. First, it can change because of the flows where the independent variables $\{t_{2k+1}\}_0^\infty$ evolve but the functional form of τ remains fixed. For example, the two-soliton solution (4.47)

$$\tau = 1 + e^{\theta_1} + e^{\theta_2} + e^{\theta_1+\theta_2+A_{12}},$$

$$\theta_j = \sum_0^\infty (-1)^k \eta_j^{2k+1} t_{2k+1}, \qquad e^{A_{12}} = \left(\frac{\eta_1-\eta_2}{\eta_1+\eta_2}\right)^2,$$

evolves under the flows but still remains a two-soliton solution. Second, as we have just learned in the previous section, the τ-function can change its functional form via a Bäcklund transformation while keeping the sequence of independent variables $\{t_{2k+1}\}_0^\infty$ fixed. Each of the changes is a group action and, in each case, the new τ also satisfies all the equations of the KdV family, either the infinite sequence of quadratic Hirota equations or the sequence

$$q_{t_{2k+1}} = \frac{\partial}{\partial x} L^k q, \qquad L^k q = 2 \frac{\partial^2}{\partial t_1 \, \partial t_{2k+1}} \ln \tau. \tag{4.115}$$

Therefore, the solution space of the KdV family is mapped out by the joint action of the flows and Bäcklund transformations. The infinitesimals of these actions, the infinitesimal symmetries, form an infinite dimensional graded Lie algebra (a Kac–Moody algebra) which is isomorphic to the central extension of the loop algebra of $sl(2, C)$, the latter denoted by $\widetilde{sl}(2, C)$. This is the point of view developed by Date, Jimbo, Kashiwara and Miwa in [39]. It contrasts with the point of view developed in Chapter 5, in which part of $\widetilde{sl}(2, C)$ is used as the phase space. The transition from the latter picture in which solutions are curves in a Lie algebra to the former in which solutions are points in a representation space, i.e. $\tau(t_1, t_3, \ldots)$, has not yet been worked out in a logical, Lie theoretic way but I will try to tie the two points of view together with some suggestive formulae in Section 5j.

We now turn to the business of identifying and finding a representation for the infinitesimal symmetries corresponding to the flows and Bäcklund transformations. First, observe that the action of the flows on $\tau(t_1, t_3, \ldots)$ is simply the action of translating the arguments and can be represented by

$$\exp\left(\sum_0^\infty a_{2k+1} \frac{\partial}{\partial t_{2k+1}}\right) \tau(t_1, \ldots, t_{2k+1}, \ldots) = \tau(t_1 + a_1, \ldots, t_{2k+1} + a_{2k+1}, \ldots)$$
$$\tag{4.116}$$

for arbitrary values of $a_1, \ldots, a_{2k+1}$. The infinitesimals of this group action are represented by the sequence $\{\partial/\partial t_{2k+1}\}_0^\infty$. We also observe that since all the quantities of interest

$$L^k q = 2 \frac{\partial^2}{\partial t_1 \, \partial t_{2k+1}} \ln \tau \tag{4.117}$$

are second log derivatives of the τ function, one can multiply τ by an exponential whose argument is linear in the times $\{t_{2k+1}\}_0^\infty$, i.e. by $\exp \sum_0^\infty b_{2k+1} t_{2k+1}$ for arbitrary $b_1, \ldots, b_{2k+1}, \ldots$. The infinitesimals of this class of symmetries are represented by $\{t_{2k+1}\}_0^\infty$. The two sets of elements $\{\partial/\partial t_{2k+1}\}_0^\infty$ and $\{t_{2k+1}\}_0^\infty$ generate a Heisenberg algebra

$$\left[\frac{\partial}{\partial t_{2k+1}}, t_{2j+1}\right] = \delta_{kj},$$

$$\left[\frac{\partial}{\partial t_{2k+1}}, \frac{\partial}{\partial t_{2j+1}}\right] = 0, \qquad [t_{2k+1}, t_{2j+1}] = 0. \tag{4.118}$$

Again, I remind the reader that this algebra is derived from the symmetries arising from the flows and from the fact that there is an equivalence class of τ-functions each of which corresponds to the same solution q of the KdV family. It is useful, at this point, to think in terms of inverse scattering theory language. We know that the *solution type* is specified by the "initial" scattering data

$$S(0) = \{R(\xi, 0), \xi \text{ real}; (\zeta_j = i\eta_j, b_j(0) = e^{2\eta_j \bar{x}_j})_1^N\}. \tag{4.119}$$

The flows change the scattering data merely by changing the phase of the reflection coefficient and the position coordinates $\bar{x}_j$ of the solitons linearly in the times. That is,

$$S(t) = \left\{ R(\xi, 0) \exp\left(2i \sum_0^\infty \xi^{2k+1} t_{2k+1} \right) \xi \text{ real};\right.$$

$$\left. \left(\zeta_j = i\eta_j, b_j(t) = b_j(0) \exp\left(2i \sum_0^\infty \zeta_j^{2k+1} t_{2k+1} \right) \right)_1^N \right\}. \tag{4.120}$$

(*Digression.* We have pointed out already that the inverse scattering transform is a canonical transformation which carries us from the old coordinates $q(x)$ to the new ones, the action-angle variables (see [13], [70], [75])

$$\mathbf{p} = \left\{ p_j = -2\eta_j^2, p(\xi) = -\frac{2\xi}{\pi} \ln\left(1 - |R|^2\right) \right\}, \tag{4.121}$$

$$\mathbf{q} = \{ q_j = \ln b_j, q(\xi) = \operatorname{Arg} b(\xi) \},$$

whose Poisson brackets also form a Heisenberg algebra,

$$\{p_j, q_k\} = \delta_{jk}, \qquad \{p(\xi), q(\xi')\} = \delta(\xi - \xi')$$

and all other brackets are zero. I do not yet know a Lie theoretic way of identifying these two Heisenberg algebras, or if indeed, one is the manifestation of the other.)

The flows preserve the solution type. On the other hand, the Bäcklund transformation changes the solution type in the sense that it adds new components to the scattering data. For example, starting with the vacuum state

$$S = \{R(\xi, 0) \equiv 0, \xi \text{ real}; N = 0\},$$

one can build the one-soliton state

$$S = \left\{ R(\xi, 0) \equiv 0; \xi \text{ real}; \zeta_1 = i\eta, \right.$$

$$\left. b_1 = \exp\left(2\eta x_0 + \sum_0^\infty (-1)^{k+1} \eta^{2k+1} t_{2k+1} \right) \right\}$$

by applying the Bäcklund transformation

$$\tau_{\text{new}} = (AX(\zeta) + BX(-\zeta))\tau_{\text{old}} \tag{4.122}$$

with $\tau_{\text{old}} = 1$ and $X(\zeta)$ the operator given by (4.18). In principle, one can build all the solutions out of the vacuum state by Bäcklund transformations, but in practice only the multisoliton solutions are tractable. The Kyoto group like to think of the multisoliton solutions as being dense in the space of all solutions, but they do not make clear in what sense this is so. Nevertheless, for the purpose of continuing this discussion, let us accept this. For the multisoliton solutions, we can rewrite (4.122) as

$$\tau_{\text{new}} = (1 + \beta Y(\zeta))\tau_{\text{old}}, \tag{4.123}$$

where $\beta = B/A = e^{-2\eta x_0}$ locates the initial position of the soliton we are about to add and

$$Y(\zeta) = \exp\left(-2i \sum_0^\infty \zeta^{2k+1} t_{2k+1}\right) \exp\left(\sum_0^\infty \frac{2}{i(2k+1)\zeta^{2k+1}} \frac{\partial}{\partial t_{2k+1}}\right). \tag{4.124}$$

Equation (4.123) holds because one can always divide out of τ the exponential factor whose argument is linear in the times. The reader should prove that

$$Y(\zeta) \cdot Y(\zeta')1 = \left(\frac{\zeta - \zeta'}{\zeta + \zeta'}\right)^2 \exp\left(-2i \sum_0^\infty (\zeta^{2k+1} + \zeta'^{2k+1}) t_{2k+1}\right). \tag{4.125}$$

Note, in particular, that $Y^2(\zeta)1 = 0$ and that one can then write (4.123) as

$$\tau_{\text{new}} = \exp(\beta Y(\zeta))\tau_{\text{old}}. \tag{4.126}$$

The infinitesimal action (as the initial point $x_0 \to +\infty$, β becomes progressively smaller) is given by the *vertex operator* $Y(\zeta)$. Formally $Y(\zeta)$ can be expressed as an infinite Laurent series

$$Y(\zeta) = \sum_{-\infty}^\infty Y_{2k+1}\zeta^{2k+1}.$$

Because the operators $Y(\zeta)$, $Y(\zeta')$ do not commute when $\zeta + \zeta' = 0$ (because of the factor $(\zeta + \zeta')^2$ in the denominator of (4.125)), the coefficients Y_{2k+1} obey a nontrivial set of commutator relations [39].

The important fact is this. Lepowsky and Wilson [102] have shown that these relations, together with the Heisenberg algebra (4.118), are isomorphic to the infinite dimensional graded Lie algebra (Kac–Moody algebra) $\widetilde{sl}(2, C) \oplus Z$, denoted by $A_1^{(1)}$, the central extension of the loop algebra of $sl(2, C)$. Each term in $\widetilde{sl}(2, C)$ is the product of a grading parameter λ times an element of $sl(2, C)$ which can be written in a matrix representation as $hH + eE + fF$ with the basis vectors

$$H = \begin{pmatrix} 1 & 0 \\ 0 & -1 \end{pmatrix}, \qquad E = \begin{pmatrix} 0 & 1 \\ 0 & 0 \end{pmatrix}, \qquad F = \begin{pmatrix} 0 & 0 \\ 1 & 0 \end{pmatrix}.$$

We therefore have one answer to the question:

"What does $sl(2, C)$ have to do with KdV?"

Solutions of soliton equations on the KdV family form an orbit (the set of all $\tau(t_1, t_3, \ldots))$ of the highest weight vector (which corresponds to $\tau = 1$) in the basic representation of $\tilde{s}l(2, C) \oplus Z$. The algebra acts on solutions as an algebra of symmetries. The alternative point of view in which the algebra is used as a phase space and the connections between the two points of view are given in Chapter 5.

Historical note. In examining the equations of motion for the general top, Kowalevski observed that in the two special cases (the Euler and Lagrange tops) where the system was known to be integrable, the solutions involved elliptic Θ functions and had no singular points other than poles for finite complex values of time. She wondered whether this (what we now call Painlevé) property might in fact be also valid for the general top. She found the answer to be negative but in the process discovered a new choice of parameter relations (between the moments of inertia, etc.) for which the property does hold and for which the top equations are integrable. Kowalevski therefore was the first person to utilize the Painlevé property. The reader should consult S. Kowalevski, Acta. Math., 12 (1889), pp. 177ff.; 14 (1890), pp. 81ff. See also the article by H. Yoshida in the volume cited in [39].

CHAPTER 5

Connecting Links Among the Miracles of Soliton Mathematics

5a. Overview. In this chapter, we investigate the mathematical structure of soliton equations and attempt to develop a viewpoint from which most or all of the miracles of soliton mathematics appear to be natural consequences. At the very least, we should like to see some common thread which ties them together. The miracles include:

1. An infinite number of local conservation laws and symmetries; membership in an infinite family of commuting flows; Hamiltonian structure (sometimes structures).

2. An equivalent statement of the nonlinear equations in bilinear form (the Hirota equations); the τ-function, which when considered as a function of infinitely many independent variables, contains so much information about the solution manifold; the Painlevé property.

3. The association with a linear eigenvalue problem; inverse scattering; isospectral, iso-Riemann surface and isomonodromic deformations; the Riemann–Hilbert problem.

4. Bäcklund and Schlesinger transformations; vertex operators.

5. The ideas of Wahlquist–Estabrook and the appearance of a rich algebraic structure; "what does $sl(2, C)$ have to do with NLS and KdV?"; the notion of reduction and the connection with the Zakharov–Shabat "dressing" scheme.

The key elements of our new perspective are that the appropriate phase space in which the (one spatial dimension) soliton equations live is a Kac–Moody algebra (an infinite dimensional, graded Lie algebra) and that it is important to think of each dependent variable as a function of an infinite number of independent variables (the flow times t_k) with no one distinguished from another. The only time that it is important to distinguish one variable is when we attach some global behavior to the dependent variables as function of one of the independent variables, as for example when we wish to solve the equation of interest as an initial-boundary value problem. All the work described in this chapter is joint with my colleagues Hermann Flaschka and Tudor Ratiu of the University of Arizona and has appeared (or is about to) in a series of papers in Physica D [38].

It would also be desirable to understand the fascinating connections between soliton mathematics and integrable nonlinear partial differential equations and other solvable models in statistical physics such as the nearest neighbour Ising model. I will give some references to these latter developments [103] but will not pursue them further in this set of lectures. Neither will anything be said about soliton equation hierarchies with spatial dimension greater than one but the interested reader should be aware of reference [39] in which the Kyoto group discuss the KP (Kadomtsev–Petviashvili) hierarchy. (See also Exercise 3b(5)).

The outline of the chapter is as follows. In Section 5b we show how, beginning with an equation or set of equations, one determines if it is integrable and, if so, how one finds the algebraic structure of the phase space in which the solutions live. The method used is a variant of the Wahlquist–Estabrook method in that it follows their basic ideas while avoiding the differential geometry terminology. In particular, we show that the algebraic structure of the AKNS hierarchy is isomorphic to (a subalgebra of) $\widetilde{sl}(2, C)$, the loop algebra $(\sum_{-\infty}^{\infty} X_{-i}\zeta^i, \ X_{-i} \in sl(2, C))$ associated with $sl(2, C)$. From the properties of many computational results, it would appear that the augmented Kac–Moody algebra $\hat{A}_1^{(1)}$, consisting of $A_1^{(1)}$ $(\widetilde{sl}(2, C) \oplus Z$ with the addition of a derivative element, may be more relevant. However, there are difficulties associated with working with $\hat{A}_1^{(1)}$ which I will discuss in Section 5l.

The Wahlquist–Estabrook ideas provide the motivation for the choice of phase space. Once established, and once it is realized that the phase space G is the direct sum of two algebras K, N in which the orthogonal complement $K^{\perp}$ of one is the dual N^* of the other with respect to a suitably defined inner product, there is a natural way to define Poisson brackets and Hamiltonian vector fields on $K^{\perp} = N^*$. When the latter are generated by functions of a special class, the so-called ad-invariant functions Φ_k, one immediately obtains an infinite family of commuting flows in Lax form $Q_{t_k} = [Q^{(k)}, Q]$ where Q is the general element in the phase space (the dual of one of the subalgebras) and $Q^{(k)} = \pi_N \nabla \Phi_k$, where ∇ is the gradient and π_N the projection into the subalgebra N. All the commutability properties are automatic consequences of very general theorems. This material is contained in Sections 5c. Also contained in this section is a discussion of how one relates the flows and Hamiltonian structure introduced in the manner described above with the flows and Hamiltonian structures which arise when x is a distinguished variable as, for example, when we begin with a distinguished eigenvalue problem $V_x = PV$, with P polynomial in ζ. The reader will recall from Section 3c that if P is of degree one (a gauge transformation, see Section 5g, can always bring it to the form (3.31)), then the AKNS hierarchy results; if it is of degree two, then the DNLS hierarchy [78] obtains.

At the end of Section 5c, I go through several exercises in which examples of Hamiltonian vector fields on duals of subalgebras are given in other contexts. The first two examples give rise to the equations for the simple harmonic oscillator and the Toda lattice with free ends. The third example shows how to include the KdV and MKdV families in a Lie-algebraic setting. The Miura transformation relating the solutions of the two families is an immediate consequence.

In Section 5d, we take advantage of the particular shape of the Lax equations and write down straightaway all the conservation laws

$$\frac{\partial}{\partial t} \text{conserved density} = \frac{\partial}{\partial x} \text{flux},$$

where the x and t are any two members of the infinite string of independent variables $\{t_k\}$; moreover, we can give explicit expressions for all the conserved densities and fluxes. These formulae are new.

At this stage, one can proceed in either of two directions, each one associated with the miracles listed under (2) and (3) above.

First, we note that the conservation laws have a structure which expresses the fact that the curls of three infinite dimensional vectors are zero. This immediately invites the introduction of *potentials*. These potentials are the set of Hirota τ-functions which for $sl(2, C)$ consist of a triplet $\{\tau, \sigma, \rho\}$. When the equations of motion are rewritten with these potentials as the dependent variables, one obtains the *Hirota bilinear equations*. As we have already mentioned in Chapter 4, the conditions for multisoliton solutions are equivalent to the conditions which ensure the Painlevé property.

Second, one observes that the Lax equations can be solved formally by the introduction of an auxiliary matrix V, $Q = V Q_0 V^{-1}$, where Q_0 is constant in all the times. The V then satisfies $V_{t_k} = Q^{(k)} V$ which is the sequence of auxiliary equations, the integrability conditions for which the equations in all the hierarchies are associated with $sl(2, C)$. Any one of these can be chosen to be the "eigenvalue" problem by imposing global constraints on the behavior of the dependent variables Q with respect to that single independent variable. For example, one might demand that all the nonconstant entries in Q approach zero as the special independent variable tends to $\pm\infty$. All the other independent variables then play time-like roles and are treated in the initial value problem sense. For example, in the AKNS hierarchy, t_1 is the special variable; in the DNLS hierarchy, of which the derivative nonlinear Schrödinger equation and the massive Thirring model are members, t_2 is the special variable. Associated with the constraints in the special variable is a certain corresponding analytic behavior of a suitably normalized fundamental solution matrix V, when considered as a function of ζ, the grading parameter in the Kac–Moody algebraic structure. In particular, the notion of iso-spectral flows can be introduced. This idea, together with a discussion of iso-Riemann surface and isomonodromic flows is given in Section 5f.

The connection between this second direction and the Hirota functions is reestablished when we examine in Section 5e, just as we did in Chapter 4 for the KdV equation, the formal asymptotic behavior in ζ of V. One finds that the asymptotic series, which first is expressed in terms of the entries in Q, can be reexpressed in terms of the potentials and gives formal relations between V and $\{\tau, \sigma, \rho\}$ with the aid of suitably defined "vertex" operators.

In Section 5g, we introduce Bäcklund transformations. Our approach is very general. We simply ask: what transformations on V keeps the form of the Lax equation $Q_{t_k} = [Q^{(k)}, Q]$ invariant? The resulting gauge transformations

$$V_{\text{new}} = R V_{\text{old}} S,$$

in which R plays the dominant role, induce a Bäcklund transformation on Q; indeed the relation between the new and old Q's has a very simple form. It is algebraic,

$$Q_{\text{new}} = R Q_{\text{old}} R^{-1}.$$

Several examples are discussed and two types of Bäcklund transformation are introduced. The first may be familiar to the reader; it is the one that adds solitons. Like the time flows, these Bäcklund transformations are continuous symmetries; namely, new solutions can be built as continuous deformations from old ones. The second, called a Bäcklund–Schlesinger transformation, is somewhat novel. It is designed so as to change the monodromy of the fundamental solution matrix at $\zeta = \infty$ and, when applied sequentially, becomes a difference equation. This transformation corresponds to a discrete symmetry of the equation family and adds a new integer-valued variable n to the list of independent variables. We will see how, as functions of n and t_1, the dependent variables of the AKNS hierarchy satisfy the differential-difference equation of the Toda lattice.

Moreover, the Bäcklund transformations which add solitons can be rewritten in terms of the "vertex" operators acting on the τ-functions $\{\tau, \rho, \sigma\}$. It turns out that the auxiliary τ-functions ρ and σ can be obtained by applying Bäcklund–Schlesinger transformations to the main τ-function τ. Indeed, repeated application of the Bäcklund–Schlesinger transformation gives the succession of τ-functions $\tau(n, t_1, t_2, t_3, \ldots)$ for the Toda lattice family; namely $\tau(45, t_1)$ is the time t_1 history of the τ-function for the Toda lattice (from neighboring pairs of which the displacements may be computed) at the 45th lattice site.

In Section 5h, we introduce the notion of grading. The basic idea is that there is more than one way that one can decompose a given algebra G into two subalgebras K and N. Each independent decomposition leads to a different set of flows. The different decompositions can be found through a process called *grading* by which one assigns different weights, consistent with all the commutation relations, to the basis vectors and grading parameter. In the case of $\widetilde{sl}(2, C)$, the basis vectors are the H, E, F defined in (5.40) and equivalent to the Pauli spin matrices, the grading parameter is ζ and the general element is $X = \sum_{-\infty}^{\infty} X_{-j} \zeta^j$, $X_{-j} = h_{-j} H + e_{-j} E + f_{-j} F$. The number of independent gradings is related to the number of independent automorphisms of finite order of the underlying algebra. For $\widetilde{sl}(2, C)$, whose elements we decompose into elements of two subalgebras K and N, with degrees less than zero and greater than or equal to zero respectively, there are two. The first, called the homogeneous grading, gives rise to the AKNS flows, the second, called the principal grading, to the KdV and MKdV families.

In Section 5i, we introduce a second Hamiltonian structure which arises by changing the definition of the inner product on the algebra. It turns out that this structure is more convenient in Section 5j in which we show how the Lax equations $Q_{t_k} = \lceil Q^{(k)}, Q \rceil$ are a reduction of a much simpler flow in a larger phase space. The reduction is achieved by taking advantage of the symmetries enjoyed by the system of equations. The basic idea is not new. It has long been known that if an m-dimensional Hamiltonian system has $n \leqq m$ constants of the motion in involution (equivalent by Noether's theory to n symmetries), then the phase space can be reduced from $2m$ dependent variables to $2(m - n)$. If $n = m$, the system is said to be completely integrable. In Section 4c, I showed how the motion of a mass spring system on the plane (of dimension two) can be reduced to a one-dimensional system (described by a second order ordinary differential equation) by using angular momentum conservation which corresponds to the rotational symmetry inherent in the problem. For the Korteweg–deVries and nonlinear Schrödinger equations, some of the infinite number of motion constants admit simple physical interpretations, like conservation of mass, momentum, energy, number density, current density and so on, but most do not. They are then called hidden. However, their identity is known once we have identified the Lie algebra G in which their solutions live. If $\bar{G}$ is the corresponding Lie group and $\bar{K}$, $\bar{N}$ the subgroups corresponding to the subalgebras K, N, then the following is true. The large symplectic manifold on which the flows are simple (just like action-angle variables, half the variables are constants, the other half move linearly with time) is $T^*\bar{G}$, the cotangent bundle of $\bar{G}$. The symmetry group by which we reduce the phase space $T^*\bar{G}$ is $\bar{K}$ (the abstract analogue of the classical reduction theorem is due to Marsden and Weinstein [88]) followed by a (trivial) reduction by $\bar{N}$. The reduced phase space is N^* and it is on this that the solutions Q to the Lax equations $Q_{t_k} = [Q^{(k)}, Q]$ live.

Furthermore, the reduction process gives us, in principle, a means by which we can solve the Lax equations. The key step is one in which an element g of $\bar{G}$ is factored into $k^{-1}n$, with the left and right factors belonging to $\bar{K}$ and $\bar{N}$ respectively. This step is the algebraic equivalent of the Riemann–Hilbert problem. The factorization process gives us another opportunity to define the τ-function. Here it arises as an infinite dimensional determinant. (You will recall that it makes its first definitional appearance as a potential; see Sections 4b, 5d.) All this is done in Section 5j. At the end of this section, we show how the formal solution of the Lax equations also leads to an algorithm by which one solution type is transformed into another and in particular how multisoliton solutions are built out of the vacuum state. This algorithm, which is equivalent to a Bäcklund transformation, turns out to be completely analogous to the "dressing" scheme proposed by Zakharov and Shabat. At this time, we also discuss how the phase space is mapped out by the joint action of the flows and Bäcklund transformations, which are continuous symmetries of the equation family, and Bäcklund–Schlesinger transformations which are discrete symmetries. For the principal grading, in which the KdV and MKdV families arise and for which soliton equations there is only one τ-function, there are no Bäcklund–Schlesinger transformations. In that case, as we discussed in Section 4g, the connection between the appearance of the Kac–Moody algebra as the phase space on the one hand and as an algebra of symmetries on the other can be made. For the homogeneous grading, in which the discrete symmetries are present, the complete connection is more difficult to make. In Section 5k, we attach to the flows $Q_{t_k} = [Q^{(k)}, Q]$, $k \geq 0$, the ones corresponding to the negative times t_k, $k < 0$. The most familiar examples of these are the sine-Gordon and massive Thirring model equations. This material is new.

Finally, in Section 5l, we discuss the changes which are necessary if we take as the phase space an extension of the loop algebra, the additional elements being a center and a derivation. There is much evidence that it is useful to include the extra elements. For example, certain formulae make much more sense when they are used. However, there is a major difficulty. Whereas the Lax equations themselves still appear to hold, the notion of the ad-invariant function, so important in Section 5c, is lost. In the new algebra, such functions may not even exist. We certainly have not been able to identify any. Another disappointing feature of the present theory is that it still does not allow for a Lie algebraic definition of the τ-function, nor do we yet have any idea of the space in which it lives. It makes its appearance in a fairly natural way as a potential and as the determinant of the coefficient matrix in an infinite set of linear equations; it also can be formally defined in terms of the V, the auxiliary function by which the Lax equations is solved, and its "ζ" derivative (which corresponds to the action of the derivative element in the extended algebra). But these are computational facts of life, rather than Lie algebraic necessities and clearly a deeper understanding of this remarkable beast is needed.

As a final note in this overview, I include a diagram, Fig. 6, which attempts to bring a visual perspective to the various soliton miracles.

5b. The Wahlquist–Estabrook approach[7] [37], [77], [97].

(i) *Introduction.* The goal of this section will be to answer the question: is a given equation integrable and, if it is, what is the natural setting in which its integrability is readily apparent? In answering this question, two guiding principles, gained from a decade of experience, are kept in mind. The first is that soliton equations arise as the integrability condition of linear systems and the second is that each soliton equation is a member of an infinite family of commuting flows. Our first goal, therefore, is to try to write the nonlinear equation as the integrability condition of the pair of linear systems $V_x = PV$, $V_t = QV$ by an appropriate choice of P and Q. This step will lead to expressions for the dependence of P and Q on the dependent variable of the nonlinear equation and its derivatives together with an open-ended set of commutator relations. Further restrictions on P and Q arise by insisting that the infinite sequence of possible Q's commute. This second requirement reads directly to the choice of Kac–Moody algebra as phase space. In the case of the AKNS hierarchy, the coefficient matrices turn out to be elements of $\tilde{sl}(2, C)$, an infinite dimensional loop algebra in which each basis vector can be written as the product of one of the basis vectors

$$\bar{H} = \begin{pmatrix} 1 & 0 \\ 0 & -1 \end{pmatrix}, \quad E = \begin{pmatrix} 0 & 1 \\ 0 & 0 \end{pmatrix}, \quad F = \begin{pmatrix} 0 & 0 \\ 1 & 0 \end{pmatrix}$$

---

[7] Section 5b is rather lengthy; on first reading, you may prefer to accept the conclusions and pass on to the later sections. Please read also the comments in the Note, p. 234.

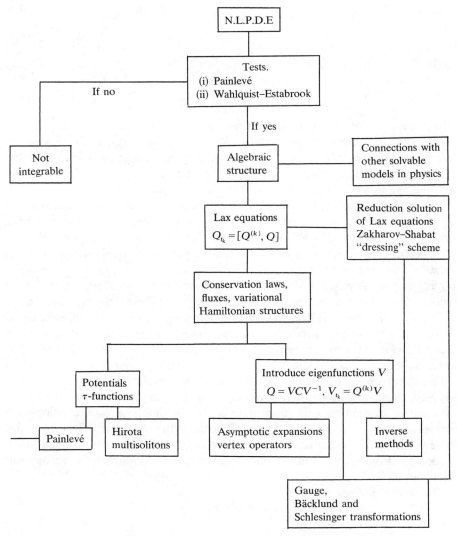

FIG. 6

of $sl(2)$ and a complex valued parameter ζ raised to an integer power. The full Kac–Moody algebra $\widehat{sl}(2) = \widetilde{sl}(2) + \mathbb{C}Z + \mathbb{C}D$ contains a center and derivative term whose roles in the theory will be discussed in Section 5l.

To our knowledge, this is the first time that the open-ended algebra generated by the Wahlquist–Estabrook method has been interpreted in a Kac–Moody framework.

(ii) *The nonlinear Schrödinger equations.* Consider the first pair of nontrivial equations of the AKNS hierarchy,

$$q_t = i/2(q_{xx} - 2q^2 r),$$
$$r_t = -i/2(r_{xx} - 2qr^2). \tag{5.1}$$

We attempt to write these equations as the integrability condition

$$P_t - Q_x + [P, Q] = 0 \tag{5.2}$$

of the pair of linear equations

$$V_x = PV, \tag{5.3}$$

$$V_t = QV. \tag{5.4}$$

Here we think of P and Q as matrices of arbitrary order whose coefficients depend on q, r and their derivatives. We could have taken (5.3), (5.4) to be nonlinear, $V_x = F(V)$, $V_t = G(V)$ in which case the commutator in (5.2) would be the general Lie bracket but in all cases to date it has been more convenient to take the linear representation of the underlying algebra.

We begin by making the simplest assumption that P depends only on q and r. If this is so, it must depend linearly on these variables as the following argument shows. If $P = P(q, r)$, from (5.1) we have $P_t = (i/2)P_q(q_{xx} - 2q^2r) - (i/2) \, P_r(r_{xx} - 2qr^2)$ where subscripts denote partial derivatives. In order to balance this term in (5.2), Q must depend on q_x, r_x, q, r and $Q_x = Q_q q_x + Q_r r_x + Q_{q_x} q_{xx} + Q_{r_x} r_{xx}$. A balance of the q_{xx}, r_{xx} terms shows $Q_{q_x} = (i/2)P_q$, $Q_{r_x} = -(i/2)P_r$ which after integration gives $Q = (i/2)P_q q_x - (i/2)P_r r_x + \tilde{Q}(q, r)$. Now $Q_x = (i/2)P_q q_{xx} - (i/2)P_r r_{xx} + (i/2)P_{qq}q_x^2 - (i/2)P_{rr}r_x^2$ plus terms which are at most linear in q_x, r_x. Since the commutator has terms proportional at most to q_x, r_x, we must have $P_{qq} = P_{rr} = 0$ which means that $P(q, r)$ can have the form $-iH + qE + rF + qrG$. However, as can readily be verified by a parallel of the ensuing analysis, G commutes with all other elements (because the coefficients of q^2r_x, r^2q_x and $rq_x - qr_x$ must vanish) and therefore belongs to the center. Therefore it plays no role in the commutator portion of (5.2) but is simply a manifestation of a conservation law $P_t = Q_x$ (equation (5.2) without the commutator) which in this case reads $(qr)_t G = (i/2)(rq_x - r_xq)_x G$. Since this information is contained in the equations anyway, we will, without loss of generality, simply leave out this element.

Therefore, we set

$$P = -iH + qE + rF \tag{5.5}$$

and from (5.1), (5.2) reads

$$Q_x + [Q, P] = \frac{i}{2}(q_{xx} - 2q^2r)E - \frac{i}{2}(r_{xx} - 2qr^2)F. \tag{5.6}$$

Now, solve (5.6) for Q; first write $Q = (i/2)q_xE - (i/2)r_xF + \bar{Q}(q, r)$. Then collecting perfect x derivatives from the commutator, we find $\bar{Q} = qE_1 + rF_1 - (i/2)qrH_0 - i\tilde{H}_2$ where we have defined H_0, E_1, F_1 by

$$[E, F] = H_0, \quad [H, E] = 2E_1, \quad [H, F] = -2F_1 \tag{5.7}$$

and $-i\tilde{H}_2$ is simply a constant matrix of integration. Therefore,

$$Q = -i\tilde{H}_2 + qE_1 + rF_1 + \frac{i}{2}q_xE - \frac{i}{2}r_xF - \frac{i}{2}qrH_0, \tag{5.8}$$

provided the relation

$$[\bar{Q}, P] = -iq^2 rE + iqr^2 F \tag{5.9}$$

is satisfied. We find by equating the coefficients of $q^2 r$, qr^2, q^2, r^2, qr, q, r, 1 that

$q^2 r$: $[H_0, E] = 2E,$ (5.10a)

qr^2: $[H_0, F] = -2F,$ (5.10b)

q^2: $[E, E_1] = 0,$ (5.10c)

r^2: $[F, F_1] = 0,$ (5.10d)

qr: $\frac{1}{2}[H_0, H] + [E, F_1] - [E_1, F] = 0,$ (5.10e)

q: $[\tilde{H}_2, E] = [H, E_1]$ (5.10f)

r: $[\tilde{H}_2, F] = [H, F_1],$ (5.10g)

1: $[\tilde{H}_2, H] = 0.$ (5.10h)

A few remarks are now in order.

1. The set of commutator relations is unclosed. The table will be given in Table 2.

2. The set (5.10) contains the closed subalgebra $sl(2)$ H_0, E, F (see (5.7) and (5.10a,b)).

3. Note the sense in which soliton equations involve a balance of nonlinearity (here represented by the terms $q^2 r$, qr^2) and dispersion (q_{xx}, r_{xx}). Equations (5.10a,b) arise because of the balance $[-(i/2)qrH_0, qE + rF]$ with $-iq^2 rE + iqr^2 F$. But the latter term arises directly from the nonlinearity in the equation while the former is a result of the integration of the product of the integration of the linear terms $q_{xx}E$ and $r_{xx}F$ with $qE + rF$. Had the nonlinear terms been $q^3 r^2$ and $q^2 r^3$, no such balances would have been possible and the only possible P, Q pair would have been the trivial ones $P \propto qrG$, $Q \propto (rq_x - r_x q)G$ expressing the existence of a (single) conservation law. This is one of the real advantages of the Wahlquist–Estabrook method. Nonsoliton equations show their inadequacies, quickly!!

4. From the Jacobi identity, (5.10e) implies

$$[H, H_0] = 0 \tag{5.10i}$$

and

$$[E, F_1] = [E_1, F], \tag{5.10j}$$

and we define

$$[E, F_1] = H_1. \tag{5.10k}$$

5. The last three equations (5.10f, g, h) define $\tilde{H}_2$, the arbitrary constant of integration; they do not give any information about what we consider to be the basic elements H, E, F from which all other elements are generated. For examples, $[H_1, E] = 2E_1$, and, as we shall see when we construct the table for (5.10), $[H, E_1] = 2E_2$ and so on. Nevertheless, one can use $\tilde{H}_2$ to effect an artificial closure on the commutator relations. Satisfy (5.10h) by letting $\tilde{H}_2 = \zeta H$ where ζ

is an arbitrary constant. Then $[\bar{H}_2, E] = \zeta[H, E] = 2E_1$ and (5.10f) and (5.10g) give $[H, E_1] = 2\zeta E_1$ and $[H, F_1] = -2\zeta F_1$ respectively. We obtain Table 1.

TABLE 1

	H_0	E	F	H	H_1	E_1	F_1
H_0	0	$2E$	$-2F$	0	0	$2E_1$	$-2F_1$
E		0	H_0	$-2E_1$	$-2E_1$	0	H_1
F			0	$2F_1$	$2F_1$	$-H_1$	0
H				0	0	$2\zeta E_1$	$-2\zeta F_1$
H_1					0	$2\zeta E_1$	$-2\zeta F_1$
E_1						0	ζH_1
F_1							0

This algebra admits the well-known representation

$$H = H_1 = \zeta H_0, \quad E_1 = \zeta E, \quad F_1 = \zeta F,$$

$$H_0 = \begin{pmatrix} 1 & 0 \\ 0 & -1 \end{pmatrix}, \quad E = \begin{pmatrix} 0 & 1 \\ 0 & 0 \end{pmatrix}, \quad F = \begin{pmatrix} 0 & 0 \\ 1 & 0 \end{pmatrix}. \tag{5.11}$$

The artificial closure is the means by which Wahlquist and Estabrook arrived at the form of P and Q in the context of the Korteweg–deVries equation and it is the way in which succeeding authors have proceeded [97]. We will now reexamine the table without making any closure assumptions. In Table 2, the integers in the right-hand corners of the boxes mean: 1, by definition; 2, direct deduction from (5.10); 3, a consequence of (5.10) and the Jacobi identities.

As an exercise in the application of the Jacobi identity, I invite the reader to complete the table. Observe that, if the elements

$$[H, H_j] = 2X_j, \quad j = 1, 2, \ldots \tag{5.12}$$

are zero, then the elements H_p, E_q, F_r obey the relations

$$[H_p, E_q] = 2E_{p+q}, \quad [H_p, F_r] = -2H_{p+r}, \quad [E_q, F_r] = H_{q+r},$$
$$[H_p, H_q] = [E_p, E_q] = [F_q, F_r] = 0 \tag{5.13}$$

(in Table 2 we associate the index 1 with H). The algebra defined by (5.13) is what we call $\widetilde{sl}(2)$, namely each term can be represented by a product $H_p = \zeta^p \bar{H}$, $E = \zeta^q \bar{E}$, $F_r = \zeta^r \bar{F}$ where $\bar{H}, \bar{E}, \bar{F}$ are a basis for $sl(2)$. However, the nonlinear Schrödinger equations do not in and of themselves force $X_j = 0$, $j = 1, 2, \ldots$. It is clear that such a choice is consistent with (5.13) but is not imposed. The reason that $X_j = 0$, $j = 1, 2, \ldots$ is connected with the second guiding principle, namely the requirement that equation (5.1) is a member of an infinite sequence of commuting flows. Before we impose this condition,

TABLE 2

	H_0	E	F	H	H_1	E_1	F_1	H_2	E_2	F_2	$\tilde{H}_2$
H_0		$^2\,2E$	$^2\,-2F$	$^2\,0$	$^3\,0$	$^3\,2E_1$	$^3\,-2F_1$	$^3\,0$	$^3\,2E_2$	$^3\,-2F_2$	$^3\,-4X_1$
E			$^1\,H_0$	$^1\,-2E_1$	$^3\,-2E_1$	$^2\,0$	$^1\,H_1$	$^3\,-2E_2$ $-2[EX_1]$	$^3\,0$	$^1\,H_2$	$^2\,-2E_2$
F				$^1\,2F_1$	$^3\,2F_1$	$^2\,-H_1$	$^2\,0$	$^3\,2F_2$ $+2[F_2X_1]$	$^3\,-H_2$ $-2X_1$	$^1\,0$	$^2\,2F_2$
H					$^3\,2X_1$	$^1\,2E_2$	$^1\,-2F_2$	$^1\,2X_2$	$^1\,2E_3$	$^1\,-2F_3$	$^2\,0$
H_1						$^3\,2E_2$ $+[EX_1]$	$^3\,-2F_2$ $-[FX_1]$		$^3\,2E_3$ $+[E_1X_1]$	$^3\,-2F_3$ $-[F_1X_1]$	
E_1						$^3\,H_2$ $+X_1$					
F_1											
H_2											
E_2											
F_2											
$\tilde{H}_2$											

however, we will examine the algebra generated by the nth equation of the AKNS hierarchy.

(iii) *The nth equation pair of the* AKNS *hierarchy.* The nth equation pair of the AKNS hierarchy can be written

$$q_{t_n} = b_{nx} - 2a_n q,$$
$$r_{t_n} = c_{nx} - 2a_n r \tag{5.14}$$

where a_n, b_n and c_n are determined iteratively from

$$-2ib_{s+1} = b_{sx} - 2a_s q, \tag{5.15a}$$
$$2ic_{s+1} = c_{sx} + 2a_s r, \tag{5.15b}$$
$$a_{sx} = rb_s - qc_s, \qquad s = 0, 1, \ldots, n-1, \tag{5.15c}$$
$$a_0 = -i, \qquad b_0 = c_0 = 0. \tag{5.15d}$$

The constant of integration in a_s is made zero. We assume and verify a posteriori that

1. qb_s, rc_s, qc_s, qa_s, ra_s and a_s are functionally independent; $rb_s = qc_s + a_{sx}$.
2. qb_s, rc_s are not perfect x derivatives except when $s = 2$, in which case

$$qb_2 = \left(\frac{iq^2}{4}\right)_x, \qquad rc_2 = \left(-\frac{ir^2}{4}\right)_x.$$

We want to choose P, $Q^{(n)}$ such that (5.14) is the integrability condition

$$Q_x^{(n)} + [Q^{(n)}, P] = P_{t_n} \tag{5.16}$$

of the equation pair

$$V_x = PV, \qquad V_{t_n} = Q^{(n)}V. \tag{5.17}$$

Again, it can be argued that we may take

$$P = -iH + qE + rF, \tag{5.18}$$

whence (5.16) reads

$$Q_x^{(n)} + [Q^{(n)}, P] = (b_{nx} - 2a_n q)E + (c_{nx} + 2a_n r)F. \tag{5.19}$$

It is straightforward to verify that we may write

$$Q^{(n)} = -a_n H_0 + b_n E + r_n F + Q^{(n-1)}, \tag{5.20}$$

where

$$Q_x^{(n-1)} + [Q^{(n-1)}, P] = (b_{n-1x} - 2a_{n-1}q)E_1 + (c_{n-1x} + 2a_{n-1}r)F_1. \tag{5.21}$$

In order to obtain (5.21) from (5.20) we had to define

$$[E, F] = H_0, \quad [H, E] = 2E_1, \quad [H, F] = -2F_1; \tag{5.22}$$

note that $[b_n E + c_n F, qE + rF] = -a_{nx}H_0$ and use (5.15a,b) with $s = n-1$. In addition, the functional independence of $a_n q$, $a_n r$ imposed the condition

$$[H_0, E] = 2E, \qquad [H_0, F] = -2F. \tag{5.23}$$

Now, equation (5.21) is simply (5.20) with a relabelling and so we can repeat the process and find

$$Q^{(n)} = \sum_{s=n}^{3} Q_s + Q^{(2)}, \tag{5.24}$$

where

$$Q_s = -a_s H_{n-s} + b_s E_{n-s} + c_s F_{n-s} \tag{5.25}$$

and

$$Q_x^{(2)} + [Q^{(2)}, P] = (b_{2x} - 2a_2 q)E_{n-2} + (c_{2x} + 2a_2 r)F_{n-2} \tag{5.26}$$

with the following relations imposed. By definition,

$$[H, E_{n-s}] = 2E_{n-s+1}, \quad [H, F_{n-s}] = -2F_{n-s+1}, \quad [E, F_{n-s}] = H_{n-2} \tag{5.27}$$

while the coefficients of $a_{n-s}q$, $a_{n-s}r$, a_{n-s}, qb_{n-s}, rc_{n-s} give for $s = n, \dots, 3$,

$$[H_{n-s}, E] = 2E_{n-s}, \qquad [H_{n-s}, F] = -2F_{n-s},$$
$$[H, H_{n-s}] = [E, E_{n-s}] = [F, F_{n-s}] = 0. \tag{5.28}$$

From the fact that $rb_{n-s} - qc_{n-s} = a_{n-sx}$ we find,

$$[E_{n-s}, F] = [E, F_{n-s}] = H_{n-s}. \tag{5.29}$$

Using the Jacobi identities, we obtain that for all $p + q \leqq n - 3$,

$$[H_p, E_q] = 2E_{p+q}, \quad [H_p, F_q] = -2F_{p+q}, \quad [E_p, F_q] = H_{p+q}, \tag{5.30a}$$

and all other brackets of this order zero. The reason that we cannot continue the table is that at $s = 2$ we encounter the anomalous behavior described in assumption 2. Solving (5.26), we find

$$Q^{(2)} = b_2 E_{n-2} + c_2 F_{n-2} - a_2 H_{n-2} + \frac{iq^2}{4}[E, E_{n-2}]$$

$$- \frac{ir^2}{4}[F, F_{n-2}] + b_1 E_{n-1} + c_1 F_{n-1} - i\tilde{H}_n, \tag{5.30b}$$

where the new elements are subject to the constraints

q^3: $[E, [E, E_{n-2}]] = 0,$ (5.31a)

r^3: $[F, [F, F_{n-2}]] = 0,$ (5.31b)

$q^2 r$: $[H_{n-2}, E] = 2E_{n-2} - \frac{1}{2}[F, [E, E_{n-2}]],$ (5.31c)

qr^2: $[H_{n-2}, F] = -2F_{n-2} + \frac{1}{2}[E, [F, F_{n-2}]],$ (5.31d)

q^2: $[E, E_{n-1}] + \frac{1}{4}[H, [E, E_{n-2}]] = 0,$ (5.31e)

r^2: $[F, F_{n-1}] - \frac{1}{4}[H, [F, F_{n-2}]] = 0,$ (5.31f)

qr: $-\frac{1}{2}[H, H_{n-2}] + [E, F_{n-1}] - [E_{n-1}, F] = 0,$ (5.31g)

q: $[\tilde{H}_n, E] = [H, E_{n-1}],$ (5.31h)

r: $[\tilde{H}_n, F] = [H, F_{n-1}],$ (5.31i)

1: $[\tilde{H}_n, H] = 0.$ (5.31j)

Note that when $n = 2$, $E_{n-2} = E_0 = E$, $F_{n-2} = F_0 = F$ and (5.31) reduces to (5.10). As in the case $n = 2$, the Jacobi identity shows that $[H, H_{n-2}] = 0$ and that $[E, F_{n-1}] = [E_r, F_s] = [E_{n-1}, H_{n-1}]$, $r + s = n - 1$. Just as in the case of $n = 2$, we can extend this algebra indefinitely, defining new elements according to the rule

$$[H, E_r] = 2E_{r+1}, \quad [H, F_r] = -2F_{r+1}, \quad [E, F_r] = H_{r+1}, \quad r \geqq n - 1.$$

The resulting algebra is infinite dimensional, contains within it the structure of $\widetilde{sl}(2)$ but leaves undetermined the elements

$$[H, H_j], j \geqq n - 1, \quad [E, E_{n-2}], \quad [F, F_{n-2}]. \tag{5.32}$$

(iv) *The imposition of commutativity.* We have seen how each equation pair in the AKNS hierarchy, taken in isolation, generates, when extended indefinitely, an infinite algebra which contains $\widetilde{sl}(2)$ as a consistent solution but which leaves the commutators (5.32) unspecified. It is easy to see that this degeneracy is removed by demanding that the flows commute with each other. If we ask that q, r solve the first two equation pairs in the AKNS hierarchy, then in addition to (5.16) with $n = 2, 3$, we must have

$$Q_{t_3}^{(2)} - Q_{t_2}^{(3)} + [Q^{(2)}, Q^{(3)}] = 0. \qquad (5.33)$$

This condition implies that the elements H_p, E_q, F_r satisfy the commutator table *common* to $n = 2, 3$; namely, we find that $[E, E_1] = [F, F_1] = 0$ (a requirement demanded when $n = 2$ but not $n = 3$) and $[H, H_1] = 0$ (imposed when $n = 3$ but not $n = 2$). Therefore, the $n = 2, 3$ tables are $\widetilde{sl}(2)$ except the $[H, H_j]$, $j \geq 2$ are unspecified. Next demand that $n = 2, 3, 4$ commute and find that $[E, E_2]$, $[F, F_2]$, each unspecified by the $n = 4$ table now are forced to be zero as is $[H, H_2]$. In general the condition that the flows commute is

$$Q_{t_k}^{(j)} - Q_{t_j}^{(k)} + [Q^{(j)}, Q^{(k)}] = 0. \qquad (5.34)$$

As we continue the process, we find at the nth stage that the inclusion of the nth equation of the AKNS hierarchy into the family of commuting flows makes $[H, H_{n-2}]$, left unspecified by the table common to $r = 2, \ldots, n-1$, zero; also $[E, E_{n-2}]$, $[F, F_{n-2}]$ left unspecified by $r = n$ are now zero because they are so for the extended tables of $r = 2, \ldots, n-1$. It is now clear that in the limit as $n \to \infty$, the elements H, H_p, E_q, F_r satisfy the $\widetilde{sl}(2)$ relations

$$[H_p, E_q] = 2E_{p+q}, \qquad [H_p, F_r] = -2F_{p+r},$$
$$[E_q, F_r] = H_{q+r}, \qquad [H_p, H_q] = [E_p, E_q] = [F_p, F_q] = 0. \qquad (5.35)$$

The commutators of the element H are the same as those of H_1.

(v) *Conclusion.* We have seen how the two guiding principles, (a) soliton equations are integrability conditions of systems of linear equations and (b) each belongs to an infinite family of commuting flows, lead us to the conclusion that the natural phase space in which the equations live is a Kac–Moody algebra. Indeed, (5.34) is the natural framework in which to express the equations. In the next section, we shall see how these equations can be recast into Lax form

$$Q_{t_k} = [Q^{(k)}, Q] \qquad (5.36)$$

by considering the evolution of $Q = \lim_{j \to \infty} \zeta^{-j} Q^{(j)}$ (where ζ is the grading parameter).

It should also be clear at this stage that the requirement that the auxiliary equations $V_{t_i} = Q^{(i)} V$ are linear can be relaxed. The linearity is simply a consequence of the fact that we can always find a linear representation of $\widetilde{sl}(2)$. In this case, it is

$$H_p = \zeta^p \begin{pmatrix} 1 & 0 \\ 0 & -1 \end{pmatrix}, \quad E_q = \zeta^q \begin{pmatrix} 0 & 1 \\ 0 & 0 \end{pmatrix}, \quad F_r = \zeta^r \begin{pmatrix} 0 & 1 \\ 0 & 0 \end{pmatrix}. \qquad (5.37)$$

We could have used another representation, for example $\zeta^p(\gamma^2(d/d\gamma), \gamma(d/d\gamma), d/d\gamma$, and obtained instead of the (2×2) linear systems $V_{t_i} = Q^{(i)} V$, a sequence of Riccati equations for $\gamma = v_1/v_2$ where $V = \begin{pmatrix} v_1 \\ v_2 \end{pmatrix}$.

Finally we remark that whereas the goal of this section has been to understand the structure of the infinitely extended algebras generated by the AKNS hierarchy, in practice it is more convenient to construct the artificial closure. However, one should realize that, having done so, one must distinguish between the elements $\zeta \begin{pmatrix} 1 & 0 \\ 0 & -1 \end{pmatrix}$ and $\begin{pmatrix} 1 & 0 \\ 0 & -1 \end{pmatrix}$ in the sense that we consider them to be linearly independent. It is only when one considers the phase space to be an infinite string of such elements (the formal power series $Q = \sum_{r=0}^{\infty} \zeta^{-r}(h_r H + e_r E + f_r F)$) that the commutator $[Q^{(k)}, Q]$ of (5.36) has a natural interpretation as a Hamiltonian vector field.

Exercise 5b.
Try the method on the equation

$$q_t + 6q^p q_x + q_{xxx} = 0.$$

Let $P = qX_1 + X_2$ (show if $P = P(q)$, that $P_{qq} = 0$) and solve for Q in (5.2),

$$Q = -q_{xx}X_1 + q_x[X_1, X_2] - \frac{6}{p+1}q^{p+1}X_1 + \frac{q^2}{2}[X_1, X_3] + q[X_2, X_3] + \bar{X}.$$

(Hint: solve first for $Q = -q_{xx}X_1 + R(q, q_x)$ and continue $R = q_x[X_1, X_2] + S(q)$ and so on.) From (5.2) we obtain, on equating powers of q, that $X_3 = 0$ if $p \neq 0, 1$ or 2. In this case X_1 and X_2 are proportional and writing the equation as the solvability condition of (5.3), (5.4) merely reflects the fact that it has the obvious conservation law. On the other hand, when $p = 1, 2$, a nontrivial algebra arises. When $p = 2$, we can solve the commutator relations by the choices,

$$\bar{X} = -4i\zeta^3 H, \quad X_1 = E - F, \quad X_2 = -iH\zeta.$$

What are the possible solutions when $p = 1$?

5c. Lax equations associated with $\widetilde{sl}(2, C)$. The material in this section is a reproduction of paper II in our sequence on Kac–Moody algebras and soliton equations. We have seen from the previous section how the natural phase space for the AKNS system is associated with an infinite dimensional Lie algebra $G = \widetilde{sl}(2, C)$ of formal series

$$X = \sum_{-\infty}^{M} X_{-i}\zeta^i, \quad M \text{ arbitrary but finite,} \tag{5.38}$$

where each element X_{-i} belongs to $sl(2, C)$.

A Lie algebra is a *vector space* equipped with a *commutator*; in this case, it is

$$[X, Y] = \sum_i \sum_{k+j=i} [X_{-j}, Y_{-k}]\zeta^i. \tag{5.39}$$

It is useful for purposes of computations to think of each X_{-i} expressed as a (complex) linear combination $h_{-i}H + e_{-i}E + f_{-i}F$ of basis elements H, E, F

which have the matrix representations

$$H = \begin{pmatrix} 1 & 0 \\ 0 & -1 \end{pmatrix}, \quad E = \begin{pmatrix} 0 & 1 \\ 0 & 0 \end{pmatrix}, \quad F = \begin{pmatrix} 0 & 0 \\ 1 & 0 \end{pmatrix}, \tag{5.40}$$

where $[H, E] = 2E$, $[H, F] = -2F$, $[E, F] = H$ and all other commutators are zero. On G we define a nondegenerate symmetric bilinear form (Killing form or inner product)

$$\langle X, Y \rangle = \text{Tr}\,(XY)_0 = \sum_{j+k=0} \text{Tr}\, X_{-j} Y_{-k}, \tag{5.41}$$

and through this pairing G can be identified with its dual G^*.

In (5.41), $\text{Tr}\,(X_{-j} Y_{-k})$ may be taken as the trace of the product of X_{-j} and Y_{-k} in their matrix representations. (Otherwise one can define it as the trace of the matrices representing the adjoint actions of these two quantities. The two definitions give answers which are proportional but not equal.) One must also check the "parallelogram volume" law $\langle X, [Y, Z] \rangle = \langle Y, [Z, X] \rangle = \langle Z, [X, Y] \rangle$.

We can define the *gradient* of a complex-valued function $f(X)$ defined on $G^* = G$ as follows. Let $\delta X \in T_X G$ (which we identify with G) be an element in the tangent space to G at X. Then the directional derivative

$$\frac{d}{d\varepsilon} f(X + \varepsilon\, \delta X) \Big|_{\varepsilon=0}$$

of f at X in the direction δX is a linear functional on δX which can be written

$$\langle \nabla f(X), \delta X \rangle. \tag{5.42}$$

We call $\nabla f(X)$ the *gradient* of $f(X)$ at X. As an exercise, show that $\nabla e_j = F_{-j}$, $\nabla f_j = E_{-j}$, $\nabla h_j = \frac{1}{2} H_{-j}$, where $F_{-j} = F\zeta^i$, $E_j = E\zeta^i$ and $H_{-j} = H\zeta^i$.

We introduce the notion of an *ad-invariant* function $f(X)$ which has the property

$$[\nabla f(X), X] = 0 \tag{5.43}$$

for all $X \in G$. This equation expresses the idea that if g is an element of the Lie group associated with G, the adjoint action of g on X, gXg^{-1}, which brings one to a new element of G, leaves the value of f unchanged. A function with this property is called *ad-invariant*. This property can be expressed by the condition that for all $Y \in G$,

$$0 = \frac{d}{dt} f(e^{tY} X e^{-tY}) \big|_{t=0} = \langle \nabla f, [Y, X] \rangle.$$

Using the parallelogram volume law, this means that $\langle [\nabla f, X], Y \rangle = 0$ and the nondegeneracy of the inner product implies (5.43). Next, consider

$$K = \left\{ \sum_{-\infty}^{-1} X_{-j} \zeta^i \right\}, \quad N = \left\{ \sum_{0}^{M} X_{-j} \zeta^i \right\}, \tag{5.44}$$

of which G is the direct sum. The dual N^* of N with respect to the inner product (5.41) (the minimal set of all elements whose inner product with any member of N is nonzero) is also the orthogonal complement $K^\perp$ of K (the set of all elements whose inner product with any element of K is zero),

$$K^\perp = N^* = \left\{ \sum_{-\infty}^{0} X_{-i}\zeta^i \right\}.$$

This will be our phase space and I will write a typical element of $K^\perp$ as

$$Q = \sum_{0}^{\infty} (h_r H_r + e_r E_r + f_r F_r), \qquad (5.45)$$

where $H_r = \zeta^{-r}H$, $E_r = \zeta^{-r}E$ and $F_r = \zeta^{-r}F$.

Now on $K^\perp$, considered as dual to N, there is a natural Poisson bracket. For two functions, $f(Q)$, $g(Q)$ on $K^\perp$, it is defined to be

$$\{f, g\}(Q) = -\langle [\pi_N \nabla f, \pi_N \nabla g], Q \rangle \qquad (5.46)$$

where π_N is the projection of $\nabla f(Q)$ into N. The reader should check that the two properties, anticommutativity $\{f, g\} = -\{g, f\}$ and the Jacobi identity, $\{\{f, g\}, h\} + \{\{h, f\}, g\} + \{\{g, h\}, f\} = 0$, are satisfied. Moreover, to each function $f(Q)$, $Q \in K^\perp$, there is associated a *Hamiltonian vector field*

$$x_f = -\pi_{K^\perp} [\pi_N \nabla f(Q), Q]. \qquad (5.47)$$

This means that the directional derivative of any function $g(Q)$ at Q in the direction x_f is given by $\{f, g\}(Q)$.

Therefore to each function $f(Q)$, there is a flow or vector field. The extra ingredient we need in order to make the system completely integrable is a set of functions $\{f_i\}$ which give rise to commuting vector fields $\{f_j, f_k\} = 0$. Then, every member of the set of functions $\{f_i\}$ is conserved along the vector field associated with any one of them.

Candidates for membership in this class are functions which are ad-invariant. The Adler–Kostant–Symes theorem tells us that if $f(X)$ and $g(X)$ are ad-invariant, then it follows that

(i) $\{f, g\} = 0$ on $K^\perp$,

(ii) the vector fields x_f, x_g commute.

The AKNS hierarchy arises via this theory from the simplest ad-invariant functions

$$-\Phi_k(Q) = \tfrac{1}{2}\langle S^k Q, Q \rangle, \qquad j = 0, \dots, \infty, \qquad (5.48)$$

where S^k is the shift,

$$S^k : \sum X_{-i}\zeta^i \to \sum X_{-i}\zeta^{i+k},$$

i.e., multiplication by ζ^k. To see this, note that for $X \in G$,

$$\nabla \Phi_k(X) = -S^k X \qquad (5.49)$$

and it is clear that

$$[\nabla \Phi_k(X), X] = 0. \qquad (5.50)$$

Next, observe that for $Q \in K^{\perp}$,

$$Q_{t_k} = -\pi_{K^{\perp}}[\pi_N \nabla \Phi_k(Q), Q] = \pi_{K^{\perp}}[\pi_K \nabla \Phi_k(Q), Q] \tag{5.51}$$

from (5.50),

$$= [\pi_K \nabla \Phi_k(Q), Q]$$

because $[\pi_K \nabla \Phi, Q] \in K^{\perp}$ already,

$$= -[\pi_N \nabla \Phi_k(Q), Q] = [Q^{(k)}, Q] \tag{5.52}$$

where

$$Q^{(k)} = \pi_N S^k Q = \zeta^k \left(Q_0 + \cdots + \frac{Q_k}{\zeta^k} \right) \tag{5.53a}$$

with

$$Q_r = h_r H + e_r E + f_r F. \tag{5.53b}$$

These are exactly the flows (3.49) in the AKNS hierarchy. The property of the ad-invariance of f_k was important in removing the projection ($\pi_{K^{\perp}}$) operator from the outside of the commutator.

The following points are emphasized:

(i) A priori, there is no favorite t_k which we need to call x. In previous analyses, t_1 has played the special role; it is the independent variable in the "eigenvalue" problem. When this choice is made, the AKNS hierarchy results. But if t_2 is chosen as the special x variable, a different hierarchy results.

A special case of this "new" hierarchy associated with t_2 is what sometimes is known as the DNLS hierarchy; it includes the derivative nonlinear Schrödinger equation. On the inverse side, which I will discuss in Section 5k, it includes the massive Thirring model in that same way that the AKNS hierarchy includes the sine-Gordon equation. But it is not really a new hierarchy at all; the equation (5.52), which is written in component form in (5.55), does not change. The hierarchies are all part of the greater hierarchy connected with $\widetilde{sl}(2, C)$: to paraphrase Rudyard Kipling;

> The AKNS thing
> And the DNLS string
> Are sisters under the skin.

(ii) Although the choice of $\widetilde{sl}(2, C)$ was motivated by writing the equation of interest as an integrability condition, once we begin with a given G and a decomposition into two subalgebras K and N, the flows arise naturally without reference to any concept such as an isospectral deformation. All iso- this and iso- that will follow as natural consequences when additional structure of an analytic nature is imposed. *So far, everything is purely algebraic.*

(iii) In developing the AKNS hierarchy, the various Q_k were expressed in terms of q, r (here e_1, f_1) and their x derivatives. (5.52) is a set of equations for the simply infinite set of variables $\{h_r, e_r, f_r\}$ as functions of an infinite number of times $\{t_1, t_2, \ldots\}$.

(iv) The result that all the flows of the AKNS hierarchy commute follows trivially from general theory.

The reader should verify for himself the following facts:

(a) The $\Phi_k(Q)$ are the coefficients of ζ^{-k} in the series $-\frac{1}{2} \operatorname{Tr} Q^2 = (h^2 + ef)$ where $h, e, f = \sum_0^\infty (h_r, e_r, f_r)\zeta^{-r}$.

(b) The Poisson brackets of h_r, e_p, f_q are

$$\{h_r, e_s\} = e_{r+s}, \quad \{h_r, f_s\} = -f_{r+s}, \quad \{e_r, f_s\} = 2h_{r+s} \tag{5.54}$$

and all other brackets are zero.

(c) The equations for h_r, e_p, f_q can be derived either from (5.54) or (5.52) and are

$$e_{j,t_k} = 2 \sum_0^{\min(j-1,k)} (h_r e_{j+k-r} - e_r h_{j+k-r}),$$

$$f_{j,t_k} = -2 \sum_0^{\min(j-1,k)} (h_r f_{j+k-r} - f_r h_{j+k-r}),$$

$$h_{j,t_k} = \sum_0^{\min(j-1,k)} (e_r f_{j+k-r} - f_r e_{j+k-r}). \tag{5.55}$$

As immediate corollaries, we have that h_0, e_0, f_0, h_1 are independent of all t_k. We choose $h_0 = -i$, $e_0 = f_0 = h_1 = 0$ as defining the canonical equations. Note also that $h^2 + ef$ is independent of t_k and, consistent with our choice of h_0, e_0, f_0 we will take this constant to be -1. From this, the h_k are determined as linear combinations of products of e's and f's. The reader should also show that all the $Q_k = h_k H + e_k E + f_k F$ can be written as functions of $e_1(q)$ and $f_1(r)$ and their x derivatives. For example, $e_{1,t_1} = -2ie_2$, $e_{2,t_1} = -2ie_3 - 2h_2 e_1$, $h_2 = -(i/2)e_1 f_1$ whence $e_2 = (i/2)e_{1,t_1}$, $e_3 = -\frac{1}{4}(e_{1,t_1 t_1} - 2e_1^2 f_1)$. Also note that equations (3.42a) (the generalized NLS equations) are simply $e_{1,t_2} = -2ie_3$ and $f_{1,t_2} = -2if_3$. Moreover, we can also write all Q_k, $k \geq 3$ as functions of e_1, f_1, e_2, f_2 and their t_2 derivatives. As a further exercise, the reader should write equations for e_1, f_1, e_2, f_2 as partial differential equations in t_2 and t_4. Can you find a consistent reduction $(e_2 = f_2 = 0, f_1 = \pm e_1^*)$ which gives the derivative NLS equation

$$u_{t_4} = iu_{t_2 t_2} \pm (u^2 u^*)_{t_2} \quad ?$$

(Hint: you will need to make a transformation of the form $e_1 \propto u \exp(i\alpha \int uu^* \, dt_2)$.)

(d) The connection between the Lie-algebraic framework and the variational Hamiltonian structure.

Once a special x is chosen, say t_1, then we may consider the phase space to be the differential algebra consisting of polynomials in $e_1 = q$, $f_1 = r$ and their x derivatives to arbitrary order, together with the symbol $\partial/\partial x$ which carries q to q_x, q_x to q_{xx} and so on. This is the phase space which has been most frequently studied [40], [106] and it admits the following Hamiltonian description. Consider

$$H_k[q, r] = \frac{4}{k+1} \int h_{k+2}(q, r, q_x, r_x, \ldots) \, dx \tag{5.56a}$$

and its variational or Fréchet gradient

$$\nabla H_k = \frac{\delta H_k}{\delta q}, \frac{\delta H_k}{\delta r},$$

where

$$\frac{\delta H_k}{\delta q} = \frac{4}{k+1} \sum_{0}^{\infty} (-1)^s \frac{\partial^s}{\partial x^s} \frac{\partial h_{k+2}}{\partial q^{(s)}}, \qquad (5.56b)$$

and $\delta H_k / \delta r$ is defined similarily. $q^{(s)}$ is $\partial^s q / \partial x^s$. The reader will recognize $\delta H_k / \delta q$ as the partial variational derivative; i.e.

$$\lim_{\varepsilon \to 0} \frac{1}{\varepsilon} H_k[q + \varepsilon \, \delta q, r] = \int \frac{\delta H_k}{\delta q} \delta q \, dx.$$

The flows (5.52) can be written as

$$\begin{pmatrix} q \\ r \end{pmatrix}_{t_k} = J \nabla H_k \qquad (5.56c)$$

with $J = \begin{pmatrix} 0 & 1 \\ -1 & 0 \end{pmatrix}$. For a proof of this, see [75]. The Poisson bracket of two functionals $F[q, r]$, $G[q, r]$ is

$$\{F, G\} = \int J \nabla F \cdot \nabla G \, dx, \qquad (5.56d)$$

where ∇ is the variational gradient. The proof that the H_k, defined by (5.56a), are in involution under this bracket is also given in [70], [75]. In Section 5d, I will tell you what the Hamiltonians and conjugate variables are if the special x is chosen to be t_j, $j > 1$.

For now, I want to emphasize two points. The first is that the soliton flows generated in the Lie-algebraic setting by the sequence of ad-invariant functions Φ_k are special if one also demands that the differential algebra interpretation is meaningful. The reason for this is that if x is to be special in the sense that we can think of all quantities in $K^\perp$ like e_1, f_1 etc. as functions of the independent variable x, then the only vector fields we can allow on $K^\perp$ (which corresponds to a choice of the other independent variables t_k, $t_k \neq x$) are those which commute with $x = t_1$. Therefore, if we want the freedom of choosing any t_j as the distinguished x, we must choose as flows in $K^\perp$ only those vector fields that commute with the vector field x_{Φ_j} generated by the Hamiltonian Φ_j. In [38], we prove a theorem which says that the only vector fields which satisfy this condition are those generated by taking a linear combination of the Φ_k as Hamiltonian.

The second point is that in the Lie-algebraic framework, the values of the Hamiltonians Φ_k (which, for the canonical equations set, we have chosen to be $\Phi_0 = 1$, $\Phi_k = 0$, $k \geq 1$) have no significance. On the other hand, the values of the Hamiltonians H_k in the differential algebra framework are important. They are the integrals of the conserved densities, the constants of the motion which can be directly related to the scattering data (the soliton and radiation data parameters) by the trace formulae (see (3.70), (3.71), (3.72) and [75]).

Exercises and examples. I now discuss three examples using the algebraic setting. The first is the harmonic oscillator, the second is the finite Toda lattice and the third is a new way of decomposing $\widetilde{sl}(2, C)$ so that the KdV and MKdV families arise naturally without the necessity of making $f_1 = -1$ or $f_1 = \pm e_1$. I will show in Section 5h how this new decomposition arises naturally when one considers alternative gradings of $\widetilde{sl}(2, C)$.

Exercises 5c.

1. Take $G = sl(2, C)$, $X = hH + eE + fF$, $\langle X, Y \rangle = \text{Tr } XY$. Find a decomposition $G = K + N$ such that the general element Q of $K^{\perp}$ is $hH + eE$. The only ad-invariant function Φ up to scalar multiplication is $-(h^2 + ef)$ and $-\nabla\Phi(x) = hH + eE + fF$. Show that the Hamiltonian vector field on $K^{\perp}$

$$Q^{\cdot} = -\pi_{K^{\perp}}[\pi_N \nabla\Phi, Q] = -[\pi_N \nabla\Phi, Q],$$

which implies that

$$h^{\cdot} = 0, \qquad e^{\cdot} = 2he.$$

For an imaginary h and complex e, this is the equation for the harmonic oscillator.

2. Consider the Lie algebra of trace-free $n \times n$ matrices

$$X \cong \begin{pmatrix} b_1 & a_1 & d_1 & & & \\ c_1 & b_2 & a_2 & \cdot & & \\ e_1 & c_2 & \cdot & \cdot & \cdot & \\ & \cdot & \cdot & \cdot & & a_{n-1} \\ & & \cdot & \cdot & c_{n-1} & b_n \end{pmatrix}$$

with the usual matrix commutator and $\langle X, Y \rangle = \text{Tr } (XY)$. Take the decomposition

$$X = \begin{pmatrix} 0 & a_1 & d_1 & & & \\ -a_1 & 0 & a_2 & d_2 & & \\ -d_1 & -a_2 & & \cdot & \cdot & \\ & & \cdot & \cdot & \cdot & a_{n-1} \\ & & & & -a_{n-1} & 0 \end{pmatrix} + \begin{pmatrix} b_1 & & & & \\ c_1+a_1 & b_2 & & & \\ d_1+e_1 & c_2+a_2 & & & \\ & & \cdot & \cdot & \\ & & & c_{n-1}+a_{n-1} & b_n \end{pmatrix}$$

$$= \qquad\qquad K \qquad\qquad + \qquad\qquad N,$$

where K is skew symmetric and N is lower triangular. X may also be decomposed into $K^\perp + N^\perp$, where $K^\perp$ is the set of symmetric matrices

$$Q = \begin{pmatrix} b_1 & a_1 & d_1 & & & \\ a_1 & b_2 & a_2 & & & \\ d_1 & a_2 & & \cdot & \cdot & \\ & & \cdot & \cdot & \cdot & \\ & & \cdot & \cdot & & a_{n-1} \\ & & & a_{n-1} & b_n \end{pmatrix}$$

and $N^\perp$ is strictly lower triangular

$$\begin{pmatrix} 0 & & & & \\ c_1 - a_1 & & & & \\ e_1 - d_1 & c_2 - a_2 & \cdot & \cdot & \\ & & \cdot & \cdot & \\ & & c_{n-1} - a_{n-1} & 0 \end{pmatrix}.$$

While K and N are subalgebras, $K^\perp$ is not. The ad-invariant function

$$\Phi(\mathbf{a}, \mathbf{b}, \mathbf{c}, \mathbf{d}, \mathbf{e}, \ldots) = \tfrac{1}{2} \operatorname{Tr} X^2$$

has its gradient $\nabla \Phi(X) = X$. The Hamiltonian vector field on $K^\perp$ is given by

$$Q^\cdot = -\pi_{K^\perp}[\pi_N \nabla\Phi(Q), Q] = \pi_{K^\perp}[\pi_K \nabla\Phi(Q), Q] = [\pi_K \nabla\Phi(Q), Q].$$

Thus the equation of motion is $Q^\cdot = [B, Q]$, where

$$B = \begin{pmatrix} 0 & a_1 & d_1 & & & \\ -a_1 & 0 & a_2 & & \cdot & \\ -d_1 & & \cdot & \cdot & \cdot & \cdot \\ & & \cdot & \cdot & & a_{n-1} \\ & & \cdot & & & \\ & & & -a_{n-1} & 0 \end{pmatrix}.$$

Note in particular that if we restrict X to be tridiagonal, namely $d = e = \cdots = 0$, Q is symmetric and tridiagonal. Further, $B = \pi_K \nabla\Phi(Q)$ is also tridiagonal and so is the commutator $[B, Q]$. Hence the flow produced by the Hamiltonian Φ preserves the tridiagonal nature of Q. Therefore we can make the consistent reduction to tridiagonal form and obtain a simpler set of equations. In component form, it reads

$$b_r^\cdot = 2a_r^2 - 2a_{r-1}^2, \quad r = 1, \ldots, n, \qquad a_0 = a_n = 0,$$
$$a_r^\cdot = a_r(b_{r+1} - b_r), \qquad r = 1, \ldots, n-1.$$

Let

$$a_r = \tfrac{1}{2} \exp \tfrac{1}{2}(u_r - u_{r+1}), \qquad b_r = -\tfrac{1}{2} u_r^\cdot,$$

and the equations become

$$u_r^{\cdot\cdot} = e^{u_{r-1} - u_r} - e^{u_r - u_{r+1}}, \quad r = 1, \ldots, n, \quad u_0 = -\infty, \quad u_{n+1} = \infty.$$

These equations describe the finite Toda lattice (see Exercise 2b(iv) in Chapter 2) with the 0th and $(n+1)$st particles stretched out to the two infinities. The solution of this system was given by Moser [98] and its Lie algebraic setting has been discussed in considerable detail by Kostant [99].

3. Consider a different decomposition of $\widetilde{sl}(2, C)$, $G = K + N$ s.t.

$$N = \sum_1^M X_{-i}\lambda^i + e_0 E + h_0 H,$$

$$K = \sum_{j=1}^\infty (h_j H + e_j E + f_j F) \cdot \lambda^{-j} + f_0 F$$

with the inner product defined[8] as in (5.41). Then an element Q in $K^\perp$ has the general form,

$$Q = h_0 H + f_0 F + \sum_{j=1}^\infty (h_j H + e_j E + f_j E)\lambda^{-j}.$$

To the set of ad-invariant functions

$$\Phi_k = -(h^2 + ef)_k,$$

where h, e, f each refer to the infinite strings and the subscript to the λ^{-k} component, we assign the values $\Phi_1 = 1$, $\Phi_k = 0$, $k \neq 1$. This means that we make the consistent choices $h_0 = 0$, $f_0 = 1$, $e_1 = 1$. These choices are analogous to the choice $h_0 = -i$, $e_0 = f_0 = 0$ made earlier in this section. The quantities

$$Q^{(k)} = -\pi_N \nabla\Phi_k(Q),$$

have the form

$$k = 1, \quad -\lambda F + h_1 H + E,$$
$$k = 2, \quad -\lambda^2 F + \lambda(h_1 H + E + f_1 F) + h_2 H + e_2 E,$$
$$\vdots$$
$$\vdots$$
$$k = k, \quad -\lambda^k F + \lambda^{k-1}(h_1 H + E + f_1 F) + \cdots + \lambda^{k-r}(h_r H + e_r E + f_r F)$$
$$+ \cdots + h_k H + e_k E. \qquad (5.57)$$

Show that the equations

$$Q_{t_k} = [Q^{(k)}, Q] \qquad (5.58a)$$

are

$$h_{j,t_k} = e_{j+k} + \sum_{r=1}^{(j-1,k)} (e_r f_{j+k-r} - f_r e_{j+k-r}) + e_j f_k, \qquad (5.58b)$$

$$e_{j,t_k} = 2 \sum_{r=1}^{j-1,k} (h_r e_{j+k-r} - e_r h_{j+k-r}), \qquad (5.58c)$$

$$f_{j,t_k} = -2h_{j+k} - 2 \sum_{r=1}^{(j-1,k)} (h_r f_{j+k-r} - f_r h_{j+k-r}) - 2h_j f_k, \qquad (5.58d)$$

where the notation $(j-1, k)$ means that we may sum to $j-1$ or k.

[8] We will see in Section 5h that this decomposition is entirely natural and corresponds to the principal grading of $\widetilde{sl}(2, C)$.

I will ask you to observe, and in some cases prove, the following.
(i) The sequence e_j derives from a potential since

$$\frac{\partial e_{j+1}}{\partial t_k} = \frac{\partial e_{k+1}}{\partial t_j}.$$

In anticipation of what is to come, I will write

$$e_j = -\frac{\partial}{\partial t_{j-1}} \frac{\partial \ln \tau}{\partial t_1}, \qquad j = 2, \dots.$$

(ii) $e_2 = \frac{1}{2} h_{1,t_1} + \frac{1}{2} h_1^2$, $f_1 = \frac{1}{2} h_{1,t_1} - \frac{1}{2} h_1^2$.

Before we begin the proof, look at the first relation. Do you recognize it at all?

Proof. From (5.58b,d)

$$h_{1,t_k} = e_{k+1} + e_1 f_k, \qquad f_{1,t_k} = -2h_{k+1} - 2h_1 f_k,$$

which gives us that

$$2h_1 h_{1,t_k} + e_1 f_{1,t_k} = 2h_1 e_{k+1} - 2e_1 h_{k+1}.$$

A little further calculation show that

$$h_{1,t_k t_1} = 2h_1 e_{k+1} - 2e_1 h_{k+1} + e_1 f_{1,t_k},$$

which when subtracted from the line above gives

$$2h_1 h_{1,t_k} - h_{1,t_k t_1} = -2e_1 f_{1,t_k} = -2f_{1,t_k}, \quad \text{since } e_1 = 1.$$

Hence $f_1 = \frac{1}{2} h_{1,t_1} - \frac{1}{2} h_1^2$ and from $e_{k+1} = h_{1,t_k} - f_k$ we have $e_2 = \frac{1}{2} h_{1,t_1} + \frac{1}{2} h_1^2$.

(iii) h_1, as a function of t_1, t_2, satisfies the modified Korteweg–deVries equation

$$h_{1,t_2} = e_3 + f_2 = h_{2,t_1} = \left(-\frac{1}{4} h_{1,t_1 t_1} + \frac{1}{2} h_1^3 \right)_{t_1},$$

since

$$h_2 = -\frac{1}{2} f_{1,t_1} - h_1 f_1 = -\frac{1}{4} h_{1,t_1 t_1} - \frac{1}{2} h_1^3.$$

(iv) *Corollary.* As a function of t_1, t_2, e_2 satisfies the KdV equation and $e_2 = \frac{1}{2} h_{1,t_1} + \frac{1}{2} h_1^2$ is the Miura transformation.

(v) I will leave it to you to prove

$$e_{j,t_k} = e_1 e_{k+j-1x} + \cdots + e_k e_{jx} - e_{1x} e_{k+j-1} - \cdots - e_{kx} e_j.$$

Look at (4.13)!

(vi) $Ne_{j+1} = -\frac{1}{4} Me_j = NLe_j$, where L, M and N are defined in (3.6), (3.12). This means that the string

$$e = \sum_{j=1}^{\infty} \frac{e_j}{\lambda^j} = \sum_{j=1}^{\infty} \frac{L^j e_1}{\lambda^j}, \qquad \text{with } e_1 = 1.$$

Thus $e_j = -B_{j-1}$ (recall $B_0 = -1$, $B_1 = q/2, \dots B_{r+1} = \frac{1}{2} L' q$) and from this and (4.6), we see that the potential τ defined by $e_j = -(\partial/\partial t_{j-1})((\partial/\partial t_1) \ln \tau)$ is indeed the τ-function for the KdV family.

(vii) Note that the solution of (5.58a) is $Q = VCV^{-1}$, C a constant, with V satisfying

$$V_{t_k} = Q^{(k)}V. \tag{5.59}$$

From (5.57), we can read off $Q^{(k)}$ and find for $V = (v_1, v_2)^T$,

$$v_{1x} = h_1 v_1 + e_1 v_2, \quad e_1 = 1, \quad v_{2x} = -\lambda v_1 - h_1 v_2$$

and

$$\begin{pmatrix} v_1 \\ v_2 \end{pmatrix}_{t_k} = \begin{pmatrix} h^{(k)} & e^{(k)} \\ f^{(k)} & -h^{(k)} \end{pmatrix} \begin{pmatrix} v_1 \\ v_2 \end{pmatrix},$$

where

$$h^{(k)} = \lambda^k \sum_0^k \frac{h_j}{\lambda^j}, \quad e^{(k)} = \lambda^k \sum_1^k \frac{e_j}{\lambda^j}, \quad f^{(k)} = \lambda^k \sum_0^{k-1} \frac{f_j}{\lambda^j}.$$

Note that both v_1 and v_2 satisfy the Schrödinger equation with potential $q = -2e_2 = -h_{1x} - h_1^2$ and $h_{1x} - h_1^2$ respectively. Let us rewrite these equations for

$$v = v_1, \quad v_x = v_2 + h_1 v_1,$$

whereby we obtain

$$\begin{pmatrix} v \\ v_x \end{pmatrix}_x = \begin{pmatrix} 0 & 1 \\ -\lambda + h_{1x} + h_1^2 & 0 \end{pmatrix} \begin{pmatrix} v \\ v_x \end{pmatrix}$$

and

$$\begin{pmatrix} v \\ v_x \end{pmatrix}_{t_k} = \begin{pmatrix} h^{(k)} - h_1 e^{(k)} & e^{(k)} \\ f^{(k)} + 2h_1 h^{(k)} + h_{1,t_k} - h_1^2 e^{(k)} & h_1 e^{(k)} - h^{(k)} \end{pmatrix} \begin{pmatrix} v \\ v_x \end{pmatrix}.$$

Notice that $h^{(k)} - h_1 e^{(k)} = -\frac{1}{2} e_x^{(k)}$ and that these equations are precisely the matrix form of (3.3).

5d. Conservation laws, fluxes, potentials and the Hirota equations.

A further and immediate consequence of (5.55) is that

$$\frac{\partial h_{j+1}}{\partial t_k} = \frac{\partial h_{k+1}}{\partial t_j}, \quad \frac{\partial e_{j+1}}{\partial t_k} = \frac{\partial e_{k+1}}{\partial t_j}, \quad \frac{\partial f_{j+1}}{\partial t_k} = \frac{\partial f_{k+1}}{\partial t_j}. \tag{5.60}$$

This suggests that the infinite vector strings h, e, f can be written in terms of three potentials. In particular, we also know from the equations themselves that since

$$e_{1,t_k} = -2ie_{k+1}, \quad f_{1,t_k} = 2if_{k+1}$$

e_1 and f_1 act as potentials for all members in each of their respective strings. Also we know from previous experience that $\int h_k \, dx$ are the conserved densities and so we expect that h_{j+1} will not only be the t_j derivative of a potential but also a t_1 derivative. Therefore, we define a new potential $\tau(t_1, t_2, \ldots)$ by

$$h_{k+1} = \frac{\partial}{\partial t_k} \left(\frac{i}{2} \frac{\partial \ln \tau}{\partial t_1} \right). \tag{5.61}$$

What we now want to show, of course, is that

$$F_{jk} = \frac{\partial^2 \ln \tau}{\partial t_j \, \partial t_k} \tag{5.62}$$

is a local function of (h_r, e_p, f_q). It is reasonable to call F_{jk} the *flux tensor*. In reference [38], we obtain an expression for this quantity. It is

$$F_{jk} = \frac{1}{2} \operatorname{Tr}\left[\sum_{r=0}^{j} (j-r) Q_r Q_{k+j-r} \right] + \frac{1}{2}\left[\sum_{r=0}^{k} (r-k) Q_r Q_{k+j-r} \right]$$

$$= \left\langle \zeta \frac{d}{d\zeta} Q^{(j)}, \zeta^k Q \right\rangle_0 . \tag{5.63}$$

The principal idea in the proof of this is to use (5.55) in order to write

$$\frac{\partial h_{j+1}}{\partial t_k} = \sum_{1}^{k} (e_r f_{k+j+1-r} - f_r e_{j+k+1-r})$$

as a t_1 derivative. For example, note that $f_{k+i+1-r} = -(i/2)(f_{k+i-r,t_1} - 2h_{k+i-r}f_1)$.

Each of the two expressions in (5.63) is equal to the other if we normalize the h, e, f series in such a way that $h^2 + ef = -1$. Otherwise, they differ by an amount which depends on the Hamiltonian $\Phi_k(Q)$. In order to ensure symmetry, I will define F_{jk} as the symmetric sum but in all calculations we keep $h^2 + ef = -1$ and therefore we can calculate the flux tensor by just using one of the expressions.

For the first time in the literature, we actually have expressions for the fluxes of all conserved densities with respect to all the flows:

$$\frac{\partial h_{j+1}}{\partial t_k} = \frac{\partial}{\partial t_1} \frac{i}{2} F_{jk}. \tag{5.64}$$

It is interesting to observe that the fluxes are most naturally expressed as derivatives with respect to *all* the time variables of a single function $\ln \tau(t_1, t_2, \ldots)$. In other words, the flux of the conserved density h_4 (whose integral is the varational Hamiltonian for the NLS equation) with respect to the NLS flow time t_2 is best expressed as the second order partial with respect to t_2 and t_3, the time for an altogether different flow, rather than t_2 and x; i.e. $(\partial/\partial t_2)h_4 = (\partial/\partial t_1)i/2 \cdot \partial^2 \ln \tau/\partial t_2 \, \partial t_3$. It is still possible, of course, to express $\partial^2 \ln \tau/\partial t_2 \, \partial t_3$ in terms of e_1, f_1 and their t_1 derivatives but these expressions are without natural structure and extremely clumsy.

Further, suppose we think of a different t_k as our special x, say t_2. This would be appropriate if we were studying the derivative nonlinear Schrödinger equation, for example. In that case, the conservation laws are

$$\frac{\partial}{\partial t_j} F_{k2} = \frac{\partial}{\partial t_2} F_{jk},$$

the conserved densities are F_{k2} and the appropriate Hamiltonians proportional to $\int F_{k2} \, dt_2$. For those situations when we take t_j as the special x (in which case we write e_{j+r}, f_{j+r}, $r > 0$, as functions of $e_1, e_2, \ldots, e_j, f_1, \ldots, f_j$ and their $x = t_j$ derivatives), the conserved densities are F_{kj} and the Hamiltonian functions are

$$H_k^{(j)} = \frac{2i}{k+1} \int F_{k+1\,j}\, dt_j.$$

I mention here, but do not prove (the interested reader should refer to reference [38]) that the conjugate variables in this case are not (e_1, f_j), $(e_2, f_{j-1}), \ldots, (e_j, f_1)$ but rather $(\tilde{e}_1, \tilde{f}_1), \ldots, (\tilde{e}_j, \tilde{f}_1)$ where $\tilde{e}_r$, $\tilde{f}_r$ are the coefficients of ζ^{-r} in the strings generated by the formal expansions of $e/\sqrt{i-h}$ and $f/\sqrt{i-h}$ about $\zeta = \infty$. For $j = 1$ and 2, the conjugate variables are (e_1, f_1) and (e_1, f_2), (e_2, f_1) respectively but after that they change.

Finally, it turns out to be more convenient to use σ and ρ defined as τe_1 and τf_1 respectively as the scalar potentials of the e and f strings.

We have seen how we can replace the triply infinite set of equations satisfying (5.55) by three scalar equations for the potentials τ, σ and ρ. It is natural to ask what equations they satisfy. This is a straightforward calculation. From (5.55)

$$-2ie_{k+1} = \frac{\partial e_1}{\partial t_k} = \frac{\partial e_k}{\partial t_1} + 2h_k e_1,$$

which means

$$\frac{\partial}{\partial t_k}\left(\frac{\sigma}{\tau}\right) = \frac{i}{2}\frac{\partial^2}{\partial t_1\,\partial t_{k-1}}\left(\frac{\sigma}{\tau}\right) + ie_1\frac{\partial^2 \ln \tau}{\partial t_1\,\partial t_{k-1}}.$$

This gives

$$\sigma_k\tau - \sigma\tau_k = \frac{i}{2}(\sigma_{1\,k-1}\tau - \sigma_1\tau_{k-1} - \sigma_{k-1}\tau_1 + \sigma\tau_{1\,k-1}) \tag{5.65}$$

where the subscript j denotes the partial derivative with respect to t_j. Now recall Section 4c (4.36), (4.37) in which the Hirota operator is defined. Equation (5.65) is simply

$$\left(D_{t_k} - \frac{i}{2}D_{t_1}D_{t_{k-1}}\right)\sigma \cdot \tau = 0. \tag{5.66}$$

In a similar way,

$$\left(D_{t_k} - \frac{i}{2}D_{t_1}D_{t_{k-1}}\right)\rho \cdot \tau = 0. \tag{5.67}$$

The third equation is obtained by noting that

$$h_2 = \frac{i}{2}\frac{\partial^2}{\partial t_1^2}\ln \tau = -\frac{i}{2}e_1 f_1$$

becomes

$$D_{t_1}^2 \tau \cdot \tau = -2\sigma\rho. \tag{5.68}$$

From these equations, one can readily calculate the multisoliton solutions. In order to do this, it is convenient to make a change of independent variables

$$2it_k \rightarrow t_k,$$

(note the "new" time is pure imaginary) so that the Hirota equations become

$$(D_{t_k} - D_{t_1}D_{t_{k-1}})\sigma \cdot \tau = 0, \qquad (5.69a)$$

$$(D_{t_k} - D_{t_1}D_{t_{k-1}})\rho \cdot \tau = 0, \qquad (5.69b)$$

$$D_{t_1}^2 \tau \cdot \tau = \tfrac{1}{2}\sigma\rho. \qquad (5.69c)$$

We seek solutions in the form

$$\tau, \sigma, \rho = \sum_{\mu_r,\nu_s=0,1} D_{\tau,\sigma,\rho}(\mu, \nu) \exp\left(\sum_1^N \mu_r H_r + \sum_1^N \nu_r \bar{H}_r + \sum_{1\leq r<s\leq N} A_{rs}\mu_r\mu_s\right.$$

$$\left. + \sum_{1\leq r<s\leq N} \bar{A}_{rs}\nu_r\nu_s + \sum_{r,s=1}^N B_{rs}\mu_r\nu_s\right)$$

where

$$H_r = \sum \zeta_r^k t_k, \qquad \bar{H}_r = -\sum \bar{\zeta}_r^k t_k,$$

$$D_\tau = \begin{cases} 1 & \text{if } \sum \mu_r = \sum \nu_r, \\ 0 & \text{otherwise,} \end{cases}$$

$$D_\rho = \begin{cases} 1 & \text{if } \sum \mu_r = \sum \nu_r + 1, \\ 0 & \text{otherwise,} \end{cases}$$

$$D_\sigma = \begin{cases} -1 & \text{if } \sum \mu_r + 1 = \sum \nu_r, \\ 0 & \text{otherwise,} \end{cases}$$

and

$$e^{B_{rs}} = -\frac{1}{4(\zeta_r - \bar{\zeta}_s)^2}, \quad e^{A_{rs}} = -4(\zeta_r - \zeta_s)^2, \quad e^{\bar{A}_{rs}} = -4(\bar{\zeta}_r - \bar{\zeta}_s)^2. \qquad (5.70)$$

For general $e_1, f_1, \bar{\zeta}_k$ bears no relation to ζ_k. However, when $f_1 = -e_1^*$, then $\bar{\zeta}_k = \zeta_k^*$, the complex conjugate of ζ_k.

Each of the Hirota equations (5.69) is expressed in the form

$$P(D_{t_k})f \cdot g = 0 \qquad (5.71)$$

where P has the properties that

$$P(0) = 0, \qquad (5.72a)$$

$$P(\zeta^k) = 0. \qquad (5.72b)$$

In deriving (5.70), the rule

$$P(D_{t_i})\exp\left(\sum a_1^{(k)}t_k\right) \cdot \exp\left(\sum a_2^{(k)}t_k\right)$$

$$= P(a_1^{(j)} - a_2^{(j)})\exp\left(\sum (a_1^{(k)} + a_2^{(k)})t_k\right) \qquad (5.73)$$

is used extensively.

As before, the phase shifts A_{rs}, $\bar{A}_{rs}$ and B_{rs} are the same for every member of the $\widetilde{sl}(2, C)$ hierarchy. We can therefore, in a manner similar to that discussed in Chapter 4, determine all polynomials in this hierarchy once we are given the phase shifts. At each level (the level is the sum of the indices in the Hirota equations; for example in (5.69a), the level is k), there are many (can you compute how many?) Hirota equations which are compatible with each other.

Exercise 5d.
Compute the one-soliton solution when

$$f_1 = -e_1^*;$$

$$N = 1, \qquad \bar{\zeta}_1 = \zeta_1^* = \xi - i\eta;$$

$$H_1 = 2i \sum \zeta_1^k t_k, \qquad \bar{H}_1 = -2i \sum \zeta_1^{*k} t_k;$$

$$\tau = 1 + \exp{(H_1 + \bar{H}_1 + B_{11})}, \quad \sigma = -\exp{\bar{H}_1}, \quad \exp\left(-\frac{B_{11}}{2}\right) = -4\eta;$$

$$q(t_k) = \frac{\sigma}{\tau} = 2\eta \exp\left(\frac{\bar{H}_1 - H_1}{2}\right) \operatorname{sech} \left(\frac{H_1 + \bar{H}_1 + B_{11}}{2}\right),$$

$$\frac{1}{2}(H_1 + \bar{H}_1 + B_{11}) = i \sum (\zeta_1^k - \zeta_1^{*k}) t_k + B_{11}/2,$$

$$\frac{1}{2}(\bar{H}_1 - H_1) = -i \sum (\zeta_1^k + \zeta_1^{*k}) t_k.$$

5e.   The eigenvalue problem, asymptotic expansions and vertex operators.
The goal in this section is to introduce the eigenfunction $V(t_k; \zeta)$ and derive certain of its properties:
  (i) its asymptotic expansion about $\zeta = \infty$;
  (ii) the fact that the phase space $Q(t_k)$ is an orbit through the element $-iH$;
  (iii) an expansion for $\zeta \, \partial V/\partial \zeta$;
  (iv) the connection between V and the Hirota equations.
  The introduction of $V(t_k; \zeta)$. We derived the $\widetilde{sl}(2, C)$ hierarchy without the introduction of any auxiliary variables. How do these enter the picture? The answer is very simple. An equation in Lax form

$$Q_{t_k} = [Q^{(k)}, Q] \tag{5.74}$$

immediately invites solution by

$$Q = V\tilde{Q}_0 V^{-1}. \tag{5.75}$$

Substitution into (5.74) gives us that $\tilde{Q}_0$ is independent of all t_k and

$$V_{t_k} = Q^{(k)} V. \tag{5.76}$$

Conversely, of course, as we showed in Section 3c, the integrability condition of the set of equations (5.76) is (5.74). Whereas (5.75) does not solve (5.74) explicitly, its very form is, nevertheless, useful. It provides the connection with the idea that the flows (5.74) are reductions of simpler flows on a larger manifold. We will discuss this idea in detail in Section 5j.

Asymptotic expansion of V about $\zeta = \infty$. Just as in the case of the KdV hierarchy, it is useful to seek formal asymptotic expansions for $V(t_1, t_2, \ldots ; \zeta)$

$$V \sim \left(I - \frac{i}{2} \sum_1^\infty \frac{b_r E - c_r F}{\zeta^r} \right) \exp \left(\left(-i \sum \zeta^k t_k + \psi \right) H + \phi \right) \tag{5.77}$$

where ψ and ϕ are written as an inverse power series in ζ. The following results hold:

$$e_r = b_r + \frac{i}{2} \sum_{\substack{m+n=r \\ m \neq 0}} h_m b_n, \tag{5.78a}$$

$$f_r = c_r + \frac{i}{2} \sum_{\substack{m+n=r \\ m \neq 0}} h_m c_n. \tag{5.78b}$$

These can be rewritten in compact form,

$$2ie = b(i - h), \tag{5.79a}$$

$$2if = c(i - h) \tag{5.79b}$$

where $(h_0 = -i, h_1 = 0)$,

$$h = \sum_0^\infty \frac{h_r}{\zeta^r}, \quad e = \sum_1^\infty \frac{e_r}{\zeta^r}, \quad f = \sum_1^\infty \frac{f_r}{r}, \quad b = \sum_1^\infty \frac{b_r}{\zeta^r}, \quad c = \sum_1^\infty \frac{c_r}{\zeta^r}. \tag{5.80}$$

I also leave it as an exercise to show that

$$b(t_k, \zeta) = \frac{1}{\zeta} e_1 \left(t_k + \frac{i}{2k\zeta^k} \right) \equiv \frac{1}{\zeta} e_{1+}, \tag{5.81a}$$

$$c(t_k, \zeta) = \frac{1}{\zeta} f_1 \left(t_k - \frac{i}{2k\zeta^k} \right) \equiv \frac{1}{\zeta} f_{1-}, \quad k = 1, 2, \ldots . \tag{5.81b}$$

The following results hold:

$$\phi + \psi = \ln \tau_- - \ln \tau, \tag{5.82a}$$

$$\phi - \psi = \ln \tau_+ - \ln \tau, \tag{5.82b}$$

where

$$\tau_+ = \tau \left(t_k + \frac{i}{2k\zeta^k} \right), \quad \tau_- = \tau \left(t_k - \frac{i}{2k\zeta^k} \right), \quad k = 1, 2, \ldots . \tag{5.83}$$

One can also show that

$$e^{2\phi} = \frac{\tau_+ \tau_-}{\tau^2} = \frac{i - h}{2i} = \left(1 - \frac{bc}{4} \right)^{-1}. \tag{5.84}$$

We introduce the operators

$$X_+(\zeta) = \exp \left(i \sum \zeta^k t_k \right) \exp \left(\sum \frac{i}{2k\zeta^k} \frac{\partial}{\partial t_k} \right), \tag{5.85a}$$

$$X_-(\zeta) = \exp \left(-i \sum \zeta^k t_k \right) \exp \left(-\sum \frac{i}{2k\zeta^k} \frac{\partial}{\partial t_k} \right). \tag{5.85b}$$

Remark. We call these operators, "vertex" operators, by analogy with the similar looking operator $Y(\zeta)$, introduced in (4.124) in connection with the τ-function of the KdV family. In that context, however, when the operator X (or a linear combination of $X(\zeta)$ and $X(-\zeta)$) was applied to τ or when the exponential of $Y(\zeta)$ was, applied to τ, one recovered another τ function. Although it is straightforward to use $X_+(\zeta)$, $X_-(\zeta)$ in a computational manner in order to construct multisoliton solutions (we do this in Section 5g), it is difficult to identify the space of functions on which the operators $X_+(\zeta)$ and $X_-(\zeta)$ naturally apply.

We shall refer to the parts of these operators without the $\exp(\pm i \sum \zeta^k t_k)$ factors as $\tilde{X}_+(\zeta)$ and $\tilde{X}_-(\zeta)$ respectively. Now, using these results,

$$V \sim \left(I - \frac{i}{2\zeta} e_{1+}E + \frac{i}{2\zeta} f_{1-}F\right)\left(\frac{I+HX_-\tau}{2} \frac{1}{\tau} + \frac{I-HX_+\tau}{2}\frac{1}{\tau}\right).$$

But

$$e_{1+}X_+\tau = X_+e_1\tau = X_+\sigma \quad \text{and} \quad f_{1-}X_-\tau = X_-f_1\tau = X_-\rho.$$

Therefore

$$V \sim \frac{1}{\tau}\begin{pmatrix} X_-\tau & -\dfrac{i}{2\zeta}X_+\sigma \\ \dfrac{i}{2\zeta}X_-\rho & X_+\tau \end{pmatrix}. \tag{5.86a}$$

Its determinant is unity because

$$\det V = \frac{1}{\tau^2}\left(\tau_+\tau_- - \frac{1}{4\zeta^2}\sigma_+\rho_-\right) = \frac{\tau_+\tau_-}{\tau^2}\left(1 - \frac{bc}{4}\right) = 1$$

from (5.84). Its inverse, therefore, is

$$V^{-1} = \frac{1}{\tau}\begin{pmatrix} X_+\tau & \dfrac{i}{2\zeta}X_+\sigma \\ -\dfrac{i}{2\zeta}X_-\rho & X_-\tau \end{pmatrix}. \tag{5.86b}$$

Some commentary on how these results are found is warranted. First, one substitutes the ansatz (5.77) into (5.76) and equates the coefficients of ζ^r, $-\infty < r < k$. For $r \geq 0$, (5.78) follows rather easily. What must be proved, in addition, is that it is also true for $r < 0$ at which stage the derivatives of ψ and ϕ enter the picture. For this, one needs to use the equations satisfied by h_r, e_r and f_r. Of course, it must be true that the expressions (5.78) are independent of which equations (which t_k) we use. Two equations which are useful are

$$h_r = -\frac{i}{4}\sum_{m+n=r} e_m c_n + f_m b_n, \tag{5.87a}$$

or

$$h = -i - \frac{i}{4}(ec + fb)$$

and

$$\sum_{m+n=r} (e_m c_n - f_m b_n) = 0, \tag{5.87b}$$

which can be written

$$ec = fb.$$

Q is on an orbit through $-iH$. Next, we show that a typical element of the phase space Q lies on an orbit through $-iH$.

We wish to show that

$$\hat{V}(-iH)\hat{V}^{-1} = Q, \tag{5.88a}$$

or equivalently that

$$Q\hat{V} = \hat{V}(-iH), \tag{5.88b}$$

where

$$\hat{V} = \begin{pmatrix} 1 & -\dfrac{1}{2}b \\ \dfrac{i}{2}c & 1 \end{pmatrix} e^{\psi H + \phi}, \tag{5.89a}$$

is the left factor of V in its asymptotic expansion about $\zeta = \infty$; i.e. formally,

$$V = \hat{V} \exp\left(-i \sum \zeta^k t_k H\right). \tag{5.89b}$$

Since the right-hand factors in (5.89a,b) commute with H, it is sufficient to show that

$$\begin{pmatrix} h & e \\ f & -h \end{pmatrix} \begin{pmatrix} 1 & -\dfrac{1}{2}b \\ \dfrac{i}{2}c & 1 \end{pmatrix} = \begin{pmatrix} 1 & -\dfrac{i}{2}b \\ \dfrac{i}{2}c & 1 \end{pmatrix} \begin{pmatrix} -i & \\ & i \end{pmatrix}.$$

This follows from (5.79) and (5.87).

Notice the equation which $\hat{V}$ satisfies. From (5.76) and (5.89b), it is

$$\hat{V}_{t_k} = Q^{(k)}\hat{V} + \hat{V}(iH\zeta^k). \tag{5.89c}$$

We will meet this equation again in Section 5j.

I also point out that if one takes the distinguished x to be t_j, then the phase space is the differential algebra of polynomials on $e_1, e_2, \ldots, e_j, f_1, \ldots, f_j$ and their $x = t_j$ derivatives to arbitrary order. The set $\{e_1, \ldots, e_j, f_1, \ldots, f_j\}$ when considered as functions of $x = t_j$ generates the phase space. Since these are exactly the elements contained in $Q^{(j)} = h^{(j)}H + e^{(j)}E + f^{(j)}F$, with $h^{(j)} = \zeta^j \sum_0^j h_r \zeta^{-r}$ and $e^{(j)}, f^{(j)}$ defined similarily, we can consider the phase space to be $Q^{(j)}$. From (5.89c) then, the typical phase space element $Q^{(j)}$ is written,

$$Q^{(j)} = \hat{V}(-iH\zeta^j - \partial_{t_j})\hat{V}^{-1} + \hat{V}_{t_j}\hat{V}^{-1}. \tag{5.89d}$$

Remark. The use of the word orbit is deliberate and meant to be suggestive. If every element X of the phase space, which is the dual G^* of a Lie algebra G, can be generated by the coadjoint action $X = gX_0g^{-1}$ (in matrix representation form) of elements g belonging to the Lie group $\bar{G}$ through an element X_0 of G^*, then we know that the phase space is a symplectic manifold and has a nondegenerate two-form. For example, if we take $x = t_1$ to be a distinguished element and take the phase space to be the function space of the pair $e_1(x; t_k)$, $f_1(x; t_k)$, then the phase space is a symplectic manifold with two-form $\int_{-\infty}^{\infty} \delta e_1 \wedge \delta f_1\, dx$ (see [75]). For distinguished element $x = t_j$, the phase space consists of the conjugate function pairs $(\tilde{e}_1, \tilde{f}_j), \ldots (\tilde{e}_j, \tilde{f}_1)$ defined in Section 5d and the two-form is

$$\int_{-\infty}^{\infty} (\delta\tilde{e}_1 \wedge \delta\tilde{f}_j + \ldots + \delta\tilde{e}_j \wedge \delta\tilde{f}_1)\, dt_j.$$

One can interpret (5.89d) (A. Reyman; private communication) as a statement that the phase space is a coadjoint orbit through the element $-iH\zeta^i - \partial/\partial t_j$

Exercise 5e. (*Important.*)
Consider the Lax equations (5.58a) of Exercise 5c(3). They, too, are solved by introducing $V(t_k, \lambda)$ by setting

$$\lambda Q = VCV^{-1}, \tag{5.90a}$$

where C is independent of t_k and V satisfies

$$V_{t_k} = Q^{(k)}V. \tag{5.90b}$$

The $Q^{(k)}$ are given in (5.57). The C in (5.90a) is $(-iH\zeta)$ if we take

$$V = \frac{1}{\sqrt{2}} \begin{pmatrix} v(t_k, -\zeta) & \dfrac{-i}{\zeta} v(t_k, \zeta) \\[2ex] v_x(t_k, -\zeta) & \dfrac{-i}{\zeta} v_x(t_k, \zeta) \end{pmatrix} \tag{5.90c}$$

where $\zeta^2 = \lambda$, $v(t_k, \zeta)$ is the v satisfying (3.1), (3.9) (in Section 3b, the time labelled t_{2k-1} corresponds to the time labelled t_k in (5.58a) and above) with asymptotic expansion given by exercise 3b(iii).

$$v(t_k, \zeta) \sim \exp\left(i\zeta \sum_1^{\infty} \lambda^{k-1} t_k\right) \tau\left(x - \frac{1}{i\zeta}, t_2 - \frac{1}{3i\zeta^3}, \ldots\right). \tag{5.90d}$$

$\tau(t_1, t_2, \ldots)$ is the τ-functions of the KdV family. The reader might check (5.90a) in the vacuum case where $q = 0$. Now (5.90b) may be written as

$$V = \hat{V} \exp\left(-i \sum_1^{\infty} \zeta^{2k-1} t_k H\right), \tag{5.90e}$$

where

$$\hat{V}_{t_k} = Q^{(k)}\hat{V} + \hat{V}(iH\zeta^{2k-1}) \tag{5.90f}$$

or, solving for $Q^{(k)}$,

$$Q^{(k)} = \hat{V}(-iH\zeta^{2k-1})\hat{V}^{-1} + \hat{V}_{t_k}\hat{V}^{-1}. \tag{5.90g}$$

The reader should also note that, if we take

$$\hat{V} = \frac{1}{\sqrt{2}}\,\hat{U}\begin{pmatrix} 1 & -i/\zeta \\ -i\zeta & 1 \end{pmatrix},$$

equations (5.90f,g) for $\hat{U}$ change only by replacing $-iH\zeta^{2k-1}$ by $X\lambda^k$ where $X = -F + E/\lambda$.

Further remark. Formulae (5.88a), (5.90a) are important as they appear to be one means through which the link between the two roles of the Kac–Moody algebra $A_1^{(1)}$ can be established. The reader will recall that in Section 4g we discussed how $A_1^{(1)}$ was an algebra of symmetries and that solutions $\tau(t_1, t_3, \ldots)$ of soliton equations of the KdV family formed an orbit through the highest weight vector ($\tau = 1$) in some basic representation of $A_1^{(1)}$, an orbit defined by the quadratic Hirota equations. On the other hand, in the present chapter, the algebra is the phase space. We will see in 5j that formulae (5.88a), (5.90a) arise very naturally as the coadjoint action of the Kac–Moody "group" on a special element ε (which is either $-iH$ or $X = -F + (E/\zeta)$). We will also see in the subsection after next how (5.88a) contains the Hirota equations.

An expression for $\zeta(\partial V/\partial\zeta)$. In Section 5l, we indicate why the operator $D = \zeta(\partial/\partial\zeta)$ is an important element in the whole theory. Therefore it is worth calculating

$$DVV^{-1} = \left(\sum k\zeta^k t_k - \frac{1}{4\zeta^k}\frac{\partial}{\partial t_k} - \frac{1}{2\zeta^k}\frac{\partial \ln \tau}{\partial t_k}\right)Q + \frac{i}{2}(eE - fF). \tag{5.91a}$$

From (5.91a), it is a straightforward exercise to show that

$$\langle \hat{V}^{-1}D\hat{V}, -i\zeta^i H\rangle_0 = \frac{\partial \ln \tau}{\partial t_j}. \tag{5.91b}$$

The term $-i\zeta^i H$ is $\nabla\Phi_j(-iH)$ which is the gradient of the Hamiltonian Φ_j calculated at the point $-iH$. Equation (5.91b) is the formula by which the Kyoto group define the τ-function [39]. In proving (5.91), we use the facts that

$$\zeta\frac{\partial}{\partial\zeta}f\left(t_k + \frac{i}{2k\zeta^k}\right) = \left(\sum_1^\infty \frac{-i}{2\zeta^i}\frac{\partial}{\partial t_j}\right)f\left(t_k + \frac{i}{2k\zeta^k}\right),$$

$$Db = -e_+ = -e\left(t_k + \frac{i}{2k\zeta^k}\right),$$

$$Dc = -f_- = -f\left(t_k - \frac{i}{2k\zeta^k}\right),$$

$$1 + \frac{bc}{4} = -\frac{2h}{i-h},$$

and also (5.79) and (5.84).

The connection with Hirota. Let us take another look at (5.89) and write it as

$$
\begin{pmatrix} h & e \\ f & -h \end{pmatrix}
\begin{pmatrix} X_{-\tau} & -\dfrac{i}{2\zeta}X_{+}\sigma \\ \dfrac{i}{2}X_{-\rho} & X_{+\tau} \end{pmatrix}
=
\begin{pmatrix} X_{-\tau} & -\dfrac{i}{2}X_{+}\sigma \\ \dfrac{i}{2}X_{-\rho} & X_{+\tau} \end{pmatrix}
\begin{pmatrix} -i & 0 \\ 0 & i \end{pmatrix},
$$

which leads to the four relations

$$(i+h)X_{-\tau} = -\frac{i}{2\zeta}\,eX_{-\rho}, \tag{5.92a}$$

$$\tfrac{1}{2}(1+ih)X_{+}\sigma = \zeta eX_{+\tau}, \tag{5.92b}$$

$$\tfrac{1}{2}(1+ih)X_{-}\rho = \zeta eX_{-\tau}, \tag{5.92c}$$

$$(i+h)X_{+\tau} = -\frac{i}{2\zeta}\,fX_{+}\sigma. \tag{5.92d}$$

Let us again prove (5.92b) in a way that helps improve one's manipulation with vertex operators. Since $\sigma = e_1\tau$, then it is also true that $\tilde{X}_{+}\sigma = \tilde{X}_{+}e_1\tau = \tilde{X}_{+}e_1\tilde{X}_{+}\tau$ (an obvious property of the shift part of the vertex operator) $= \zeta b\tilde{X}_{+}\tau = (2ie\zeta/(i-h))\tilde{X}_{+}\tau$. Multiplying by the exponential factor $\exp{(i\sum \zeta^k t_k)}$ gives $1/2(1+ih)X_{+}\sigma = \zeta eX_{+}\tau$, which is (5.92b). But these are also the Hirota equations. Expand (5.92b) using

$$\frac{1}{2}(1+ih) = 1 - \frac{1}{4}\sum \frac{1}{\zeta^k}\frac{\partial^2}{\partial t_1\,\partial t_{k-1}}\ln\tau,$$

$$\zeta e = e_1 + \frac{i}{2}\sum_2 \frac{1}{\zeta^{k-1}}\frac{\partial e_1}{\partial t_{k-1}} = \frac{\sigma}{t} + \frac{i}{2}\sum_2 \frac{1}{\zeta^{k-1}}\frac{\sigma_{k-1}\tau - \sigma\tau_{k-1}}{\tau^2},$$

and

$$\tilde{X}_{+}\sigma = \sigma + \frac{i}{2\zeta}\sigma_1 + \frac{1}{\zeta^2}\left(\frac{i}{4}\sigma_2 - \tfrac{1}{8}\sigma_{11}\right) + \frac{1}{\zeta^3}\left(\frac{i}{6}\sigma_3 - \tfrac{1}{8}\sigma_{12} - \frac{i}{48}\sigma_{111}\right) + \cdots.$$

We remind the reader that subscripts on τ, σ, ρ denote partial derivatives. The coefficient of ζ^{-2} is

$$\left(D_{t_2} - \frac{i}{2}D_{t_1}^2\right)\sigma\cdot\tau = 0.$$

The coefficient of ζ^{-3} includes the equations

$$\left(D_{t_3} - \frac{i}{2}D_{t_1}D_{t_2}\right)\sigma\cdot\tau = 0$$

and

$$(D_{t_3} + \tfrac{1}{4}D_{t_1}^3)\sigma\cdot\tau = 0.$$

The higher order balances appear to include all the equations

$$P(D_{t_k})\sigma\cdot\tau = 0$$

in the Hirota hierarchy. I do not have a proof of this.

Another more powerful way of deriving all the Hirota equations is to use the following identity,

$$\frac{1}{2\pi i}\int_C V(\mathbf{x}+\mathbf{y};\zeta)V^{-1}(\mathbf{x}-\mathbf{y};\zeta)\,d\zeta = I, \qquad (5.93)$$

where C is the circle at $\zeta=\infty$ in the anticlockwise direction and $\mathbf{x}=(x_1, x_2, x_3, \ldots)$, $(x_j = t_j)$. This form was suggested by a similar expression used by Date, Jimbo, Kashiwara, Miwa [39] for members of the KP hierarchy. I find their proof too intricate to understand. The best way to think of (5.93) is to appeal to analytic arguments and think of (5.93) as being an expression of the completeness of the eigenstates of $V_X = Q^{(1)}V$. For a discussion of this property, the reader should consult the third reference in [23], appendix 6.

In any event, (5.93) tells us that the ζ^{-1} component of $V(\mathbf{x}+\mathbf{y};\zeta)V^{-1}(\mathbf{x}-\mathbf{y};\zeta)$ is the identity. Using (5.86), (5.87),

$$V(\mathbf{x}+\mathbf{y})=\frac{1}{\tau^+}\begin{pmatrix} X_-^+\tau & -\dfrac{i}{2\zeta}X_+^+\sigma \\[2mm] \dfrac{i}{2\zeta}X_-^+\rho & X_+^+\tau \end{pmatrix}, \quad V^{-1}(\mathbf{x}-\mathbf{y})=\frac{1}{\tau^-}\begin{pmatrix} X_+^-\tau & \dfrac{i}{2\zeta}X_+^-\sigma \\[2mm] -\dfrac{i}{2\zeta}X_-^-\rho & X_-^-\tau \end{pmatrix},$$

where the superscript on the vertex operator means that the argument has been displaced by plus or minus $\mathbf{y}$; i.e.

$$X_-^+\tau = \exp\left(-i\zeta(x_1+y_1)-i\zeta^2(x_2+y_2)\ldots\right)\tau\left(x_1+y_1-\frac{i}{2\zeta}, x_2+y_2-\frac{i}{4\zeta^2}, \ldots\right).$$

Thus,

$$\tau^+\tau^- V(\mathbf{x}+\mathbf{y},\zeta)V^{-1}(\mathbf{x}-\mathbf{y};\zeta)$$

$$=\begin{pmatrix} X_-^+\tau X_+^-\tau-\dfrac{1}{4\zeta^2}X_+^+\sigma X_-^-\rho & -\dfrac{i}{2\zeta}(X_+^+\sigma X_-^-\tau-X_+^-\sigma X_-^+\tau) \\[3mm] \dfrac{i}{2\zeta}(X_-^+\rho X_+^-\tau-X_+^+\tau X_-^-\rho) & X_+^+\tau X_-^-\tau-\dfrac{1}{4\zeta^2}X_-^+\rho X_+^-\sigma \end{pmatrix}.$$

I invite you to expand the $1/\zeta$ of each of these terms in a Taylor series in $\mathbf{y}=(y_1, y_2, \ldots)$. You will find that the coefficients of y_2, y_3, y_1^3 are nontrivial and equating them to zero gives the Hirota equations.

I will expand the (1, 2) element as our example. It is

$$\exp\left(2i\sum \zeta^k y_k\right)\sigma\left(x_k+y_k+\frac{i}{2k\zeta^k}\right)\tau\left(x_k-y_k-\frac{i}{2k\zeta^k}\right)$$

$$-\exp\left(-2i\sum \zeta^k y_k\right)\sigma\left(x_k-y_k+\frac{i}{2k\zeta^k}\right)\tau\left(x_k+y_k-\frac{i}{2k\zeta^k}\right)$$

and we seek the ζ^0 component. The second term is obtained from the first by changing $y_k \to -y_k$. Therefore all terms even in $\mathbf{y}$ automatically vanish (y_1^2,

$y_1 y_2$, $y_1 y_3$ etc.). First, a little calculation shows that

$$\sigma\left(x_k + y_k + \frac{i}{2k\zeta^k}\right)\tau\left(x_k - y_k - \frac{i}{2ky^k}\right)$$

$$= \sigma(x_k)\tau(y_k) + \sum_k \left(y_k + \frac{i}{2k\zeta^k}\right)D_k\sigma \cdot \tau$$

$$+ \frac{1}{2!}\sum \left(y_k + \frac{i}{2k\zeta^k}\right)\left(y_j + \frac{i}{2j\zeta^j}\right)D_k D_j\sigma \cdot \tau$$

$$+ \frac{1}{3!}\sum \left(y_k + \frac{i}{2k\zeta^k}\right)\left(y_j + \frac{i}{2j\zeta^j}\right)\left(y_l + \frac{i}{2l\zeta^l}\right)D_k D_j D_l\sigma \cdot \tau$$

$$+ \cdots,$$

where D_k is the Hirota operator (4.36), (4.37). Multiply this expression by $\exp\left(2i \sum \zeta^k y_k\right)$, take the ζ^0 component, then set $y_k \to -y_k$ and subtract the latter expression from the former. One finds

y_1: $\quad D_1 - D_1$,

y_2: $\quad \dfrac{1}{2}D_2 - \dfrac{i}{4}D_1^2$,

y_3: $\quad \dfrac{2}{3}D_3 - \dfrac{i}{4}D_1 D_2 + \dfrac{1}{24}D_1^3$,

y_1^3: $\quad \dfrac{2}{9}D_3 - \dfrac{3i}{9}D_1 D_2 - \dfrac{1}{9}D_1^3$.

Setting these to zero gives us

$$\left(D_2 - \frac{i}{2}D_1^2\right)\sigma \cdot \tau = 0,$$

$$\left(D_3 - \frac{i}{2}D_1 D_2\right)\sigma \cdot \tau = 0,$$

$$(D_3 + \tfrac{1}{4}D_1^3)\sigma \cdot \tau = 0.$$

The reader can develop a more fancy notation and obtain expressions for the Hirota polynomials in terms of the Schur functions $p_n(x_1, x_2, \ldots)$

$$e^{kx_1 + k^2 x_2 + \cdots} = \sum_0^\infty k^n p_n(x_1, x_2, \ldots).$$

One can also count the number of Hirota equations at each level [39]. At level n, there are the same number N_n of equations as there are decompositions of the integer n into a sum of odd integers $n_1 + n_2 + \cdots + n_r$, each $n_s \leq n$.

5f. Iso-spectral, iso-Riemann surfaces and iso-monodromic deformations.
(i) *Iso-spectral deformations.* Up to this point in the present chapter, all considerations have been local in the sense that we interpret the equations

(5.52) as describing the evolution of the infinite dimensional point Q in infinite dimensional time $\mathbf{t}(t_1, t_2, \ldots)$.

In this section, we revert to the more traditional way of thinking by imagining that one of the independent variables is special. If we choose it to be t_1 and call $t_1 = x$, then the resulting equations

$$e_{1,t_j} = e_{1,t_j}(e_1, f_1, e_{1x}, f_{1x}, \ldots), \tag{5.94a}$$

$$f_{1,t_j} = f_{1,t_j}(e_1, f_1, e_{1x}, f_{1x}, \ldots) \tag{5.94b}$$

are the AKNS hierarchy. If we choose it to be t_2, then the equations e_{1,t_j}, e_{2,t_j}, f_{1,t_j}, f_{2,t_j} are the DNLS hierarchy. Whichever independent variable is declared special, the problem is then defined as an initial-boundary value problem on $-\infty < x < \infty$ with boundary conditions $e(x, 0)$, $f(x, 0)$ given and behaving in such a way that all the quantities which enter the inverse scattering method are suitably defined. The periodic problem on a finite interval can also be posed. Consider (5.94) on $-\infty < x < \infty$ and assume that $e_1(x, t_j)$, $f_1(x, t_j) \to 0$ as $x \to \pm\infty$. Let V be the fundamental solution matrix of

$$V_x = Q^{(1)}V, \tag{5.95}$$

and since $e_1, f_1 \to 0$ as $x \to \pm\infty$, we will normalize V so that

$$V \to V_0 = \begin{pmatrix} e^{-i\zeta x} & 0 \\ 0 & e^{i\zeta x} \end{pmatrix}$$

as $x \to -\infty$. This particular V we call Φ with columns $\phi(\phi_1, \phi_2)^T$ and $-\bar{\phi}(-\bar{\phi}_1, -\bar{\phi}_2)$.[9] This normalization causes the time evolution to change to

$$\Phi_{t_j} = Q^{(j)}\Phi - \Phi(V_0^{-1}Q_0^{(j)}V_0), \qquad j \geq 2, \tag{5.96}$$

where $Q_0^{(j)}$ is the asymptotic behavior of $Q^{(j)}$ as $x \to \pm\infty$. It is an easy matter to show that the extra term on the right-hand side does not change the integrability conditions. When one cross-differentiates (5.96) and (5.95) or any two members of (5.96), the last terms in (5.96) automatically cancel.

Now, if Ψ is the fundamental matrix solution of (5.95) such that $\Psi \to V_0$ when $x \to +\infty$, then Φ and Ψ are related by a constant (in x) matrix

$$A(\zeta, t_j) = \begin{pmatrix} a(\zeta, t_j) & -\bar{b}(\zeta, t_j) \\ b(\zeta, t_j) & \bar{a}(\zeta, t_j) \end{pmatrix}, \qquad a\bar{a} + b\bar{b} = 1, \tag{5.97a}$$

and equal to $\Psi^{-1}\Phi$. As $x \to +\infty$

$$\Phi \to V_0 A(\zeta, t_j). \tag{5.97b}$$

We call $A(\zeta, t_j)$ the *scattering matrix*.

In this and what follows, I will use the notation of reference [23] and the reader should refer to that paper for all the details. Strictly, the matrix which is called the scattering matrix relates the states $\phi(x, \zeta)$ (the first column of Φ) and

---

[9] $\bar{\phi}_1$ does *not* mean the complex conjugate of ϕ_1.

$\psi(x, \zeta)$ (the second column of Ψ) with the states $-\bar{\phi}$, $\bar{\psi}$, the second and first columns of Φ and Ψ respectively. The reason for this is that the former can be associated with outgoing radiation and the latter with incoming radiation. As an exercise show that

$$(\bar{\psi}, -\bar{\phi}) = (\phi, \psi)S, \quad \text{where} \quad S = \begin{pmatrix} \dfrac{1}{a} & \dfrac{\bar{b}}{a} \\ \dfrac{b}{a} & \dfrac{1}{a} \end{pmatrix}.$$

The time dependence of $A(\zeta, t_j)$ is found by substitution (5.97b) in (5.96)

$$A_{t_j} = [V_0^{-1}Q_0^{(j)}V_0, A]. \tag{5.98}$$

In the hierarchies we consider,

$$V_0^{-1}Q_0^{(j)}V_0 = -i\zeta^j H.$$

In particular, (5.98) has Lax form and *the diagonal elements of $a(\zeta)$ and $\bar{a}(\zeta)$ of A are constants of the motion.* Now it can be shown given $\int_{-\infty}^{\infty}(|e_1|, |f_1|)\, dx < \infty$, that

$$a(\zeta) = W(\phi, \psi), \qquad \bar{a}(\zeta) = W(\bar{\phi}, \bar{\psi})$$

can be analytically continued into the upper and lower half ζ planes respectively. But if Im $\zeta > 0$, then $\phi(\sim \binom{1}{0}e^{-i\zeta x})$ and $\psi(\sim \binom{0}{1}e^{i\zeta x})$ are the only solutions which decay at $-\infty$ and $+\infty$ respectively. Thus if ζ_j is such that $\phi(x, \zeta_j) \to 0$ at $x = +\infty$, then

$$\phi(x, \zeta_j) = b_j\psi(x, \zeta_j)$$

and $a(\zeta_j) = 0$. Similarly the zeros $\bar{\zeta}_j$ of $\bar{a}(\zeta)$, Im $\bar{\zeta}_j < 0$ give rise to the bound states $\bar{\phi}(x, \bar{\zeta}_j)$ and $\bar{\psi}(x, \bar{\zeta}_j)$; $\bar{\phi}(x, \bar{\zeta}_j) = \bar{b}_j\bar{\psi}(x, \bar{\zeta}_j)$.

Therefore, the *spectrum of* (5.95) is preserved under each of the flows (5.52) in the hierarchy. As in the case of the KdV equations studied earlier, the eigenvalues ζ_j, $\bar{\zeta}_j$ are associated with the soliton component of the solution. I also point out a fact which is repeatedly used in developing soliton solutions (as in Section 3h or in the following section on Bäcklund transformations). It is that at an eigenvalue ζ_j, the columns (ϕ, ψ) of a fundamental solution matrix are proportional.

In the cases where $f_1 = -e_1^*$, $f_1 = -e_1$ ($r = -q^*$, $-q$ in [23]), certain simplifications occur. In particular, in the former $\bar{\zeta}_j = \zeta_j^*$ and in the latter $\bar{\zeta}_j = -\zeta_j$. In paticular, if $f_1 = -e_1$ are real, then the eigenvalues are purely imaginary $\zeta_j = i\eta_j$ (solitons, kinks) or arise in pairs $(\zeta_j, -\zeta_j^*)$ (breathers or bions).

The inverse problem is straightforward and we give a brief summary of the results here. For more details the reader should refer to [23]. What we want to do is reconstruct Φ or Ψ from the scattering data. A knowledge of these functions will give us the $Q^{(j)}$ and in particular $Q_1 = e_1E + f_1F$. The easiest way to obtain Q_1 from Φ, Ψ is to calculate the $1/\zeta$ term in their asymptotic expansions about $\zeta = \infty$ (see previous section).

Consider the function $(\phi(\xi)/a(\xi))e^{i\xi x}$, meromorphic in Im $\xi > 0$ with asymptotic behavior $\binom{1}{0}$ as $\xi \to \infty$ there. It is connected to the function $\bar{\psi}(\xi)e^{i\xi x}$ analytic in $\xi < 0$ and having asymptotic behavior $\binom{1}{0}$ as $|\xi| \to \infty$ by $(b(\xi)/a(\xi))\psi(\xi)e^{i\xi x}$ (simply write down the equation for the first column of $\Phi = \Psi A$) across the real axis. We therefore want to solve the *Riemann–Hilbert problem* of finding a function, analytic for all ξ except at a finite number of poles with prescribed behavior as $\xi \to \infty$, given that it possesses a prescribed jump across the real ξ axis. The construction is accomplished by considering for Im $\zeta < 0$,

$$\int_{-\infty}^{\infty} \frac{\phi(\xi)e^{i\xi x}}{a(\xi)(\xi - \zeta)} \, d\xi$$

and computing its value twice, first by closing the contour on the semicircle at $\xi = \infty$, Im $\xi > 0$ and then by replacing ϕ by $a\bar{\psi} + b\psi$ and computing the first integral by closing in the lower half plane. One obtains

$$\bar{\psi}(\zeta, x)e^{i\zeta x} = \binom{1}{0} + \sum_{1}^{N} \frac{1}{\zeta - \zeta_j} \gamma_j \psi_j e^{i\zeta_j x} + \frac{1}{2\pi i} \int_{-\infty}^{\infty} \frac{b\psi e^{i\xi x}}{a(\xi - \zeta)} \, d\xi,$$

where $\psi_j = \psi(\zeta_j)$, $\gamma_j = b_j(a_j')^{-1}$, $a_j' = (da/d\zeta)$ at $\zeta = \zeta_j$, $j = 1, \ldots, N$, the N zeros of $a(\zeta)$. Following a similar prescription, we can find an equation for $\psi(\zeta, x)e^{-i\zeta x}$ (Im $\zeta > 0$) linear in $\bar{\psi}$. It is the fact that the jumps across the real ζ axis are linear in ψ and $\bar{\psi}$ which makes the inverse problem linear.

If we are dealing with reflectionless potentials, $b = \bar{b} \equiv 0$ and the equations for ψ, $\bar{\psi}$, ϕ, $\bar{\phi}$ are

$$\bar{\psi}(\zeta, x)e^{i\zeta x} = \binom{1}{0} + \sum_{1}^{N} \frac{\gamma_j \psi_j e^{i\zeta_j x}}{\zeta - \zeta_j}, \qquad \text{Im } \zeta < 0, \qquad (5.99a)$$

$$\psi(\zeta, x)e^{-i\zeta x} = \binom{0}{1} - \sum_{1}^{N} \frac{\bar{\gamma}_j \bar{\psi}_j e^{-i\bar{\zeta}_j x}}{\zeta - \bar{\zeta}_j}, \qquad \text{Im } \zeta > 0, \qquad (5.99b)$$

$$\bar{\phi}(\zeta, x)e^{-i\zeta x} = \binom{0}{-1} - \sum_{1}^{N} \frac{\beta_j \phi_j e^{-i\zeta_j x}}{\zeta - \zeta_j}, \qquad \text{Im } \zeta < 0, \qquad (5.99c)$$

$$\phi(\zeta, x)e^{i\zeta x} = \binom{1}{0} + \sum_{1}^{N} \frac{\bar{\beta}_j \bar{\phi}_j e^{i\bar{\zeta}_j x}}{\zeta - \bar{\zeta}_j}, \qquad \text{Im } \zeta > 0, \qquad (5.99d)$$

where $\beta_j = \bar{b}_j(a_j')^{-1}$, $\bar{\beta}_j = b_j(\bar{a}_j')$. Let $\zeta = \bar{\zeta}_k$ in (5.99a,c), and $\zeta = \zeta_k$ in (5.99b,d), and then the resulting linear equations for ψ_j, $\bar{\psi}_j$ are easily solved. The determinant of the coefficient matrix is the τ function up to multiplication by exponential factors with phases linear in t_k.

The reader should solve the one-soliton case corresponding to the pair ζ_1, $\bar{\zeta}_1$. If $r = -q^*$, then $\bar{\zeta}_1 = \zeta_1^*$ and the solutions are

$$e_1 = 2\eta \, \text{sech} \, \theta \exp i\phi, \qquad f_1 = -e_1^*,$$

with

$$\theta = i \sum (\zeta_1^k - \zeta_1^{*k})t_k + 2\eta x_0,$$

$$\phi = -\sum (\zeta_1^k + \zeta_1^{*k})t_k + \phi_0,$$

where the position parameters x_0, ϕ_0 are related to the coefficients b_j, $\bar{b}_j = b_j^*$.

The property I want you to remember, however, is the structure of the fundamental solution matrix. Notice that each of the columns $\bar{\psi}$ and ψ has a finite number of poles independent of x, t_k, $k = 2, \ldots$. We are free to renormalize the fundamental solution matrix by multiplying the solution $\bar{\psi}$ by $\zeta^{-N} \prod_1^N (\zeta - \zeta_j)$ in which case $\bar{\psi} e^{i\zeta x}$ has the form

$$\binom{1}{0} + \sum_1^n \frac{1}{\zeta^k} \mathbf{C}_k.$$

Similarly $\psi e^{-i\zeta x}$ can be renormalized to be a polynomial of degree $\bar{N}$ in inverse powers of ζ. This normalization can be accomplished by multiplication of V on the right by $\binom{F_1(\zeta)\ 0}{0\ F_2(\zeta)}$.

Conversely, just as we did in Chapter 3, one can show that if we take

$$V = (\mathbf{u}_1, \mathbf{u}_2)^T,$$

with

$$\mathbf{u}_1 = \left(\binom{1}{0} + \sum_1^N \frac{1}{\zeta^k} \mathbf{C}_{1k} \right) \exp\left(-i \sum \zeta^k t_k \right),$$

$$\mathbf{u}_2 = \left(\binom{0}{1} + \sum_1^{\bar{N}} \frac{1}{\zeta^k} \mathbf{C}_{2k} \right) \exp\left(i \sum \zeta^k t_k \right), \tag{5.100a}$$

$$\mathbf{u}_1(\alpha) = b\mathbf{u}_2(\alpha) \tag{5.100b}$$

for $\alpha = (\zeta_1, \ldots, \zeta_N, \bar{\zeta}_1, \ldots, \bar{\zeta}_{\bar{N}})$, $b = (b_1, \ldots, b_N, \bar{b}_1, \ldots, \bar{b}_{\bar{N}})$, then the column vectors of V satisfy

$$V_{t_j} = Q^{(j)} V. \tag{5.101}$$

The elements of the coefficient matrices $Q^{(j)}$ are related to $\mathbf{C}_{1k}$, $\mathbf{C}_{2k}$ and their various derivatives. For example,

$$\mathbf{C}_{21} = \left(-\frac{i}{2} e_1, -\frac{i}{2} \int^x e_1 f_1 \right), \qquad \mathbf{C}_{11} = \left(\frac{i}{2} \int^x e_1 f_1, \frac{i}{2} f_1 \right)$$

(see Section 5e). The integrability of (5.101) insures that e_1, f_1 satisfy the AKNS hierarchy.

The proof of the proposition that (5.100a,b) implies that (5.101) follows the uniqueness argument given in connection with the KdV family in Section 3h. First, note that the $2(N + \bar{N})$ equations (5.100b) determine $\mathbf{C}_{1k}$, $\mathbf{C}_{2k}$ uniquely as functions of x, $t_j \geq 2$. Next, consider the vector quantities

$$\mathbf{u}_j = (v_{j1x} + i\zeta v_{j1} - e_1 v_{j2}, v_{j2x} - i\zeta v_{j2} - f_1 v_{j1})^T,$$

with $\mathbf{v}_j = (v_{j1}, v_{j2})^T$, $j = 1, 2$. A little calculation shows they have asymptotic expansions

$$\sum_1^N \frac{1}{\zeta^k} d_k, \qquad \sum_1^{\bar{N}} \frac{1}{\zeta^k} \bar{d}_k$$

respectively and moreover $\mathbf{u}_1(\alpha) = b\mathbf{u}_2(\alpha)$. Thus the vectors $\mathbf{v}_1 + \mathbf{u}_1$, $\mathbf{v}_2 + \mathbf{u}_2$ satisfy all the conditions (5.100), (5.101). But the vectors satisfying these conditions are unique (we can calculate the $\mathbf{C}_{1k}$, $\mathbf{C}_{2k}$ explicitly) and hence

$\mathbf{u}_1 = \mathbf{u}_2 = 0$. Therefore, the vectors $\mathbf{v}_1$ and $\mathbf{v}_2$ satisfy (5.101) for $j = 1$. The proof for general t_j follows similarly.

(ii) *Iso-Riemann surface deformations.* In Section 3h I showed you how the finite-gap solutions of the KdV family were associated with a Riemann surface which does not change in time, so it will only be necessary to give a brief review of the situation here. Consider the further constraint

$$\sum_1^n u_j V_{t_j} = yV \tag{5.102}$$

added to the list (5.76). This can also be written

$$\left(\sum_1^n u_j Q^{(j)} - y\right)V = 0, \tag{5.103}$$

so that a nontrivial solution exists only when

$$\det\left(\sum_1^n u_j Q^{(j)} - yI\right) = 0. \tag{5.104}$$

Equation (5.104) is an algebraic curve (for $\widetilde{sl}(2, C)$ it is hyperelliptic)

$$y^2 = \det\left(\sum_1^n u_j Q^{(j)}\right). \tag{5.105}$$

Cross-differentiation of (5.96) and (5.103) shows that $P = \sum_1^n u_j Q^{(j)}$ satisfies the equation

$$P_{t_j} = [Q^{(j)}, P] \tag{5.106}$$

which means that P can be written

$$P = VP_0 V^{-1}, \qquad P_{0t_j} = 0. \tag{5.107}$$

Hence the characteristic polynomial of P is equal to $\det(P_0 - yI)$ and therefore independent of all the times. In addition, from the compatibility of (5.102) and (5.76) (just cross-differentiate with respect to t_k and use $Q^{(j)}_{t_k} + [Q^{(j)}, Q^{(k)}] = Q^{(k)}_{t_j}$),

$$\sum_1^n u_j Q^{(k)}_{t_j} = 0, \qquad k = 1, 2, \ldots. \tag{5.108}$$

Thus (i) implies that each $Q^{(k)}$ is only a function of $(n-1)$ linear combinations of the $t_1, \ldots, t_n$ and (ii) means that in particular $Q^{(1)}$ satisfies a nonlinear ordinary equation in $t_1 = x$ because e_{1,t_i} and f_{1,t_i} can be written as functions of e_1, f_1 and their t_1 derivatives. The finite dimensional solution manifold of this equation is left invariant by each of the flows in the whole hierarchy $Q_{t_j} = [Q^{(j)}, Q]$. This means that a solution of

$$\sum_1^n u_j Q^{(1)}_{t_j} = 0 \tag{5.109}$$

at time zero, if allowed to evolve in any of the time flows, will continue to be a solution of (5.109). Equation (5.109) is often called the *Lax–Novikov equation.*

Solutions of (5.109) as well as their time evolutions can be constructed explicitly and we demonstrated one way of doing this in Section 3h. They are Abelian functions. A method of construction which uses the theory of Riemann surfaces and the uniqueness of functions with certain properties defined on them was given by Krichever. I will not go into his ideas here but instead refer the reader to his papers [28].

(iii) *Iso-monodromic deformations.* Suppose now that instead of the constraint (5.102) one were to apply the constraint

$$\zeta V_\zeta = \sum_1^n jt_j V_{t_j} \tag{5.110}$$

to the set of equations

$$V_{t_j} = Q^{(j)} V, \qquad j = 1, \ldots, n. \tag{5.111}$$

The integrability conditions are now

$$\sum_1^n jt_j Q_{t_j}^{(k)} = \zeta Q_\zeta^{(k)} - k Q^{(k)},$$

which can be written

$$\sum_1^n jt_j P_{t_j} = \zeta P_\zeta, \tag{5.112}$$

where

$$P = \frac{Q^{(k)}}{\zeta^k} = \sum_0^k Q_r \zeta^{-r}, \qquad Q_r = h_r H + e_r E + f_r F.$$

The coefficient of $1/\zeta$ gives us that

$$(xQ_1)_x + \sum_2^n jt_j Q_{1,t_j} = 0, \tag{5.113}$$

which is the analogue of (5.109). What it means is that Q_1 is a function of the form

$$\frac{1}{x} \bar{Q}_1 \left(\frac{x}{(2t_2)^{1/2}}, \cdots, \frac{x}{(nt_n)^{1/n}} \right) \tag{5.114}$$

with $n-1$ phases.

Now, what motivates the choice of constraint (5.110)? The idea is that it reflects scaling symmetries which certain equations in the hierarchy possess in exactly the same way that the choice of finite gap constraint reflects translational symmetries of the equation (you might wish to refer to Section 3h). To make this more concrete, I will focus attention on the MKdV hierarchy which is a subset of the hierarchy of equations

$$Q_{t_k} = [Q^{(k)}, Q] \tag{5.115}$$

obtained by setting $f_1 = e_1 = q$ and disallowing any flow in the even times t_{2n}.

The first three members of the sequence are

$$q_{t_1} = q_x, \tag{5.116}$$

$$q_{t_3} = -\tfrac{1}{4}(q_{xx} - 6q^2 q_x)_x, \tag{5.117}$$

$$q_{t_5} = \tfrac{1}{16}(q_{xxxx} - 10q^2 q_{xx} - 10qq_x^2 + 6q^5)_x, \tag{5.118}$$

and corresponds to

$$Q^{(1)} = \begin{pmatrix} -i\zeta & q \\ q & i\zeta \end{pmatrix}, \tag{5.119a}$$

$$Q^{(3)} = \begin{pmatrix} -i\zeta^3 - \dfrac{iq^2\zeta}{2} & \zeta^2 q + \dfrac{i\zeta q_x}{2} - \tfrac{1}{4}(q_{xx} - 2q^3) \\ \zeta^2 q - \dfrac{i\zeta q_x}{2} - \tfrac{1}{4}(q_{xx} - 2q^3) & i\zeta^3 + \dfrac{iq^2\zeta}{2} \end{pmatrix}. \tag{5.119b}$$

I will leave it as an exercise to the reader to calculate $Q^{(5)}$. Note that (5.117) has the scaling symmetry that if $q(x, t)$ solves (5.117), then so does $\beta q(\beta x, \beta^3 t_3)$. A solution $\bar{q}(x, t)$ which is invariant under this scaling is called self-similar and it means that

$$\frac{\partial}{\partial \beta} \beta q(\beta x, \beta^3 t_3)|_{\beta=1} = 0$$

or

$$q + xq_x + 3t_3 q_{t_3} = 0, \tag{5.120}$$

which is exactly (5.113) with $n = 3$. Thus $q(x, t_3)$ has the shape

$$q(x, t_3) = \frac{1}{(3t_3)^{1/3}} f\left(X = \frac{x}{(3t_3)^{1/3}}\right). \tag{5.121}$$

It is natural to change variables in

$$V_x = Q^{(1)}V, \qquad V_{t_3} = Q^{(3)}V \tag{5.122}$$

to

$$X = \frac{x}{(3t_3)^{1/3}}, \qquad T = t_3, \tag{5.123}$$

in order to reflect the structure of the coefficients of $Q^{(1)}$ and $Q^{(3)}$. We find that $V(x, t_3; \zeta)$ rescales as $W(X, \zeta)$ where $\xi = \zeta(3t_3)^{1/3}$ and that the equations (5.122) become

$$W_X = \begin{pmatrix} -i\xi & f \\ f & i\xi \end{pmatrix} W, \tag{5.124}$$

$$\xi W_\xi = \begin{pmatrix} -i\xi^3 - if^2\dfrac{\xi}{2} - iX\xi & \xi^2 f + i\xi f_x/2 + v \\ \xi^2 f - i\xi f_x/2 + v & i\xi^3 + if^2\dfrac{\xi}{2} + iX\xi \end{pmatrix} W. \tag{5.125}$$

The integrability condition for (5.124, 125) is

$$f_{XX} = 4Xf + 2f^3 - \nu, \tag{5.126}$$

which, if q is given by (5.121), is (5.117) once integrated with respect to X, and ν is the integration constant.

Equation (5.126), which describes the self similar solution of (5.117), is the Painlevé equation of the second kind (see [35], [36] and Section 4e). It is a nonlinear, nonautonomous equation which is left invariant by the flow. For example, choose an $f(x)$ satisfying (5.126). Take $t_3 = \frac{1}{3}$. Allow $q(x, t_3)$ as given by (5.121) to evolve in (5.117) for a time $\frac{1}{3} < t_3 < t$. Then at $t_3 = t$, $q(x, t)$ has the form (5.121) where $f(X)$ again solves (5.126) except that X is now $x/(3t)^{1/3}$.

More generally we state that the class of solutions consistent with the constraint (5.110) have a multiphase self-similar structure and satisfy a nonlinear nonautonomous ordinary differential equation in x, namely (5.113), with coefficients depending on x, $t_2, \ldots, t_n$. Further, the solution manifold is left invariant by the flows Q_{1, t_j}, $j = 1, \ldots, n$ in the sense that a solution of (5.113) at $t_2^{(0)}, \ldots, t_n^{(0)}$, if allowed to evolve with the flows Q_{1, t_j}, $j = 2, \ldots, n$ till times $t_2, \ldots, t_n$ will again satisfy (5.113) with $t_2^{(0)}, \ldots, t_n^{(0)}$ replaced by the present times $t_2, \ldots, t_n$. Unlike the finite gap solutions, however, these solutions are not, repeat not, left invariant by higher flows Q_{1, t_j} $j > n$ in the hierarchy.

Now, how does one solve the initial value problem for (5.126) given f, f_X at $X = X_0$. In the inverse scattering method, you will recall we focussed our attention on the "eigenvalue" problem

$$V_x = Q^{(1)} V$$

and used the second equation in (5.122) in order to determine the time evolution of the scattering data. For solving the nonlinear autonomous ordinary differential equations associated with the finite gap solutions, we focussed our attention on the constraint

$$\left(\sum u_j Q^{(j)} - yI \right) V = 0,$$

and used the eigenvalue problem and the other equations $V_{t_i} = Q^{(i)} V$ as auxiliary equations for determining the x and t_j dependence of the μ's (see Section 3h).

Here, we again focus our attention on the constraint (5.125) and use (5.124) as an auxiliary equation. Equation (5.125) looks complicated but, when considered as function of ξ, it is really very simple as all the coefficients are rational in ξ. There are two singular points, one regular at $\xi = 0$ and the other irregular of rank three at $\xi = \infty$. It is known from the theory of ordinary differential equations that the structure of the fundamental solution matrix is completely characterized by its behavior in the neighborhood of its singular points. In particular, this behavior is measured by the *monodromy matrices* which tell one how the fundamental solution matrix changes as the solution traverses a contour surrounding the singular point.

Next $\xi = 0$, (5.125) has a solution of the form

$$\Phi(\xi; X) = \hat{\Phi}(\xi; X)\begin{pmatrix} \xi^{-\nu} & 0 \\ 0 & \xi^{\nu} \end{pmatrix} \tag{5.127}$$

where $\hat{\Phi}$ is analytic in ξ (for $\nu = $ half integer, there will, in general, be logarithms in the general solution as well). The monodromy matrix J at $\xi = 0$ is

$$\Phi(\xi e^{2\pi i}) = \Phi(\xi)J, \tag{5.128}$$

and J has the form

$$J = \begin{pmatrix} e^{-2\pi i \nu} & 0 \\ 2\pi J_1 e^{2\pi i \nu} & e^{2\pi i \nu} \end{pmatrix} \tag{5.129}$$

where J_1 is only present if ν is a half integer. For $\nu = \frac{1}{2}$, $J_1 = 2(f_X + f^2 + 2X)e^{-2u}$, $u_X = f$. Note that $J_{1X} = 0$.

Near $\xi = \infty$, (5.125) has a formal fundamental solution

$$\tilde{\Psi}(\xi; X) = \hat{\Psi}(\xi; X)\begin{pmatrix} e^{-\theta} & 0 \\ 0 & e^{\theta} \end{pmatrix} \tag{5.130}$$

where $\theta = i\xi X + i\xi^3/3$ and $\hat{\Psi} = \sum_0^\infty c_i \zeta^{-i}$ is a formal Laurent series. In each sector S_j, $(\pi/3)(j-1) \le \mathrm{Arg}\, \xi \le (\pi/3)j$, there exists a true solution Ψ_j for which (5.130) is an asymptotic expansion in S_j. However, as one traverses the point at ∞, one meets the Stokes phenomenon. Namely, as one crosses from S_1 to S_2 at $\mathrm{Arg}\, \xi = \pi/3$, the analytic continuation of the asymptotic expansion in S_1 is no longer the asymptotic expansion of the analytic continuation of the true solution. One has to multiply the true solution Ψ, by a "Stokes" matrix of the form

$$A_1 = \begin{pmatrix} 1 & 0 \\ a_1 & 1 \end{pmatrix}, \tag{5.131}$$

to get a new solution Ψ_2 whose asymptotic expansion in S_2 is (5.130). What happens is that the recessive solution in S_1 (i.e., the solution proportional to e^θ which decays exponentially) becomes the dominant solution in S_2 but a certain amount (a_1, the Stokes multiplier) of the recessive solution in S_1 must be added to the dominant solution there in order that the combination is recessive in the next sector. The fundamental solutions Ψ_j with asymptotic expansion (5.130) in the six sectors of infinity are then related by

$$\Psi_{j+1} = \Psi_j A_j \tag{5.132}$$

where the nonzero off-diagonal elements in A_j (the Stokes multipliers) alternate between the two corners. The particular details are all worked out in reference [36].

The set of matrices $J, A_1, \ldots, A_6$ together with the connection matrix A which specifies the relation between the fundamental solution Φ normalized in a certain way at $\xi = 0$ and Ψ_1, $\Phi = \Psi_1 A$ constitute the *monodromy data*.

(Because of symmetries, there are if ν is known only two independent parameters in all this data corresponding to the unknowns f and f_X.)

Now we can state a remarkable result. As $f(X)$ deforms according to (5.126), all these matrices are independent of X. Hence the term *iso-monodromic deformation*. The solution to (5.126) may then be obtained as follows. Given f, f_X at $X = X_0$, compute the monodromy data. Then at some other X, given this data and $\theta = i\xi X + i\xi^3/3$, one can reconstruct Ψ_1 and hence the coefficients in its formal asymptotic expansion, which coefficients depend on $f(X)$ and $f_X(X)$. Therefore, one can find f at any X. The details of the inversion procedure and some information about the solutions are given in [36].

I will end this section by remarking that the operator $\xi\, d/d\xi$ is very important in the whole theory and not just in connection with self-similar solutions. Some comments on its role are given in the last section.

5g. Gauge and Bäcklund transformations.

We begin with a theorem which states that under the transformations

$$Q^{(k)} \to RQ^{(k)}R^{-1} + R_{t_k}R^{-1} = \tilde{Q}^{(k)}, \qquad (5.133)$$

$$Q \to RQR^{-1} = \tilde{Q}, \qquad (5.134)$$

the equation

$$Q^{(j)}_{t_k} - Q^{(k)}_{t_j} + [Q^{(j)}, Q^{(k)}] = 0 \qquad (5.135)$$

and its limit

$$Q_{t_k} = [Q^{(k)}, Q] \qquad (5.136)$$

are invariant. The proof is straightforward. This choice of transformation is motivated by noting that (5.135) and (5.136) are the integrability conditions for the sequence of equations

$$V_{t_k} = Q^{(k)}V \qquad (5.137)$$

and (5.137) remains invariant under

$$V \to RV = W \qquad (5.138)$$

provided (5.133) holds.

In addition, it is also useful on occasion to normalize V by adding to the right-hand side of (5.137)

$$V_{t_k} = Q^{(k)}V + VN^{(k)}. \qquad (5.139)$$

The integrability condition for (5.139) is exactly (5.135) provided the curl of the vector string $N^{(k)}$ is zero; i.e.

$$N^{(k)}_{t_j} = N^{(j)}_{t_k}. \qquad (5.140)$$

(You will remember that we choose $N^{(1)} = 0$, $N^{(k)} = i\zeta^k H$ for $k \geq 2$, in order to ensure that V at $-\infty$ had asymptotic behavior

$$V_0 = \begin{pmatrix} e^{-i\zeta x} & 0 \\ 0 & e^{i\zeta x} \end{pmatrix};$$

see Section 5f(i).) The addition of a normalization $N^{(k)}$ can also be accomplished by a transformation on V,

$$V \to RVS = \tilde{V}, \tag{5.141}$$

and it is easy to show that

$$N^{(k)} = S^{-1}S_{t_k}. \tag{5.142}$$

The condition (5.140) is then satisfied if

$$[N^{(k)}, N^{(j)}] = 0. \tag{5.143}$$

Observe that the transformation (5.134) on Q does not involve S. This is because the multiplication of V on the right simply amounts to a change in basis of the column vectors in V. The transformation (5.141) with condition (5.143) together with (5.133) and (5.134) is called a *Gauge transformation*. But look at (5.134). We know that if Q satisfies (5.136), so does $\tilde{Q} = RQR^{-1}$. But equation (5.134) is just an infinite set of relations between the variables $\{h_r, e_r, f_r\}$ of Q and their corresponding counterparts $\{\tilde{h}_r, \tilde{e}_r, \tilde{f}_r\}$ in $\tilde{Q}$. Hence (5.134) is an auto Bäcklund transformation between any two members of the sequence of integrable equations associated with $\widetilde{sl}(2, C)$.

So far we have said nothing about R and S except that they should be invertible. How are they to be chosen? We have already noted that S plays no substantial role. Therefore everything depends on how we choose R.

First, note that R has the property, easily deduced from (5.133), that

$$\frac{\partial}{\partial t_k} \det R = (\text{Tr } \tilde{Q}^{(k)} - \text{Tr } Q^{(k)}) \det R = 0,$$

because $\text{Tr } Q^{(k)}$ is independent of t_k and from (5.134), so is $\text{Tr } \tilde{Q}$. Therefore $\det R$ is independent of t_k and can only be a function of ζ. Let α be a zero of $\det R$ and assume it is not a zero of $\det V$. Then $\det \tilde{V}(a) = \det R(\alpha) \det V(\alpha) = 0$ and the columns of $\tilde{V}$ are linearly dependent at $\zeta = \alpha$. Now recall from the discussion at the end of part (i) of the previous section, that this is precisely the condition that $\tilde{V}$ has a bound state at α. We know that the addition of a bound state pair at α, $\bar{\alpha}$ corresponds to the addition of a soliton. On the other hand if $\det R$ has a pole at α, then we note that the inverse transformation $V = R^{-1}\tilde{V}$ creates a new fundamental matrix V with an extra bound state parametrized by α.

In summary, then, the zeros of $\det R$ are associated with bound states of $\tilde{V}$ which are not contained in V. A bound state pair is associated with the addition of a soliton. More complicated functional forms of $\det R$ are associated with the addition of more complicated solutions which are beyond the scope of these lectures. There is, however, one other simple class of Bäcklund transformations associated with $\det R = \text{constant}$. They are transformations which change the monodromy of the fundamental solution matrix at $\zeta = \infty$ and are known as Schlesinger transformations [125]. They play a central role in the story I am about to tell you.

Let us now get down to concrete cases by illustrating these ideas with several examples. The immediate goal of the section is to write formulae for the new τ functions $\tilde{\tau}$, $\tilde{\sigma}$, $\tilde{\rho}$ in terms of the old ones τ, σ, ρ. We will find out, however, that it is better to think of each triplet as three successive entries in an infinite sequence.

The first example is well known (although you probably have not seen it approached in this manner before) and attempts to add to the solution Q one extra soliton. Again I remind you the signature of a soliton is contained in the structure of the fundamental solution matrix. Its columns become linearly dependent at values of ζ corresponding to the soliton parameters. By use of a transformation in basis (of the columns of V) via an S transformation, this criterion can be stated in a completely equivalent way by demanding that the first column of RV vanishes at $\zeta = \zeta_1$ and the second at the companion eigenvalue $\bar{\zeta}_1$. Recall the brief discussion in Section 5f of the scattering problem for the Zakharov–Shabat eigenvalue problem. For general r and q, the eigenvalues come in pairs, ζ, $\bar{\zeta}$, the zeros of $a(\zeta)$ and $\bar{a}(\zeta)$ respectively in the upper and lower half planes. If $f_1 = -e_1^*$ (or $r = -q^*$), then $\bar{\zeta} = \zeta^* = \xi - i\eta$ and η and ξ are the amplitude and speed of the soliton envelope. For the purposes of our discussion here, we will take $r = -q$ in which case if $\zeta_1 = i\eta$, $\bar{\zeta}_1 = -i\eta$. The calculations are simpler and the result is one already familiar to the reader from Section 4f. Let

$$R = \begin{pmatrix} \zeta + d + a & b \\ c & \zeta + d - a \end{pmatrix}, \qquad V = \begin{pmatrix} V_{11} & V_{12} \\ V_{21} & V_{22} \end{pmatrix}.$$

It is not hard to show that the condition that

$$R\begin{pmatrix} V_{11} \\ V_{21} \end{pmatrix} = 0 \quad \text{at } \zeta = i\eta, \qquad R\begin{pmatrix} V_{21} \\ V_{22} \end{pmatrix} = 0 \quad \text{at } \zeta = -i\eta,$$

leads to the following values for a, b, c, d:

$$a = i\eta - \frac{2i\eta}{1 + \gamma^2}, \qquad b = c = \frac{-2i\eta\gamma}{1 + \gamma^2}, \qquad d = 0, \tag{5.144}$$

where (recall the notation used in (4f))

$$\gamma = \frac{V_{21}(i\eta)}{V_{11}(i\eta)} = -\frac{V_{22}(-i\eta)}{V_{12}(-i\eta)}. \tag{5.145}$$

The latter equality is true because we know from Section 5f that if $V_1(x, \zeta)$, $V_2(x, \zeta)$ solve the Zakharov–Shabat equations so do $V_2(x, -\zeta)$, $-V_1(x, -\zeta)$ if $r = -q$. Having calculated R, let us now use the Bäcklund transformation (5.134) in order to calculate the new strings $\tilde{h}$, $\tilde{e}$, $\tilde{f}$ in terms of the old ones h, e, f.

A little calculation shows that if

$$R = \begin{pmatrix} \zeta + a & b \\ b & \zeta - a \end{pmatrix}, \qquad Q = \begin{pmatrix} h & e \\ f & -h \end{pmatrix}, \qquad \tilde{Q} = \begin{pmatrix} \tilde{h} & \tilde{e} \\ \tilde{f} & -\tilde{h} \end{pmatrix},$$

then (5.134) can be written

$$\zeta \begin{pmatrix} \tilde{h}-h \\ \tilde{e}-e \\ \tilde{f}-f \end{pmatrix} = \begin{pmatrix} a(h-\tilde{h})+b(f-\tilde{e}) \\ a(\tilde{e}+e)-b(\tilde{h}+h) \\ -a(\tilde{f}+f)+b(\tilde{h}+h) \end{pmatrix}. \tag{5.146}$$

Recall that $h_0 = \tilde{h}_0 = -i$, $e_0 = f_0 = \tilde{e}_0 = \tilde{f}_0 = h_1 = \tilde{h}_1 = 0$. Equating the ζ^0 coefficients, we obtain,

$$\tilde{e}_1 - e_1 = 2ib = \frac{4\eta\gamma}{1+\gamma^2}, \tag{5.147a}$$

$$\tilde{f}_1 - f_1 = -2ib = \frac{-4\eta\gamma}{1+\gamma^2}, \tag{5.147b}$$

which tells us that, since $f_1 = -e_1$ that $\tilde{f}_1 = -\tilde{e}_1$, that is, we remain in the solution class. Let $e_1 = -u_x/2$, $\tilde{e}_1 = -\tilde{u}_x/2$ and $\gamma = \tan(u+\tilde{u})/4$ whence (5.147a)

$$u_x - \tilde{u}_x = 4\eta \sin \frac{u+\tilde{u}}{2} \tag{5.148}$$

a distinctly familiar result (see Section 4f). The coefficients of the higher powers of ζ give us relations between $\tilde{h}_r$, $\tilde{e}_r$, and $\tilde{f}_r$ and h_s, e_s, and f_s and, recalling that $e_s = (i/2) \partial e_1/\partial t_{s-1}$, we obtain all the Bäcklund relations for the flows in the modified KdV family. In this example, we of course must keep t_{2n}, $n = 1, 2, \ldots$, fixed as otherwise we flow out of the class $f_1 = -e_1$. Therefore the expressions obtained from equating the coefficients of odd powers of ζ are satisfied automatically and do not give us new Bäcklund relations.

They have one important role. You may have asked how I was able to choose $\gamma = \tan(u+\tilde{u})/4$, above. Strictly, this does not follow at all from what I have told you. However, if you look at the coefficients of ζ in (5.146), you will find that

$$\tilde{e}_2 - e_2 = a(\tilde{e}_1 + e_1). \tag{5.149}$$

Now differentiate (5.147a) with respect to x or t_1 and multiply by $i/2$. One gets

$$\tilde{e}_2 - e_2 = i\eta \cos \phi \phi_x,$$

where we have used $e_2 = (i/2) \partial e_1/\partial t_1$ and set $\gamma = \tan(\phi/2)$. Now recall $a = -i\eta \cos \phi$ and therefore

$$\tilde{e}_1 + e_1 = -\phi_x,$$

or

$$\frac{\tilde{u}+u}{2} = \phi.$$

Therefore the introduction of $\gamma = \tan[(u+\tilde{u})/4]$ is entirely natural. There is no hocus pocus, no cheating!

The second example illustrates a Schlesinger transformation [86], [125]. This time we choose R so that the new fundamental solution matrix $\tilde{V}$ has an asymptotic expansion which is the canonical one (given by (5.77)) multiplied by the matrix

$$\begin{pmatrix} -2i\zeta & \\ & \dfrac{1}{2i\zeta} \end{pmatrix}$$

which changes its monodromy properties at $\zeta = \infty$.

First let me tell you the answer and examine its ramifications. I will tell you how to work R out later. Take

$$R = \begin{pmatrix} -2i\zeta - \dfrac{\partial}{\partial t_1} \ln e_1 & e_1 \\ \tilde{f}_1 & 0 \end{pmatrix}. \tag{5.150}$$

The canonical V is

$$V = \frac{1}{\tau} \begin{pmatrix} X_{-}\tau & -\dfrac{i}{2\zeta} X_{+}\sigma \\ \dfrac{i}{2\zeta} X_{-}\rho & X_{+}\tau \end{pmatrix} \tag{5.151}$$

and, as I have said, R has been chosen so that

$$\tilde{V} = RV = \tilde{V}_c \begin{pmatrix} -2i\zeta & \\ & \dfrac{1}{2i\zeta} \end{pmatrix}. \tag{5.152}$$

Here $\tilde{V}_c$ refers to a $\tilde{V}$ which has been renormalized through multiplication by a ζ dependent transformation so as to achieve canonical ((5.77) and (5.151) with τ, σ, ρ replaced by $\tilde{\tau}$, $\tilde{\sigma}$, $\tilde{\rho}$) form. Since $\det R = -1$, $e_1 \tilde{f}_1 = 1$.

But, from (5.152) and (5.151),

$$R = \tilde{V}_c \begin{pmatrix} -2i\zeta & \\ & \dfrac{1}{2i\zeta} \end{pmatrix} V^{-1}$$

$$= \frac{1}{\tau\tilde{\tau}} \begin{pmatrix} -2i\zeta\tilde{\tau}_{-}\tau_{+} + \dfrac{i}{8\zeta^3}\tilde{\sigma}_{+}\rho_{-} & \sigma_{+}\tilde{\tau}_{-} - \dfrac{1}{4\zeta^2}\tilde{\sigma}_{+}\tau_{-} \\ \tilde{\rho}_{-}\tau_{+} - \dfrac{1}{4\zeta^2}\tilde{\tau}_{+}\rho_{-} & \dfrac{i}{2\zeta}(\sigma_{+}\tilde{\rho}_{-} - \tilde{\tau}_{+}\tau_{-}) \end{pmatrix}, \tag{5.153}$$

where $\tau_{+} = \tau(t_k + i/2k\zeta^k)$, $\tilde{\rho}_{-} = \tilde{\rho}(t_k - i/2k\zeta^k)$. The exponential factors $e^{\mp i\Sigma\zeta^k t_k H}$ in X_{+} and X_{-} have all cancelled out. Now, comparing (5.150) and (5.153), we find

$$\sigma_{+}\tilde{\rho}_{-} = \tilde{\tau}_{+}\tau_{-}, \tag{5.154}$$

$$\tilde{\rho}_{-}\tau_{+} - \frac{1}{4\zeta^2}\tilde{\tau}_{+}\rho_{-} = \tilde{f}_1\tau\tilde{\tau} = \tilde{\rho}\tau. \tag{5.155}$$

But $\tilde{f}_1 = e_1^{-1}$ ($\det R = -1$) implies that

$$\tilde{\rho}\sigma = \tau\tilde{\tau}. \tag{5.156}$$

One finds that (recall $\tau_+\tau_- - \sigma_+\rho_-/4\zeta^2 = \tau^2$ since det $V = 1$)

$$\tilde{\tau} = \sigma, \qquad \tilde{\rho} = \tau, \tag{5.157}$$

and expanding the expression

$$\sigma_+\tilde{\tau}_- - \frac{1}{4\zeta^2}\tilde{\sigma}_+\tau_-,$$

we find

$$\tilde{\sigma} = -\frac{1}{2\tau}D_{t_1}^2\sigma \cdot \sigma, \tag{5.158}$$

where D_{t_1} is the Hirota operator. Writing (5.151) in terms of e_1, f_1, we find using the fact that $e_1 f_1 = -\tau_{11}/\tau + \tau_1^2/\tau^2$ (subscripts are $t_1 = x$ derivatives),

$$\tilde{e}_1 = -e_{1xx} + \frac{e_{1x}^2}{e_1} + e_1^2 f_1. \tag{5.159}$$

Also

$$\tilde{f}_1 = \frac{1}{e_1}. \tag{5.160}$$

Now let us imagine that we want to apply this Bäcklund–Schlesinger transformation many times. Let $(\tilde{e}_1, \tilde{f}_1) = (q_{n+1}, r_{n+1})$ and $(e_1, f_1) = (q_n, r_n)$.

Then, the successive application of the Bäcklund transformation which changes the monodromy of V by

$$\begin{pmatrix} -2i\zeta & \\ & \frac{1}{2i\zeta} \end{pmatrix}$$

gives the sequence

$$q_{n+1} = -q_{nxx} + \frac{q_{nx}^2}{q_n} + q_n^2 r_n, \tag{5.161a}$$

$$r_{n+1} = \frac{1}{q_n}. \tag{5.161b}$$

Set $q_n = e^{u_n}$ whence $r_n = u^{-u_{n-1}}$ and (5.161a) is

$$u_{nxx} = e^{u_n - u_{n-1}} - e^{u_{n+1} - u_n}. \tag{5.162}$$

If we call $x = it$, these are the equations for the Toda lattice! Thus, we have the remarkable result that the Toda lattice can be solved by applying successive Bäcklund transformations of a certain kind to the $\widetilde{sl}(2, C)$ hierarchy.

Needless to add, the Hirota equations for the Toda lattice are

$$\tau_n = \sigma_{n-1}, \qquad \rho_n = \tau_{n-1}$$

and (because $x = t_1 = it$)

$$\sigma_n = \frac{1}{2\tau_{n-1}}D_t^2\sigma_{n-1} \cdot \sigma_{n-1} = \frac{1}{2\sigma_{n-2}}D_t^2\sigma_{n-1} \cdot \sigma_{n-1}.$$

The Toda lattice analogy is useful as it allows us to visualize the effect of R in the following way. Imagine that instead of the triplet $\{\rho, \tau, \sigma\}$, we think of these three quantities as being the τ-function on the Toda lattice associated with sites $n-1$, n and $n+1$, i.e. $\tau_{n-1} = \rho$, $\tau_n = \tau$, $\tau_{n+1} = \sigma$. Under the Toda lattice rule, the τ function at the $(n+2)$nd site τ_{n+2} is given by

$$\tau_{n+2} = \frac{1}{2\tau_n} D_t^2 \tau_{n+1} \cdot \tau_{n+1}.$$

Let us apply R. Being a matrix, it does not act directly on scalars. We will therefore denote its effective action as R.

$$R \cdot \rho = \tau \Leftrightarrow R \cdot \tau_{n-1} = \tau_n,$$
$$R \cdot \tau = \sigma \Leftrightarrow R \cdot \tau_n = \tau_{n+1}.$$

Applying R to σ, we obtain

$$R \cdot \sigma = -\frac{1}{2\tau} D_{t_1}^2 \sigma \cdot \sigma$$

which is equivalent to

$$R \cdot \tau_{n+1} = \frac{1}{2\tau_n} D_t^2 \tau_{n+1} \cdot \tau_{n+1} = \tau_{n+2}.$$

The action of R on τ_{n-2} which is given by $(1/2\tau_n) D_t^2 \tau_{n-1} \cdot \tau_{n-1}$ is to shift the indices up by one and give τ_{n-1}. Thus,

$$R\{\ldots, \tau_{n-1}, \tau_n, \tau_{n+1}, \ldots\} = \{\ldots, \tau_n, \tau_{n+1}, \tau_{n+2}, \ldots\}. \tag{5.163}$$

The action of R is therefore best thought of as an action not on a triplet $\{\rho, \tau, \sigma\}$, but rather on the sequence $\{\tau_n\}_{-\infty}^{\infty}$.

Finally, let me tell you how to calculate matrices R which change the monodromy at $\zeta = \infty$. Simply take

$$R = \frac{1}{2}(I+H) \sum \frac{\alpha_r}{\zeta^r} + \frac{1}{2}(I-H) \sum \frac{\delta_r}{\zeta^r} + E \sum \frac{\beta_r}{\zeta^r} + F \sum \frac{\gamma_r}{\zeta^r}, \tag{5.164}$$

where the summations run from 0 to ∞, and write down the equations

$$RQ = \tilde{Q}R.$$

We find for the coefficient of ζ^{-n}

$$\sum_0^n \{\alpha_r(h_{n-r} - \bar{h}_{n-r}) + b_r f_{n-r} - c_r \tilde{e}_{n-r}\} = 0, \tag{5.165a}$$

$$\sum_0^n \{\delta_r(h_{n-r} - \bar{h}_{n-r}) - c_r e_{n-r} + b_r \tilde{f}_{n-r}\} = 0, \tag{5.165b}$$

$$\sum_0^n \{\alpha_r(e_{n-r} - \tilde{e}_{n-r}) = b_r(h_{n-r} + \bar{h}_{n-r})\} = 0, \tag{5.165c}$$

$$\sum_0^n \{\delta_r(f_{n-r} - \tilde{f}_{n-r}) + c_r(h_{n-r} + \bar{h}_{n-r})\} = 0. \tag{5.165d}$$

In order to change the monodromy of V by

$$\begin{pmatrix} -2i\zeta & 0 \\ 0 & \dfrac{1}{2i\zeta} \end{pmatrix},$$

look for solutions in which $\delta_r = 0$, α_r, β_r, γ_r are zero for $r \geq 2$ and $\beta_0 = \gamma_0 = 0$. Now solve the equations to find

$$\alpha_0 = \text{constant} = -2i, \qquad \alpha_1 = 2ie_2/e_1,$$
$$\beta_1 = e_1, \qquad \gamma_1 = \tilde{f}_1.$$

A little calculation shows that

$$\tilde{h}_{n-1} = h_{n-1} + \frac{i}{2} \frac{\partial^2}{\partial t_1\, \partial t_{n-2}} \ln e_1,$$

which gives

$$\tilde{\tau} = \tau e_1 = \sigma.$$

The other relations (5.157) and (5.158) also follow in a straightforward manner. In summarizing, we will adopt a more convenient notation. If

$$V \rightarrow V_+ = R_+ V \tag{5.166}$$

with

$$R_+ = \begin{pmatrix} -2i\zeta - \dfrac{\partial}{\partial t_1} \ln e_1 & e_1 \\ f_{1+} = \dfrac{1}{e_1} & 0 \end{pmatrix}, \tag{5.167}$$

then the asymptotic expansion of V_+ is

$$V_+ \sim \frac{1}{\tau_+} \begin{pmatrix} X_- \tau_+ & -\dfrac{i}{2\zeta} X_+ \sigma_+ \\ \dfrac{i}{2\zeta} X_- \rho_+ & X_+ \tau_+ \end{pmatrix} \begin{pmatrix} -2i\zeta & 0 \\ 0 & \dfrac{1}{2i\zeta} \end{pmatrix} \tag{5.168}$$

where

$$\tau_+ = \sigma, \quad \rho_+ = \tau, \quad \sigma_+ = -\frac{1}{2\tau} D_{t_1}^2 \sigma \cdot \sigma. \tag{5.169a}$$

Equivalently, in terms of the Toda lattice picture

$$\{\ldots, \tau_{n-1}, \tau_n, \tau_{n+1}, \ldots\}_+ = \{\ldots, \tau_n, \tau_{n+1}, \tau_{n+2}, \ldots\}. \tag{5.169b}$$

Note that the last equation is consistent with (5.67), the third equation in the Hirota set for τ, σ, ρ, because

$$D_{t_1}^2 \sigma \cdot \sigma = D_{t_1}^2 \tau_+ \cdot \tau_+ = -2\sigma_+ \rho_+ = -2\sigma_+ \tau.$$

Note also that the determinant of R is constant and equal to -1.

From (5.166) we find that the components (u_{1+}, u_{2+}) of a column in V_+ are related to the corresponding column in V by (recall $e_2 = i/2e_{1,t_1}$)

$$u_{1+} = \left(-2i\zeta + 2i\frac{e_2}{e_1}\right)u_1 + e_1 u_2, \qquad u_{2+} = \frac{1}{e_1}u_1. \tag{5.170}$$

The dual transformation,

$$V \to V_- = R_- V \tag{5.171}$$

with

$$R_- = \begin{pmatrix} 0 & \dfrac{1}{f_1} \\ f_1 & 2i\zeta - 2i\dfrac{f_2}{f_1} \end{pmatrix}, \tag{5.172}$$

changes the monodromy of V at $\zeta = \infty$ by the factor

$$\begin{pmatrix} -\dfrac{1}{2i\zeta} & \\ & 2i\zeta \end{pmatrix},$$

that is,

$$V_- \sim \frac{1}{\tau_-}\begin{pmatrix} X_-\tau_- & -\dfrac{i}{2\zeta}X_+\sigma_- \\ \dfrac{i}{2\zeta}X_-\rho_- & X_+\tau_- \end{pmatrix}\begin{pmatrix} -\dfrac{1}{2i\zeta} & 0 \\ 0 & 2i\zeta \end{pmatrix} \tag{5.173}$$

and

$$\tau_- = \rho, \qquad \sigma_- = \tau, \qquad \rho_- = -\frac{1}{2\tau}D_{t_1}^2\rho \cdot \rho. \tag{5.174a}$$

Equivalently, in terms of the Toda lattice picture

$$\{\dots, \tau_{n-1}, \tau_n, \tau_{n+1}, \dots\}_- = \{\dots, \tau_{n-2}, \tau_{n-1}, \tau_n, \dots\}. \tag{5.174b}$$

The columns of V_- are related to those of V by

$$u_{1-} = \frac{1}{f_1}u_2, \qquad u_{2-} = f_1 u_1 + \left(2i\zeta - 2i\frac{f_2}{f_1}\right)u_2. \tag{5.175}$$

The reader should calculate the equations analogous to (5.161).

Now we are going to use these Bäcklund–Schlesinger transformations in writing down the Bäcklund formula for adding solitons. Employing the strategy of the first example, we will ask that $R = R_L$ is chosen so that one of the two columns of $V_L = R_L V$ has a zero at $\zeta = \alpha$. If

$$R_L = \begin{pmatrix} -2i\zeta + a & b \\ c & d \end{pmatrix}, \tag{5.176}$$

with a, b, c, d independent of ζ, this means that

$$(-2i\alpha + a)u_1 + bu_2 = 0, \qquad cu_1 + du_2 = 0. \qquad (5.177)$$

From the Bäcklund transformation

$$RQ = \tilde{Q}R$$

which in component form is

$$-2i\zeta(\tilde{h} - h) = bf - \tilde{e}c, \qquad (5.178a)$$
$$(-2i\zeta + a)e = b(h + \tilde{h}) + \tilde{e}d, \qquad (5.178b)$$
$$(-2i\zeta + a)\tilde{f} = c(h + \tilde{h}) + fd, \qquad (5.178c)$$
$$ce = \tilde{f}b - d(\tilde{h} - h), \qquad (5.178d)$$

we obtain by equating coefficients of powers of ζ,

$$\begin{aligned}
\zeta^0: \quad & e_1 = b, \\
& \tilde{f}_1 = c, \\
& -2i(\tilde{h}_2 - h_2) = e_1 f_1 - \tilde{e}_1 \tilde{f}_1, \\
\zeta^{-1}: \quad & -2ie_2 + ae_1 = \tilde{e}_1 d, \\
& -2i\tilde{f}_2 + a\tilde{f}_1 = f_1 d, \\
& \tilde{e}_1 f_1 = \tilde{f}_1 e_1.
\end{aligned} \qquad (5.179)$$

From (5.177), (5.179), we have

$$\tilde{f}_1 = -d\frac{u_2}{u_1},$$

$$\tilde{e}_1 = \frac{1}{d}\left(-2ie_2 + 2i\alpha e_1 - e_1^2 \frac{u_2}{u_1}\right)$$

and

$$-2i(\tilde{h}_2 - h_2) = e_1 f_1 + \frac{u_2}{u_1}\left(-2ie_2 + 2i\alpha e_1 - e_1^2 \frac{u_2}{u_1}\right) = \frac{\partial^2}{\partial x^2}\ln u_1$$

after a small calculation using the equations satisfied by u_1, u_2; namely,

$$u_{1x} = -i\zeta u_1 + e_1 u_2, \qquad u_{2x} = i\zeta u_2 + f_1 u_1.$$

Since $h_k = (i/2)\,\partial^2 \ln \tau/\partial t_1\,\partial t_{k-1}$, this gives us that

$$\tilde{\tau} = \tau u_1. \qquad (5.180)$$

We also find

$$\tilde{\rho} = \tilde{f}_1 \tilde{\tau} = -\frac{du_2}{u_1}\cdot \tau u_1 = -d\tau u_2. \qquad (5.181)$$

We will show shortly that d must be constant; it is convenient for us to take it equal to -1. Finally

$$\tilde{\sigma} = \tilde{e}_1 \tilde{\tau} = \tau e_1 \left(-2i\alpha + 2i\frac{e_2}{e_1} + e_1\frac{u_2}{u_1} \right)$$

$$= \tilde{\tau} e_1 \frac{u_{1+}(\alpha)}{u_1} = \tau e_1 u_{1+}(\alpha) = \sigma u_{1+}(\alpha) = \tau_+ u_{1+}(\alpha). \qquad (5.182)$$

What a remarkable result! There is a close resemblance between the two formulae (5.180), (5.182). The difference is that the new σ is given in terms of the old τ, σ, ρ and u_1, u_2 which had undergone a plus Schlesinger transformation. On re-examining (5.181), we find from (5.175)

$$\tilde{\rho} = \tau u_2 = \tau f_1 u_{1-} = \rho u_{1-} = \tau_- u_{1-}. \qquad (5.183)$$

Again, the formula has the same form as (5.180) except it is applied to the (τ, σ, ρ) and (u_1, u_2) which have undergone a minus Schlesinger transformation.

The reason that d is constant is that det $\tilde{V} = -2id(\zeta - \alpha)$. Since this is also the Wronskian of the fundamental solution to the equation set $\tilde{V}_{t_k} = \tilde{Q}^{(k)}\tilde{V}$, it must be independent of all the t_k. The choice $d = -1$ renders the asymptotic expansion of $\tilde{V}$ similar to the asymptotic expansion for V except for the factor $2i(\zeta - \alpha)$ in the first column. In summary, then, the application of

$$R_L = \begin{pmatrix} -2i\zeta + 2i\alpha - e_1\dfrac{u_2}{u_1}(\alpha) & e_1 \\[2mm] \dfrac{u_2}{u_1}(\alpha) = \tilde{f}_1 & -1 \end{pmatrix} \qquad (5.184)$$

to V gives us a V_L which satisfies $V_{Lt_k} = Q_L^{(k)}V_L$, each k, with $Q_L^{(k)} = RQ^{(k)}R^{-1} + R_{t_k}R^{-1}$ and

$$R_L Q = Q_L R; \qquad (5.185)$$

also,

$$\tau_L = \tau u_1(\alpha),$$
$$\rho_L = \tau u_2(\alpha) = \tau_- u_{1-}(\alpha),$$
$$\sigma_L = \sigma u_{1+}(\alpha) = \tau_+ u_{1+}(\alpha). \qquad (5.186)$$

The vector u_1 refers to the column in V for which the corresponding column in V_L has a zero at $\zeta = \alpha$. Then u_2 is the other column. Usually, by an appropriate linear transformation, we will arrange things so that the left column in V_L has the zero; hence the use of the subscript L.

Now, since on the application of a plus (minus) Schlesinger transformation, $\tau_+ = \sigma(\tau_- = \rho)$, we can call $\sigma_L = \tau_{+L}(\rho_L = \tau_{-L})$. Then it is natural to rewrite (5.186) (calling $\tau = \tau_0$),

$$\tau_{-L} = \tau_- u_{1-}(\alpha),$$
$$\tau_{0L} = \tau_0 u_1(\alpha),$$
$$\tau_{+L} = \tau_+ u_{1+}(\alpha). \qquad (5.187)$$

Hence, a Bäcklund transformation which adds a bound state at $\zeta = \alpha$ can be expressed in a simple form, analogous to (4.99) for the KdV family for which the "eigenvalue" problem is in scalar (the Schrödinger equation) rather than in matrix form. The main τ function τ_0 transforms exactly as (4.99) and the auxiliary τ functions $\tau_-(\rho)$ and $\tau_+(\sigma)$ transform in a similar way after the application of minus and plus Bäcklund–Schlesinger transformations respectively.

It is straightforward to express (5.187) in terms of the operators X_+ and X_- acting on the old τ, σ, ρ because we know how $u(\alpha)$ is expressed in terms of these quantities. Recall from (5.86) that the canonical V has the form

$$V \sim \frac{1}{\tau} \begin{pmatrix} X_- \tau & -\dfrac{i}{2\zeta} X_+ \sigma \\ \dfrac{i}{2\zeta} X_- \rho & X_+ \tau \end{pmatrix}. \tag{5.188}$$

V_- and V_+ have asymptotic form

$$\frac{1}{\tau_-} \begin{pmatrix} -\dfrac{1}{2i\zeta} X_- \tau_- & X_+ \sigma_- \\ -\dfrac{1}{4\zeta^2} X_- \rho_- & 2i\zeta X_+ \tau_- \end{pmatrix} \tag{5.189a}$$

and

$$\frac{1}{\tau_+} \begin{pmatrix} -2i\zeta X_- \tau_+ & -\dfrac{1}{4\zeta^2} X_+ \sigma_+ \\ X_- \rho_+ & \dfrac{1}{2i\zeta} X_+ \zeta_+ \end{pmatrix} \tag{5.189b}$$

respectively. Hence when one takes (u_1, u_2) to be a linear combination of the columns of V in (5.188), one takes (u_{1-}, u_{2-}) and (u_{1+}, u_{2+}) to be the same linear combination of (5.189a) and (5.189b) respectively.

An example: the creation of a one-soliton solution. Let us apply this transformation once, starting with the trivial solution $\tau = 1$, $\sigma = \rho = 0$. Let (u_1, u_2) be the linear combination of A times column 1 and B times column 2 of V. Then

$$\rho_L = \tau_{-L} = -\frac{A}{2i\alpha} X_-(\alpha) \cdot \tau_- + BX_+(\alpha)\sigma_- = BX_+(\alpha) \cdot 1 \tag{5.190a}$$

because $\tau_- = \rho$ is zero and $\sigma_- = \tau = 1$.

$$\tau_L = \tau_{0L} = AX_-(\alpha) \cdot 1 \tag{5.190b}$$

and

$$\sigma_L = \tau_{+L} = 0. \tag{5.190c}$$

Hence $\rho_L = B \exp(i \sum \alpha^k t_k)$, $\tau_L = A \exp(-i \sum \alpha^k t_k)$, $\sigma_L = 0$. This corresponds to the solution

$$e_1 = 0, \qquad f_1 = \frac{B}{A} \exp\left(2i \sum \alpha^k t_k\right) \tag{5.191}$$

of the equation hierarchy.

In an exactly similar way we can create a zero in the determinant of RV at $\zeta = \bar{\alpha}$ by applying

$$R_R = \begin{pmatrix} -1 & \tilde{e}_1 = \dfrac{u_1}{u_2}(\bar{\alpha}) \\ f_1 & 2i\zeta - 2i\bar{\alpha} - f_1\tilde{e}_1 \end{pmatrix} \tag{5.192}$$

to V. The corresponding quantities are

$$\tau_R = \tau_{0R} = \tau_0 u_2(\bar{\alpha}),$$
$$\rho_R = \tau_{-R} = \tau_- u_{2-}(\bar{\alpha}), \tag{5.193}$$
$$\sigma_R = \tau_{+R} = \tau_+ u_{2+}(\bar{\alpha}).$$

Again, I remind the reader that if we take (u_1, u_2) to be a linear combination of the columns of the canonical V given in (5.188), then we must take the same linear combination of the columns of (5.189) for u_- and u_+.

Let us now apply (5.193) where the original τ_-, τ_0, τ_+ (ρ, τ, σ) is given by (5.190). A little calculation shows that

$$X_+(\xi) \cdot X_+(\zeta')f(t_k) = \left(1 - \frac{\zeta'}{\zeta}\right)^{1/2} \exp\left(i \sum (\zeta^k + \zeta'^k)t_k\right) f\left(t_k + \frac{i}{2k\zeta^k} + \frac{i}{2k\zeta'^k}\right),$$

$$X_+(\zeta) \cdot X_-(\zeta')f(t_k) = \left(1 - \frac{\zeta'}{\zeta}\right)^{-1/2} \exp\left(i \sum (\zeta^k - \zeta'^k)t_k\right)$$

$$\cdot f\left(t_k + \frac{i}{2k\zeta^k} - \frac{i}{2k\zeta'^k}\right) \tag{5.194}$$

where we have used the fact that

$$\exp\left(\pm \sum_1^\infty \frac{\zeta'^k}{2k\zeta^k}\right) = \left(1 - \frac{\zeta'}{\zeta}\right)^{\mp 1/2}. \tag{5.195}$$

We will take the u_2 in (5.193) to be C times the first column of V and D times the second. We find (we drop the subscripts R on τ, σ, ρ)

$$\tau = \left(1 - \frac{\alpha}{\bar{\alpha}}\right)^{1/2} DA \exp\left(-i \sum (\alpha^k - \bar{\alpha}^k)t_k\right)$$
$$\cdot \left(1 + \frac{i}{2\bar{\alpha}} \frac{CB}{DA} \exp\left(2i \sum (\alpha^k - \bar{\alpha}^k)t_k\right)\right),$$

$$\sigma = \left(1 - \frac{\alpha}{\bar{\alpha}}\right)^{1/2} CA \exp\left(-i \sum (\alpha^k - \bar{\alpha}^k)t_k\right) \exp\left(-2i \sum \bar{\alpha}^k t_k\right),$$

$$\rho = \left(1 - \frac{\alpha}{\bar{\alpha}}\right)^{1/2} 2i\bar{\alpha}BD \exp\left(-i \sum (\alpha^k - \bar{\alpha}^k)t_k\right) \exp\left(2i \sum \alpha^k t_k\right). \tag{5.196}$$

This is the one-soliton solution of the AKNS hierarchy. If we look at the special case $r = -q^*$, then $\bar{\alpha} = \alpha^*$ and with the choices

$$\frac{C}{D} = 2\alpha^* e^{2\eta x_0} e^{i\phi}, \qquad \frac{B}{A} = -i e^{2\eta x_0} e^{-i\phi},$$

the equations are

$$e_1 = -2\eta \operatorname{sech}\left(i\sum(\alpha^k - \alpha^{*k})t_k + 2\eta x_0\right)\exp\left\{-i\sum(\alpha_k + \alpha_k^*)t_k + i\left(\phi + \frac{\pi}{2}\right)\right\},$$

$$f_1 = 2\eta \operatorname{sech}\left(i\sum(\alpha^k - \alpha^{*k})t_k + 2\eta x_0\right)\exp\left\{i\sum(\alpha_k + \alpha_k^*)t_k - i\left(\phi + \frac{\pi}{2}\right)\right\}.$$

$$(5.197)$$

The reader should also verify that the formulae (5.196) are equivalent to the formulae one obtains from using Hirota's method (5.70). They are not exactly equal and differ only by the exponential factors with phases linear in the t_k which make no difference when one computes either ratios σ/τ or second log derivatives.

Note that exactly as in the case of KdV (see Section 4f), the phase shifts emerge as factors in the successive applications of the "vertex" operators.

In summary, then, the gauge transformation

$$V \to R_R(\bar{\alpha}_{\bar{N}})\cdots R_R(\bar{\alpha}_1)R_L(\alpha_N)\cdots R_L(\alpha_1)V$$

adds an $(N, \bar{N})$ bound state, which for $\bar{N} = N$ is an N-soliton state, to solutions of the AKNS hierarchy.

As the final remark in this section, let us examine the effect of repeated Bäcklund–Schlesinger transformations on an exact $(N, \bar{N})$ bound state solution. Suppose we normalize V so that it can be written in the form (5.100a). In particular the ζ^{-1} of the second component of V_2 and first component of V_1 are $(i/2)f_1$ and $-(i/2)e_1$ respectively, where the latter as functions of x have the shape of the $(N, \bar{N})$ bound state solution. Apply R_+, the plus Schlesinger transformation. This sends V as given by (5.100a) into R_+V. But we know from the definition of the plus Schlesinger transformation that its effect is to change the monodromy of V at $\zeta = \infty$ by the factor

$$\begin{pmatrix} -2i\zeta & 0 \\ 0 & \dfrac{1}{2i\zeta} \end{pmatrix}.$$

This means that the new V has the form $(-2i\zeta V_1, (1/2i\zeta)V_2)$ where V_1 and V_2 are given in (5.100a) with the vectors C_{1k}, C_{2k} given by exactly the same expressions as before with the strings e, f, h replaced by the new values $\tilde{e}, \tilde{f}, \tilde{h}$ attained under the Schlesinger transformation. The reader might check, for example, what happens to the vector

$$
\mathbf{V}_1 = \begin{pmatrix} 1 + \dfrac{i}{2\zeta} \displaystyle\int^x e_1 f_1 + \cdots + \dfrac{T_1}{\zeta^N} \\[2ex] \dfrac{i}{2\zeta} f_1 + \cdots + \dfrac{T_2}{\zeta^N} \end{pmatrix}
$$

when multiplied by

$$
R_+ = \begin{pmatrix} -2i\zeta - \dfrac{\partial}{\partial t_1} \ln e_1 & e_1 \\[2ex] \dfrac{1}{e_1} & 0 \end{pmatrix}.
$$

We find

$$
R_+ \mathbf{V}_1 = -2i\zeta \begin{pmatrix} 1 + \dfrac{i}{2\zeta} \displaystyle\int^x e_1 f_1 - \dfrac{i}{2\zeta} \dfrac{\partial}{\partial t_1} \ln e_1 + \cdots + \dfrac{T_1'}{\zeta^{N+1}} \\[2ex] \dfrac{i}{2\zeta} \dfrac{1}{e_1} + \cdots + \dfrac{T_2'}{\zeta^{N+1}} \end{pmatrix}.
$$

Recall that $\tilde{f}_1 = 1/e_1$, and check that

$$
\frac{i}{2\zeta} \int^x e_1 f_1 - \frac{i}{2\zeta} \frac{\partial}{\partial t_1} \ln e_1 = \frac{i}{2\zeta} \int^x \tilde{e}_1 \tilde{f}_1 + \frac{1}{\zeta} \int (\tilde{h}_2 - h_2)\, dx - \frac{i}{2\zeta} \frac{\partial}{\partial t_1} \ln e_1 = \frac{i}{2\zeta} \int^x \tilde{e}_1 \tilde{f}_1
$$

from the fact that $\tilde{h}_2 - h_2 = (i/2)(\partial^2/\partial t_1^2) \ln e_1$. (See the unnumbered equation after (5.165).) On the other hand

$$
R_+ \mathbf{V}_2 = -\frac{1}{2i\zeta} \begin{pmatrix} -\dfrac{i}{2\zeta} \tilde{e}_1 + \cdots + \dfrac{S_1}{\zeta^{\bar{N}-1}} \\[2ex] 1 + \cdots + \dfrac{S_2}{\zeta^{\bar{N}-1}} \end{pmatrix}.
$$

Therefore the new V corresponds to a solution $\tilde{e}_1, \tilde{f}_1$ which corresponds to an $(N+1, \bar{N}-1)$ bound state. After $\bar{N}$ applications of R_+, the second column of the new V has a second column $\binom{0}{1}$ which means that the new e_1, which we call $q_{\bar{N}}$ (the first e_1 is q, the second $\tilde{e}_1 = q_1$, and so on), is zero. But we know from (5.161) that $\bar{N}$ successive applications of R_+ solves the Toda lattice between the mass point which we label zero and the mass point we call $\bar{N}$.

Hence, if q is a $(N, \bar{N})$ bound state solution of the AKNS hierarchy, the motion of the mass point which we label zero in the lattice is given by the spatial shape of q. Further, the mass point labelled $\bar{N}$ will have solution $q_{\bar{N}} = 0$ which means that $u_{\bar{N}}$ defined by $\exp u_{\bar{N}} = q_{\bar{N}}$ is equal to $-\infty$. Therefore the successive application of a plus Schlesinger transformation on a $(N, \bar{N})$ bound state gives a sequence q_r, $0 \leq r \leq \bar{N}$, $q_0 = q$ whose shape as function of x is the time motion of the mass points labelled zero through $\bar{N}$ on a finite Toda lattice with free ends.

The set of differential-difference equations which are associated with the $sl(n + 1, C)$ flow via Schlesinger transformations has not yet been calculated.

We will again return to the subject of Bäcklund transformations at the end of 5j. There I will show how they relate to the Zakharov–Shabat "dressing" scheme and the method of reduction.

5h. The notion of grading. In general, the Kac–Moody algebra $A_1^{(1)}$ can be defined by giving six generators p_0, p_1, q_0, q_1, r_0, r_1 (for the algebra associated with $\widetilde{sl}(n + 1, C)$ we would need $3(n + 1)$) and their commutators as follows:

$$[q_i, r_j] = \delta_{ij} p_j,$$
$$[p_i, p_j] = 0,$$
$$[p_i, q_j] = A_{ij} q_j,$$
$$[p_i, r_j] = -A_{ij} r_j,$$
$$\mathrm{ad}_{q_i}^{1-A_{ij}} q_j = \mathrm{ad}_{r_i}^{1-A_{ij}} r_j = 0, \tag{5.198}$$

where the A_{ij} are the elements of the generalized Cartan matrix $\begin{pmatrix} 2 & -2 \\ -2 & 2 \end{pmatrix}$ of $A_1^{(1)}$ and no summation convention is implied. The expression $\mathrm{ad}_{q_i}^{1-A_{ij}} q_j$ means $[q_i, [q_i, \ldots, [q_i, q_j]]]$ with the commutator taken $1 - A_{ij}$ (in our case, three) times; i.e. $[q_i, [q_i, [q_i, q_j]]] = 0$, $i \neq j$. For example, consider the identification

$$
\begin{array}{cccccc}
p_0 & p_1 & q_0 & q_1 & r_0 & r_1 \\
-H + Z & H & F\zeta & E & E\zeta^{-1} & F
\end{array}. \tag{5.199}
$$

The commutator rules (5.198) are consistent with those already established for $X_j = h_j F_j + e_j E_j + f_j F_j$ with $H_j = \zeta^{-j} H$, $E_j = \zeta^{-j} E$, $F_j = \zeta^{-j} F$. Notice that $p_0 + p_1 = Z$ commutes with everything; it is called the center. Notice also how new "elements" are introduced; H_1 or ζH is produced by $[q_1, q_0]$; F_2 or $\zeta^2 F$ by $-\frac{1}{2}[[q_1, q_0], q_0]$ and so on. The reader might check that the last condition of (5.198) is satisfied.

When the center Z is added to the loop basis $\{H_j, E_j, F_j\}_{-\infty}^{\infty}$, the new set is called the *central extension* of the loop algebra $\widetilde{sl}(2, C)$.

We want to assign a weight W to each of the generators which is consistent with the commutation rules (5.198). For example, we could do this as follows:

$$
\begin{array}{cccccc}
p_0 & p_1 & q_0 & q_1 & r_0 & r_1 \\
0 & 0 & 1 & 0 & -1 & 0
\end{array}. \tag{5.200a}
$$

By adopting the rule that the weight of the commutator is the sum of the weights of its individual elements, we notice that the weighting assignment is indeed consistent; e.g. $W([q_1, r_1]) = 0 = W(p_1)$. Comparing the identification (5.199) with (5.200a), the equivalent weighting in terms of our basis is achieved by

$$H \quad E \quad F \quad \zeta$$
$$0 \quad 0 \quad 0 \quad 1 \quad . \tag{5.200b}$$

Notice that each of the blocks

$$h_j H_j + e_j E_j + f_j E_j$$

has equal weight. This is called the homogeneous grading.

But there are other possibilities. Consider the assignment

$$p_0 \quad p_1 \quad q_0 \quad q_1 \quad r_0 \quad r_1$$
$$0 \quad 0 \quad 1 \quad 1 \quad -1 \quad -1 \tag{5.201a}$$

which can be achieved by the same identification as (5.200a)

$$p_0 \quad p_1 \quad q_0 \quad q_1 \quad r_0 \quad r_1$$
$$0 \quad 0 \quad 1 \quad 1 \quad -1 \quad -1$$
$$-H+Z \quad H \quad F\zeta \quad E \quad E\zeta^{-1} \quad F \tag{5.201b}$$

except that we now assign the weights

$$W(H) = 0, \quad W(E) = 1, \quad W(F) = -1, \quad W(\zeta) = 2 \tag{5.201c}$$

to H, E, F and the grading parameter ζ.

How are the two gradings related? Consider the map which acts on the typical element $X(\zeta) = \sum_{\infty}^{-\infty} (h_j H + e_j E + f_j F)\zeta^{-j}$ of $\widetilde{sl}(2, C)$

$$X(\zeta) \rightarrow \begin{pmatrix} 1 & \\ & \lambda^{-1} \end{pmatrix} X(\lambda^2) \begin{pmatrix} 1 & \\ & \lambda \end{pmatrix} \tag{5.202a}$$

in which we identify the coefficients of h_j, e_j, f_j;

$$H\zeta^{-j} \rightarrow H\lambda^{-2j}, \tag{5.202b}$$

$$E\zeta^{-j} \rightarrow E\lambda^{-2j+1}, \tag{5.202c}$$

$$F\zeta^{-j} \rightarrow F\lambda^{-2j-1}. \tag{5.202d}$$

Let us use the identification to assign new weights to the H, E, F, ζ of the left-hand side given that, on the right-hand side, the weights are $W(H) = W(E) = W(F) = 0$, $W(\lambda) = 1$. Clearly the assignation must be (5.201c). Not surprisingly, then, there is an isomorphism between the two elements of $\widetilde{sl}(2, C)$ even though the basis vectors and grading parameter are weighted differently.

But how can this make any difference in the dynamics? The point is that different gradings induce different decompositions of the algebra. In the first grading, the terms $h_0 H$, $e_0 E$, $f_0 F$ all had the same weight, namely zero, and were therefore assigned to N. They also belonged to $N^* = K^\perp$ which was the phase space. In the second grading, the terms $e_0 E$ and $f_0 F$ belong to different subalgebras, the former has weight one and belongs to N, whereas the latter has weight -1 and must therefore be assigned to K. Now recall that this was exactly the same decomposition that we made in the third example at the end of Section 5c; namely.

$$N = \sum_{-1}^{-M} (h_j H + e_j E + f_j F)\zeta^{-j} + h_0 H + e_0 E, \qquad M \text{ arbitrary,}$$

$$K = \sum_{1}^{\infty} (h_j H + e_j E + f_j F)\zeta^{-j} + f_0 F.$$

Note that all terms in N have weights greater than or equal to zero; those in K have weights less than or equal to -1. The phase space $K^\perp$ has typical element

$$Q = h_0 H + f_0 F + \sum_{1}^{\infty} (h_j H + e_j E + f_j F)\zeta^{-j}$$

which can more conveniently be written as

$$= X_1 + \sum_{j=1}^{\infty} h_j Z_{2j} + \sum_{j=1}^{\infty} \frac{1}{2}(-f_j + e_{j+1}) X_{2j+1}$$

$$+ \sum_{j=1}^{\infty} \frac{1}{2}(f_j + e_{j+1}) Y_{2j+1}. \tag{5.203}$$

In (5.203)

$$Z_{2j} = H\zeta^{-j},$$

$$X_{2j+1} = \left(-F + \frac{E}{\zeta}\right)\zeta^{-j},$$

$$Y_{2j+1} = \left(F + \frac{E}{\zeta}\right)\zeta^{-j},$$

and the subscripts denote the inverse of the weights of each term. The reader will also recall from Section 5c(iii), that we have taken $h_0 = 0$, $-f_0 = e_1 = 1$, a choice consistent with the time flows which emerge.

Whereas the first grading leads naturally to the nonlinear Schrödinger (NLS) family of equations, the second grading leads naturally to the KdV and modified KdV families. I use the word "naturally" deliberately. It is certainly true that the NLS family contains the latter families but one has to constrain the phase space (either by $f_1 = -1$ for KdV or $f_1 = \pm e_1$ for MKdV) in order to obtain them. In the second grading, the equations simply appear without any constraints being imposed. The only constraint we did impose, which has a somewhat arbitrary look about it, was the choice of $h_0 = 0$. This is analogous to making the consistent choices $h_0 = -i$, $e_0 = f_0 = 0$ in the equations associated with the first grading. This rather minor arbitrariness can be removed by using as phase space $\varepsilon + K^\perp$ instead of $K^\perp$ where ε is distinguished element (with some constraints) in the dual K^* of K. We meet this idea in Sections 5i, 5j.

As a final comment to this section, we mention that all the independent gradings on the loop algebra $\widetilde{sl}(2, C)$ are determined by the automorphisms of finite order on $\widetilde{sl}(2, C)$. A finite order automorphism σ is a map on the algebra which preserves the Lie bracket i.e. $[\sigma(X), \sigma(Y)] = \sigma([X, Y])$, X, $Y \in \widetilde{sl}(2, C)$ such that σ^m is the identity for some integer m. All such maps are similarity transformations $\sigma(X) = aXa^{-1}$ for some a in $sl(2, C)$, $a^m = I$. For $sl(2, C)$, note that for $a = H$, σ acts as a linear transformation on the space of H, E, F and splits it into two subspaces, $\sigma(H) = H$, $\sigma(E, F) = (-E, -F)$. Note that in (5.203), the elements H and E, F appear as even and odd powers of the weighting respectively.

5i. A second Hamiltonian structure. I begin by reminding the reader that the Hamiltonian structure introduced at the beginning of Section 5c and the variational Hamiltonian structure introduced in Section 3b and by (5.56), (5.57) at the end of Section 5c are completely different. You will recall the variational Hamiltonian structure we found in Section 3b for the KdV hierarchy; namely

$$q_{t_{2n+1}} = N \frac{\delta H_{2n+1}}{\delta q} = -\frac{1}{4} M \frac{\delta H_{2n-1}}{\delta q}, \tag{5.204}$$

where N and M are given by (3.6). The two symplectic structures N and M are local (although also degenerate) in the sense that the application of either to a member of the phase space (the differential algebra consisting of q and all its x derivatives) keeps it there. On the other hand, the variational Hamiltonian structure for the AKNS hierarchy does not appear to admit two local structures. In Section 5d, we showed that $(q = e_1, r = f_1)$

$$\binom{q}{r}_{t_n} = -2iL^n \binom{q}{-r} \tag{5.205}$$

and, in [75], it is shown that this can be written as $J \nabla H_n$ where

$$\nabla = \left(\frac{\delta}{\delta q}, \frac{\delta}{\delta r} \right), \qquad J = \begin{pmatrix} 0 & 1 \\ -1 & 0 \end{pmatrix}$$

and H_n is a constant of the motion proportional to the ζ^{-n} coefficient in the asymptotic expansion of $\ln a(\zeta)$ (see Section 5f(i)) about $\zeta = \infty$. The operator L is given in Section 5d, and even though one might write (5.205) as

$$\binom{q}{r}_{t_n} = JL \nabla H_{n-1} = JL^2 \nabla H_{n-2}, \quad \text{etc.},$$

the symplectic operators JL, JL^2, etc. are no longer local and take us out of the original phase space which in the variational Hamiltonian structure consists of q, r and their x derivatives to all orders.

It is therefore interesting that the Hamiltonian framework associated with the algebraic approach of Section 5c naturally admits two local structures. The second one is obtained by defining the Killing form or inner product $\langle X, Y \rangle$ on G to be the ζ^{-1} coefficient, rather than the ζ^0 coefficient, in the trace of the product XY. We call it $\langle X, Y \rangle_{-1}$. Now, a little thought shows us that $K^{\perp} = K$, (also $N^{\perp} = N$). As before, there is a natural Hamiltonian structure on $K^{\perp}$ and, by translation, there will also be one on $K^{\perp} + \varepsilon$ where ε is a fixed element of G. We take ε to be in the orthogonal complements of both $[K, K]$ and $[N, N]$. It is clearly in the latter, for by writing $\varepsilon + K^{\perp}$, we imply that ε does not belong to $K^{\perp}$. Therefore it must belong to $N^{\perp}$ and the orthogonal complement of $[N, N]$. This will only be the case if $\varepsilon = X\zeta^0$, since $[K, K]$ contains only terms of $\zeta^p, p \leq -2$ and $[N, N]$ contains only terms $\zeta^p, p \geq 0$; the (new) inner product of ε with either is zero.

If ε satisfies this condition and if Φ is ad-invariant on G, the Hamiltonian vector field is given by

$$x_{\Phi}(X + \varepsilon) = [\pi_K \nabla \phi(X + \varepsilon), X + \varepsilon]. \tag{5.206}$$

Moreover, if Φ, Ψ are two ad-invariant functions on G, they are in involution with respect to the Poisson bracket on $K^{\perp} + \varepsilon$. For our purposes, we take

$$\varepsilon = -iH^-, \quad \Phi_k = -\tfrac{1}{2} \langle S^k X, X \rangle_0 = -\tfrac{1}{2} \langle S^{k-1} X, X \rangle_{-1} \tag{5.207}$$

where the subscripts 0 and -1 refer to which inner product we take. The Hamiltonian which gives rise to the t_k flow is then Φ_{k+1}. Then for $P \in K^{\perp} = K$, the flows are given by

$$P_{t_k} = -[\pi_k \nabla \Phi_{k+1}(\varepsilon + P), \varepsilon + P] = [\pi_N S^k(\varepsilon + P), \varepsilon + P] = [Q^{(k)}, \varepsilon + P]. \tag{5.208}$$

These are exactly the same Lax equations as given in (5.52) as $\varepsilon + P = Q$.

The major differences between the two approaches associated with the two Hamiltonian structures are (i) the Hamiltonians are shifted; and (ii) the ζ^0 element, which was constant as a consequence of the flows in the first approach, is fixed once and for all in the second. In our first grading $\varepsilon = -iH$, whereas in the second grading introduced in the Section 5h, $\varepsilon = -F + E/\zeta$. The choice of the second Hamiltonian structure allows us to avoid what appear to be rather arbitrary choices such as setting $e_0 = f_0 = 0$ in Section 5c and $h_0 = 0$ in Section 5h. I want to stress, however, that there is no essential difference between the two structures nor is there a significant advantage to be gained by using one over the other.

In the next section in which we discuss the reduction procedure, we use the first structure. There we have $K^{\perp} = N^*$ ($X = \sum_0^{\infty} X_j \zeta^j$), $K^* = N^{\perp}$ ($X = \sum_1^{\infty} X_j \zeta^j$) and we will take ε, which belongs to K^* to be $-iH\zeta$. The typical element in our phase space will then be simply ζ times the Q we used earlier,

$$-iH\zeta + Q_1 + \frac{Q_2}{\zeta} + \cdots = Q\zeta$$

where Q, Q_1, Q_2, etc. are exactly as before.

5j. Inverse scattering and the Riemann–Hilbert problem, algebraic style.
By this stage, I hope you have been convinced that the special qualities enjoyed by integrable systems are algebraic in character. It is therefore

reasonable to ask if there is an algebraic analogue to the inverse scattering transform and the Riemann–Hilbert problem. There is. The main idea is that the Lax equation (5.52) is a reduction of a simpler flow on a larger manifold, the reduction being achieved by using constants of the motion of the simpler flow and their corresponding symmetries to obtain a smaller phase space. The price one pays for the smaller phase space is that the simple flow does not look so simple any more.

In Section 4c, I pointed out that to each symmetry of a Hamiltonian system, there corresponds a constant of the motion (Noether's theorem) and vice versa. The well-known symmetries, like the invariance of the Hamiltonian under translation or rotation, give rise to linear and angular momentum conservation. To a group of symmetries, there corresponds a vector of momenta which are constants of the motion and which are in involution with each other under the Poisson bracket associated with the flow manifold. It is a theorem that if there are n independent symmetries and thereby n constants of the motion in involution, then $2n$ of the $2m$ variables in the phase space can be eliminated. If $n = m$, the phase space reduces to a single point and the motion is exactly integrable. Such a system is called *completely integrable*. This classical theorem has been put in a general framework by Marsden and Weinstein [88] and the process of eliminating variables through use of the symmetries is called *reduction*. Roughly speaking the method consists of the following steps. For a complete account the reader should consult Abraham and Marsden [84], Arnold [105] and the forthcoming book by Marsden, Ratiu, Weinstein, Schmidt and Spencer.

Suppose we are given a symplectic manifold P with a group of symmetries $\bar{G}$ acting on P by canonical transformations. The two-form ω on P is thereby preserved by the action of $\bar{G}$. Let G be the Lie algebra of $\bar{G}$ and G^* be its dual. Then the *momentum map* J assigns to a point in P a value in G^*; namely J is a vector belonging to G^* listing all the constants of the motion. The level set $J^{-1}(\mu)$, namely the set of points p in P for which $J(p) = \mu$ is a manifold and invariant under the isotropy subgroup $\bar{G}_\mu$ of $\bar{G}$ under the coadjoint action; namely $\bar{G}_\mu$ is the set of $g \in \bar{G}$ such that (in matrix representation) $g\mu g^{-1} = \mu$. Then the Marsden–Weinstein theorem states that the quotient $P_\mu = J^{-1}(\mu)/\bar{G}_\mu$ is a symplectic manifold with its symplectic form ω_μ induced by ω. P_μ is the reduced phase space. Moreover, if Φ is a Hamiltonian function on P, invariant under the action of $\bar{G}$, then it induces a Hamiltonian Φ_μ on P_μ. If F_t is the flow of the Hamiltonian vector field corresponding to Φ, then by the Ad*-equivariance of J (Ad*-equivariance means that $J(g \cdot p) = \text{Ad}^*_{g^{-1}} J(p) = g\mu g^{-1}$ where $g \cdot p$ denotes the action of $\bar{G}$ on P), the level set $J^{-1}(\mu)$ is invariant under F_t and F_t induces a flow F_t^μ of symplectic diffeomorphisms on the reduced manifold P_μ. The second part of the Marsden–Weinstein theorem states that F_t^μ is the flow of the Hamiltonian vector field corresponding to Φ_μ.

We shall also be using a second type of reduction, a Poisson reduction. Given a symplectic manifold P with a group of canonical symmetries $\bar{G}$, the quotient manifold $P/\bar{G}$ has a natural Poisson bracket induced by the one on P. Moreover the Hamiltonian vector field corresponding to Φ on P and to $\hat{\Phi}$ on

$P/\bar{G}$, $\hat{\Phi}([p]) = \Phi(p)$ ([p] is the class of $p \in P$ in $P/\bar{G}$) are related by the projection map $P \to P/\bar{G}$ which, if canonical, preserves Poisson brackets.

In our case, we take the original manifold to the $T^*\bar{G}$, the cotangent bundle of a group $\bar{G}$, the Lie group corresponding to the loop algebra $G = \widetilde{sl}(2, C)$. (I mean by this that every element in $\bar{G}$ is the exponential $\exp t\xi$ of some element ξ in the algebra. Later considerations, which have to do with the Bäcklund–Schlesinger transformations of Section 5g and which will be discussed in the summary of this section, suggest that it is probably appropriate to augment $\bar{G}$ to include certain discrete symmetries in addition to the continuous ones, analogous to including the action of reflection when constructing $O(3)$ from $so(3)$. But the present definition will suffice for now.) The manifold $T^*\bar{G}$ is naturally symplectic and consists of a base space $\bar{G}$ at every point g of which a fibre, the dual to the tangent space at g is attached. In the vernacular of classical mechanics, $\bar{G}$ is the set of position coordinates (the q's), the tangent space at g is the set of velocities (the q''s) and the fiber is the space of momenta (the p's). Every member g of $\bar{G}$ can be uniquely factored (the analogue of the Riemann–Hilbert problem) into a product

$$g = k^{-1}n \tag{5.209}$$

where k and n are exponentials of elements in the subalgebras K and N respectively. The choice of writing the left factor as an inverse is merely one of convenience. The group of symmetries with respect to which we perform the Marsden–Weinstein reduction on the phase space $T^*\bar{G}$ will be $\bar{K}$, the subgroup corresponding to K. The reduced manifold will be $\bar{N} \times (\varepsilon + N^*)$, where ε is a single, distinguished element in K^* which will shortly be identified. Then the trivial application of the Poisson reduction by the $\bar{N}$ action leaves us with $\varepsilon + N^*$, the phase space of Section 5c. The time flows in $\varepsilon + N^*$ induced by the Hamiltonians Φ_j defined in (5.48) can be "integrated" to give

$$Q(t_j) = k(t_j)(-iH)k^{-1}(t_j) \tag{5.210}$$

where k^{-1} is the left factor of $g(t_j)$. We will see that $g(t_j)$ evolves very simply in time

$$g(t_j) = \exp\left(-i\sum \zeta^i t_j H\right) g_0 \tag{5.211}$$

where $g_0 = k^{-1}(0)$ and $Q(0) = k(0)(-iH)k^{-1}(0)$.

We will elaborate this procedure in five steps. First, we will trivialize $T^*\bar{G}$ as $\bar{G} \times G^*$ and thereby endow it with coordinates. Second, we will take our original Hamiltonians Φ_j of Section 5c, which are Ad*-invariant on G^*, and extend their definitions to all of $T^*\bar{G}$. We will then use the natural symplectic structure on $T^*\bar{G}$ to find the flow induced by Φ_j. The flow on $T^*\bar{G}$ will look very like that satisfied by the scattering matrix (5.98). Third, we will find the flow on the reduced manifold $\bar{N} \times (\varepsilon + N^*)$ ($\bar{N}$ is the subgroup of $\bar{G}$ associated with the subalgebra N and N^* is the dual of N) with the canonical symplectic structure. The Poisson reduction to $\varepsilon + N^*$ is then trivial. The typical element of the second component in $\bar{N} \times (\varepsilon + N^*)$, which will be ζ times the Q of (5.45), will evolve according to (5.210) and satisfy the Lax equation (5.52). In the fourth step, we will identify $k(t_j)$ of (5.210), the inverse of the left factor of $g(t_j)$, with $\hat{V}(t_j)$ defined in 5e.

Finally, in step 5, we discuss how to solve the Lax equation (5.52) in an algebraic way. The most important step as we shall see is the one in which we factor a group element g into $k^{-1}n$. This factorization is the algebraic analogue of the Riemann–Hilbert problem which you will recall was the principal step in the construction of the fundamental solution matrix Φ in Section 5f(i) from the scattering data. Having found k, we will then have the solution to the Lax equation. Because these steps require the introduction of a lot of new mathematical ideas and notation, I will attempt to discuss the results in the language already familiar to the reader of this book. The reader interested in studying these details further is referred to paper IV in our series "Kac–Moody Lie algebras and soliton equations" [38].

I want to point out that the notion that the AKNS flows are reductions of simpler flows in a larger manifold is not new with us. I refer the reader to the papers of Reyman and Semenov–Tian–Shansky [106]; also the ideas are very similar to those used by Kostant, Kazhdan, Sternberg [100], Moser [107] in connection with the Toda lattice, the Calogero and the Moser–Sutherland systems; they are also closely related to the "dressing" scheme of Zakharov and Shabat [108]. What is new with us (Flaschka, Ratiu and the present author) is the incorporation of the flows corresponding to the times t_k, $k < 0$, (the "sine-Gordon" flows) into the general picture. I discuss this in the following section. At this time, I want to elaborate further on steps one through five.

Step 1. We (right) trivialize $T^*\bar{G}$ by identifying it with $\bar{G} \times G^*$ where G^* is the dual of the Lie algebra $G = T_e\bar{G}$, the tangent space to $\bar{G}$ at the identity. The identification proceeds as follows. Take a curve $e^{t\xi}$, $\xi \in G$, through the identity of $\bar{G}$ and take its tangent vector there, $(d/dt)e^{t\xi}|_{t=0}$. Right translate the tangent vector by g and call this element $T_eR_g\xi$. In a matrix representation, where $\xi \in sl(2, C)$ and g has determinant one, $T_eR_g\xi$ is simply ξg. Then we give $\mu_g \in T^*\bar{G}$, which lives in the fiber above g, the coordinates (g, μ), $\mu \in G^*$, where

$$\langle \mu, \xi \rangle = \langle \mu_g, T_eR_g\xi \rangle \qquad (5.212)$$

with $\langle \cdot, \cdot \rangle$ the pairing between G and G^*. The right-hand side can be written $\langle T^*R_g\mu_g, \xi \rangle$ whence $\mu = T^*R_g\mu_g$; i.e. μ is the element in G^*, the fiber at the identity, which under right translation by g takes (e, μ) to μ_g.

Step 2. Let $\Phi(\mu)$ be an Ad*-invariant function on G^*; i.e. $\Phi(g^{-1}\mu g) = \Phi(\mu)$. You will recall in Section 5c that we very quickly identified G^* with G because we could define an inner product on G itself. In that case $\Phi(X)$ was Ad*-invariant on G if $\Phi(e^{tY}Xe^{-tY}) = \Phi(X)$ for all $X, Y \in G$, t real. This condition was expressed as $[\nabla\Phi(X), X] = 0$. Here the same notion says that Φ is Ad*-invariant on G^* if $\Phi(e^{-tY}Xe^{tY}) = \Phi(X)$ for all $X \in G^*$, $Y \in G$. We extend $\Phi(\mu)$

to $T^*\bar{G}$ by

$$\bar{\Phi}(\mu_g) = \Phi(\mu). \tag{5.213}$$

The flow on $T^*\bar{G}$ can now be worked out; the details are given in [38, IV]. It is simply the straight line flow

$$\mu^{\cdot} = 0, \tag{5.214a}$$

$$g^{\cdot} = T_e R_g \frac{\delta\Phi}{\delta\mu}, \tag{5.214b}$$

which in matrix notation can be written

$$g^{\cdot} = \frac{\delta\Phi}{\delta\mu} g. \tag{5.214c}$$

Here $\delta\Phi/\delta\mu$, an element in G, is the gradient of Φ at μ; i.e. $D\Phi(\mu) = \langle \delta\Phi/\delta\mu, \mu \rangle$ where $D\Phi(\mu)$ is the Fréchet derivative of Φ at μ. Integrating (5.214), we obtain

$$\mu = \mu_0, \tag{5.215a}$$

$$g = \exp\left(t\frac{\delta\Phi}{\delta\mu_0}\right)g_0. \tag{5.215b}$$

Notice the connection with the time flow of action-angle variables. The action variables (the new momenta) are constants of the motion whereas the arguments of the new position coordinates, the angle variables, change linearly in time. This is just like the behavior of the scattering data in (5.98). Without loss of generality, we can take the matrix representation of μ_0 to be $-iH\zeta$. Then, if $\Phi(\mu)$ is Φ_{j-1}, the component of ζ^{-j+1} in the expansion of $-(h^2 + ef)$, it is Ad*-invariant and, with respect to the $\langle \cdot, \cdot \rangle_{-0}$ Killing form,

$$\frac{\delta\Phi_{j-1}}{\delta\mu_0} = -i\zeta^j H. \tag{5.216}$$

Then, as function of all the times,

$$g = \exp\left(-i\sum \zeta^i t_i H\right)g_0. \tag{5.217}$$

Step 3. We next carry out the symplectic reduction of $T^*\bar{G}$ by $\bar{K}$ at ε which belongs to $K^* = N^\perp$. It turns out to be more convenient to use the first Hamiltonian structure and the special element ε introduced there is $-iH\zeta$. I will not go through the calculations in details but simply sketch the results. The reader should refer to [38, IV] for a more detailed discussion.

As mentioned already, we trivialize $T^*\bar{G}$ as $\bar{G} \times G^*$. Let ϕ be the map that assigns the coordinates $(g, \mu = T_e^* R_g \mu_g)$ to the element μ_g in $T^*\bar{G}$. $\bar{K}$ is the group of symmetries by which we will reduce $T^*\bar{G}$. $\bar{K}$ acts on the left on $T^*\bar{G}$ and in the trivialization for $k \in \bar{K}$,

$$(k, g, \mu) \rightarrow (kg, \text{Ad}^*_{k^{-1}}\mu) \tag{5.218}$$

which in matrix representation is $(kg, k\mu k^{-1})$. The left action of $\bar{K}$ on $T^*\bar{G}$ has a *momentum map*

$$J : T^*\bar{G} \to K^*,$$

which in coordinates is

$$\mu_g \to \mu|_K \tag{5.219}$$

where $\mu|_K$ means that $\mu = T_e^* R_g \mu_g$ must lie in K^* and therefore must be restricted to K. If we identify G^* with G, K^* with $N^\perp$, this means that the inner product of $\mu|_K$ with any element of N is zero. In the trivialization, the momentum map is

$$\tilde{J} = J \cdot \phi^{-1} : (g, \mu) \to \mu|_K. \tag{5.220}$$

Now introduce the level set $\tilde{J}^{-1}(\varepsilon)$ of the element ε which belongs to $K^* = N^\perp$.

$$\tilde{J}^{-1}(\varepsilon) = \{(g, \mu) | \mu|_K = \varepsilon\}$$

which, because we have now identified G^* and G, can be written as

$$\tilde{J}^{-1}(\varepsilon) = \bar{G} \times \{\varepsilon + \nu\} \tag{5.221}$$

where ν is any element of $K^\perp$.

Next we identify the isotropy subgroup $\bar{K}_\varepsilon$ of ε. It can be shown to be $\bar{K}$. It is important that ε is chosen so that $\varepsilon|_{[K,K]} = 0$. Then in $\tilde{J}^{-1}(\varepsilon)/\bar{K}_\varepsilon$, any element $(g = k^{-1}n, \varepsilon + \nu)$ of $\tilde{J}^{-1}(\varepsilon)$ is equivalent, under the action of $\bar{K}$, to an element whose first coordinate is in $\bar{N}$; i.e.

$$(k^{-1}n, \varepsilon + \nu) \sim (n, \lambda).$$

I have told you already that it is convenient to write g as $k^{-1}n$. Now, what is λ? In order to take the k^{-1} away from $k^{-1}n$ in the first element, it is natural to act on $(k^{-1}n, \varepsilon + \nu)$ with k. But we already know this action to be

$$k(k^{-1}n, \varepsilon + \nu) = (n, \mathrm{Ad}_{k^{-1}}^*(\varepsilon + \nu)). \tag{5.222}$$

It is not difficult to show that $\mathrm{Ad}_{k^{-1}}^*(\varepsilon + \nu)$ still belongs to $\varepsilon + N^*$. In matrix notation, it is $k(\varepsilon + \nu)k^{-1}$; it is clear that both k and k^{-1} can be written as power series in ζ^{-j} beginning with the identity at $j = 0$ and since $\varepsilon + \nu$ belongs to $\varepsilon + N^*$, the product $k(\varepsilon + \nu)k^{-1}$ does also.

The final (Poisson) reduction by the right action of $\bar{N}$ simply removes the group element n from $\bar{N} \times (\varepsilon + N^*)$. Because the actions $\bar{K}$ and $\bar{N}$ commute (they act on different sides) and the $\bar{N}$ action leaves $\tilde{J}^{-1}(\varepsilon)$ invariant, this reduction is trivial. Hence the general element in the reduced phase space $\varepsilon + N^* = \varepsilon + K^\perp$ is given by $k(\varepsilon + \nu)k^{-1}$.

The time flows of k and $\varepsilon + \nu$ are given by (5.215a) and (5.217); namely $k^{-1}(t_j)$ is the left factor of $\exp(-i \sum t_j \zeta^j H)g_0$ and $\varepsilon + \nu$ is a constant of the motion which we have taken to be $\varepsilon = -iH\zeta$. Hence the motion of the point μ, which we call $\zeta Q(t_j)$ in the reduced phase spaces $\varepsilon + N^*$ is, after factoring out the ζ,

$$Q(t_j) = k(t_j)(-iH)k^{-1}(t_j). \tag{5.223}$$

Step 4. It is now useful to identify the element $k(t_j)$ with elements we have already met. A little calculation of (5.223) shows that

$$Q_{t_j} = [k_{t_j}k^{-1}, Q].$$ (5.224)

But from (5.217),

$$g_{t_j} = -i\zeta^j Hg = -k^{-1}k_{t_j}k^{-1}n + kn_{t_j},$$

which is also

$$\zeta^j k(-iH)k^{-1} = -k_{t_j}k^{-1} + n_{t_j}n^{-1}.$$ (5.225)

Each term in (5.225) is an element of G. Take the projection into N and find

$$n_{t_j}n^{-1} = \prod_N \zeta^j k(-iH)k^{-1} = \prod_N \zeta^j Q(t_j) = Q^{(j)},$$ (5.226a)

as defined in (3.48) and (5.52). Hence

$$k_{t_j} = Q^{(j)}k + k(iH\zeta^j).$$ (5.226b)

Now recall the eigenfunction V from Section 5e. In order to compute its asymptotic expansion, we wrote it as $\hat{V}\exp(-i\sum \zeta^j t_j H)$ whence $\hat{V}$ satisfies (5.226b). In fact (look at (5.88)) $Q = V(-iH)V^{-1} = \hat{V}(-iH)\hat{V}^{-1}$ because the exponential factor in V commutes with H. Therefore the element k is simply $\hat{V}$, which is the left factor of V when written as an asymptotic expansion about $\zeta = \infty$. Next look at (5.224). It is

$$Q_{t_j} = [Q^{(j)}, Q],$$ (5.227)

the Lax equation (5.52) for the $\widetilde{sl}(2, C)$ family. Thus the time flows of the element ζQ belonging to $\varepsilon + N^*$, obtained as a reduction of the linear flow on $T^*\bar{G}$ are the same as those defined by the Hamiltonian vector fields in Section 5c.

Step 5. How, then, do we "solve" (5.227)? Given $Q(0)$, calculate $k(0)$ which is defined to be

$$Ad^*_{k^{-1}(0)}(-iH) = Q(0)$$ (5.228)

i.e. $k(0)(-iH)k^{-1}(0) = Q(0)$. It is the fact that we can always find a $k(0)$ so that $\zeta Q(0)$ is similar to $-iH\zeta$ that allows us to take $\varepsilon = -iH\zeta$ with no loss of generality. We may also take $n(0) = 1$; the identity $k(0)$ defined in this way will not be unique because it can be multiplied on the right by any factor which commutes with H. This does not matter. We can take any $k(0)$ in the equivalence class because the right factor which commutes with H in $k(0)$ becomes a left factor in $k^{-1}(0)$ and therefore also commutes with the time dependence of g in (5.229) below. The factor therefore simply carries over into a constant right multiple of $k(t_j)$ and therefore disappears when we calculate $Q(t_j)$ from (5.223). The reader may have already noticed that $k(t_j)$ which we have identified as

$$\hat{V}(t_j) = \begin{pmatrix} 1 & -\dfrac{i}{2}b \\ \dfrac{i}{2}c & 1 \end{pmatrix} e^{\psi H + \phi}$$

already has a right factor which commutes with H. When we compute $Q = \hat{V}(-iH)\hat{V}^{-1}$, this factor, which is made up of nonlocal terms, disappears. Recall that these nonlocal terms are expressed as the first logarithmic derivatives of τ and therefore are integrals of the phase space coordinates h_r, e_r, f_r.

The solution to (5.227) is (5.223), where k is the inverse of the left factor in

$$g = \exp\left(-i\sum \zeta^i t_j H\right) k^{-1}(0), \tag{5.229}$$

which can be calculated. The factorization is not always an easy task and I will not give a proof that it can be done. One begins by writing (5.209) as

$$kg = n$$

and then taking the projection into $\bar{K}$

$$\prod_{\bar{K}} kg = 0. \tag{5.230}$$

These are the equations equivalent to (5.99). For the N-soliton solution, equation (5.229) is a system of nonhomogeneous linear equations of order $2N$. It turns out that the coefficient matrix is the N-soliton τ-function, the τ of (5.70). For a general solution, the system of linear equations is of infinite order and the determinant of the coefficient matrix, which is the τ-function, is of infinite order.

Therefore, we now have a second way of introducing and defining the τ-function. You will recall that we first introduced him in 5d as a *potential*. Here he is an infinite order determinant arising from the solution of the Riemann–Hilbert problem. The reader will recall from 5g that the auxiliary τ-functions σ, ρ can be constructed from τ through Bäcklund–Schlesinger transformations.

Let me illustrate that (5.230) holds in the case of a simple example. This example corresponds to the simplest form of $(N, \bar{N})$ bound state discussed in Section 5g, i.e. $N = 1$, $\bar{N} = 0$. Then from (5.99a,b) (I normalize the eigenfunctions $\bar{\psi} \sim \binom{1}{0}e^{-i\theta}$, $\psi \sim \binom{0}{1}e^{i\theta}$, $x \to +\infty$, $\theta = \sum \zeta^i t_j$)

$$\bar{\psi}(\zeta, t_j)e^{i\theta} = \binom{1}{0} + \frac{\gamma}{\zeta - \zeta_1}\psi_1 e^{i\theta_1},$$

$$\psi(\zeta, t_j)e^{-i\theta} = \binom{0}{1}, \qquad \theta_1 = \sum \zeta_1^i t_j,$$

whence $\psi_1 = \binom{0}{1}e^{i\theta_1}$. Thus,

$$k = V \exp\left(i \sum \zeta^i t_i H\right) = \begin{pmatrix} 1 & 0 \\ \dfrac{\gamma e^{2i\theta_1}}{\zeta - \zeta_1} & 1 \end{pmatrix}.$$

We can find $f_1(t_j)$ by recalling that the first term in the asymptotic expansion of the $(2, 1)$ element in k is $if_1/2\zeta$, giving

$$f_1(t_j) = -2i\gamma e^{2i\theta_1}.$$

Note $f_{1,t_j} = -2i\gamma(2i\zeta_1^j)e^{2i\theta_1} = (-i/2)^{j-1}f_{1,t_1\dots t_1}$, where $f_{1,t_1\dots t_1}$ denotes j derivatives of f_1 with respect to $t_1 = x$. e_1, and therefore e, is identically zero. These are the Lax equations in this case.

Let us check that the projection of kg into $\bar{K}$ is zero. Consider,

$$kg = k(t_j)e^{-i\theta H}k^{-1}(0) = \begin{pmatrix} e^{-i\theta} & 0 \\ \gamma e^{i\theta}\dfrac{e^{2i(\theta_1-\theta)}-1}{\zeta-\zeta_1} & e^{i\theta} \end{pmatrix}.$$

You will notice that the pole at ζ_1 is removable and that kg has a Taylor series expansion about ζ_1. It therefore belongs to $\bar{N}$. Further, let us also check

$$Q = k(t_j)(-iH)^{-1}k(t_j) = \begin{pmatrix} -i & 0 \\ \dfrac{-2i\gamma}{\zeta-\zeta_1}e^{2i\theta_1} & i \end{pmatrix} = -iH + \dfrac{f_1}{\zeta-\zeta_1}F.$$

But we know

$$f = \sum_1^\infty \frac{f_j}{\zeta^j} = \frac{1}{\zeta}f_1 - \frac{i}{2}\sum_1^\infty \frac{f_{1,t_j}}{\zeta^{j+1}},$$

and, since $f_{1,t_j} = 2i\zeta_1^j f_1$,

$$f = \frac{1}{\zeta-\zeta_1}f_1.$$

Since $e \equiv 0$, $h^2 = -1$ or $h = -i$. Thus (5.223) is

$$Q = -iH + OE + fF.$$

More on Bäcklund transformations; the Zakharov–Shabat [108] *"dressing" scheme.* It is useful to observe that the form of (5.223) is very similar to the form of (5.134),

$$\tilde{Q} = RQR^{-1},$$

for Bäcklund transformations. Also (5.226b) and (5.133) are exactly the same if we write R for k. Indeed in order to add one bound state at $\zeta = \zeta_1$ to the vacuum

$$Q(t_j) = -iH, \qquad V(t_j, \zeta) = e^{-i\theta H},$$

with $\theta = \sum_1^\infty \zeta^i t_j$, we use (5.184)

$$R_L = \begin{pmatrix} -2i(\zeta - \zeta_1) & e_1 = 0 \\ f_1 = -2i\gamma e^{2i\theta_1} & -1 \end{pmatrix},$$

which can be normalized to

$$R = \begin{pmatrix} 1 & 0 \\ \dfrac{\gamma e^{2i\theta_1}}{\zeta - \zeta_1} & 1 \end{pmatrix}.$$

Interpreting $(\zeta - \zeta_1)^{-1}$ as $\sum_0^\infty (\zeta_1^i / \zeta^{i+1})$ means that this R has exactly the form, $I + \sum_1^\infty R_i \zeta^{-i}$. Therefore it can be expressed as the exponential of an element $\sum_1^\infty X_i \zeta^{-i}$ in the subalgebra K and hence belongs to $\bar{K}$. The reader should contrast this with the structure of the R used in the Bäcklund–Schlesinger transformation which does not have this form.

A fundamental difference between (5.223) and (5.134) is, of course, that the former tells us how an initial state $Q(0)$

$$Q(0) = k(0)(-iH)k^{-1}(0) \qquad (5.231)$$

evolves under the sequence of flows $\{t_k\}_1^\infty$ whereas the latter relates two different solution types at fixed values of t_k, $k = 1, 2, \dots$. Nevertheless, they are intimately related and I will now show, with the aid of (5.231) and (5.223), that we can cast the Bäcklund transformation in the language of the reduction method.

Let us begin, for example, from the vacuum state (for which we use the subscript 0)

$$Q_0(0) = -iH, \qquad V_0(0, \zeta) = I$$

which evolves under the flows to

$$Q(t_j) = -iH, \qquad V(t_j, \zeta) = e^{-i\theta H}.$$

On the other hand, if we begin with the initial state

$$Q_1(t_j) = -iH - \frac{2i\gamma}{\zeta - \zeta_1} F,$$

then its time evolved state is

$$Q_1(t_j) = -iH - \frac{2i\gamma e^{2i\theta_1}}{\zeta - \zeta_1} F$$

where $\theta_1 = \sum_1^\infty \zeta_1^i t_j$. Its corresponding eigenmatrix V_1 is

$$V_1 = \hat{V}(t_j) e^{-i\theta H} = k(t_j) e^{-i\theta H}$$

where $k(t_j)$ is the inverse of the left factor in the decomposition of

$$e^{-i\theta H} k^{-1}(0)$$

which I denote

$$(e^{-i\theta H}k^{-1}(0))_-.$$

Notice that this is also

$$(e^{-i\theta H}k^{-1}(0)e^{i\theta H})_-$$

since $e^{i\theta H}$ belongs to $\bar{N}$ and any member of $\bar{N}$ can be multiplied on the right under the minus subscript; i.e. $(k^{-1}nn')_- = (k^{-1}n)_- = k^{-1}$.

These observations suggest an algorithm for constructing new solutions out of the vacuum state. Take an element $k^{-1}(0)$ belonging to $\bar{K}$ which, as I have mentioned, relates two solution types at time zero (here a one-bound state with the vacuum). Form the product

$$(e^{-i\theta H}k^{-1}(0)e^{i\theta H})$$

and take the inverse of its left factor

$$\{(e^{-i\theta H}k^{-1}(0)e^{i\theta H})_-\}^{-1}.$$

Then the eigenmatrix

$$V_1 = \{(e^{-i\theta H}k^{-1}(0)e^{i\theta H})_-\}^{-1}e^{-i\theta H}$$

corresponds to a one bound state solution $Q(t_j)$. This is precisely the Zakharov–Shabat "dressing" algorithm [108].

What happens if we apply this algorithm again? I will ask you to show in the exercise at the end of this section

$$\{(V_1 k_2^{-1}(0)V_1^{-1})_-\}^{-1}V_1 = \{(e^{-i\theta H}k_2^{-1}(0)k_1^{-1}(0)e^{i\theta H})_-\}^{-1}e^{-i\theta H},$$

which indicates that this action is a group action [40], [106].

Summary. In closing this section, I want to make a few important observations. First, I remind the reader that the typical element $Q(t_j)$ of the phase space is given by (5.223),

$$Q(t_j) = k(t_j)(-iH)k^{-1}(t_j),$$

where $k(t_j)$ is computed as $(\exp(-i\theta H)k^{-1}(0))_-^{-1}$. The element $k^{-1}(0)$ specifies the *solution type*, whether it is a one-soliton, two-soliton, $(1, 0)$ bound state, etc. We can think of $k^{-1}(0)$ as being analogous to the scattering data at time zero. Once the solution type is specified, the flow with respect to the sequence of times $\{t_k\}_0^\infty$ maps out a subset of the phase space. We shall see in the next section that the time flow operator $\partial/\partial t_k$ can be associated with elements $-iH\zeta^k$, $k \geq 0$ of $\widetilde{sl}(2, C)$. These elements, together with their counterparts $-iH\zeta^k$, $k < 0$, form a Heisenberg subalgebra of $\widetilde{sl}(2, C) + Z$, the loop algebra of $\widetilde{sl}(2, C)$ augmented with a center element.

On the other hand, the passage between different solution types (at some fixed values of t_k) is achieved by a Bäcklund transformation. And now we come to a very important point. It is only the Bäcklund transformations (5.134) with

$$R = I + \sum_1^\infty \frac{1}{\zeta^i}C_i$$

which fit the framework of this section. The reason for this is that if R is to be a k, it must be the exponential of an element $\sum_1^\infty D_j\zeta^{-j}$. The Bäcklund transformations which correspond to R's in this class include those which add bound states and solitons but do not include the Bäcklund–Schlesinger transformations. The latter correspond to discrete symmetries and the group elements g with which they are associated cannot be simply factored as $k^{-1}n$ but must include, in the matrix representation, a middle element which is a diagonal matrix.

Therefore much still remains to be done. First, one would like to understand how to augment the group $\bar{G}$ in order to incorporate the discrete symmetries. Second, one would like to have a theory which parallels that of the KdV family, discussed in Section 4g. There, the analogue of (5.223) is (5.90)

$$Q = kXk^{-1}, \qquad k = \hat{U} \qquad (5.232)$$

where $X = -F + E/\lambda$ and $\hat{U}$ is given in (5.90). The whole phase space is recovered from the joint action of flows and Bäcklund transformations. There are no discrete symmetries. It should not be too difficult to put a direct correspondence between formula (5.232) and the corresponding behavior of the τ function, $\tau(t_1, t_2, t_3, \ldots)$, under the action of the flows and under the action of a Bäcklund transformation (see Exercise 5j(2)).

In the present situation, the phase space is divided into discrete parts, each one labelled by the monodromy property of V at $\zeta = \infty$. In each part, flows and bound state and soliton adding Bäcklund transformations act as continuous symmetries. To move from one part to another requires a Bäcklund–Schlesinger transformation. Now recall from (5.169b) and (5.174b) that the action of the Bäcklund–Schlesinger transformation can be interpreted as a shift in a doubly infinite sequence

$$\{\ldots, \tau_{n-1}, \tau_n, \tau_{n+1}, \ldots\}$$

corresponding to the state of the Toda lattice at any given values of the times t_k. The potentials ρ, τ, σ introduced earlier can be any successive triplet in this sequence and such a sequence of three successive members will satisfy the family of Hirota equations and correspond to a solution of the AKNS hierarchy. It would appear that the analogue of the τ-function of the KdV family is not a single function or a triplet but an infinite sequence.

If this is indeed the picture, then the τ function is considered not simply as a function of infinitely many continuous time variables but also a function of n which takes on integer values. The flows change the continuous times. The Bäcklund transformations change the nature of τ from the vacuum state to a one soliton state and so on, but keep n and $\{t_k\}_0^\infty$ fixed. The Bäcklund–Schlesinger transformations change the discrete variable. What is the group connected with these actions and how are the infinitesimal actions related to the Kac–Moody algebra? How does the choice of grading indicate the need for discrete symmetries? In the principal grading of $\widehat{sl}(2, C)$ which gave us (Exercise 5c(3)) the KdV family, there was only one τ function and no discrete symmetries. In the homogeneous grading, there was a single infinity of discrete symmetries and $\tau = \tau(n, t_k)$ satisfies, as a function of n and t_1, the Toda lattice equations. The loop algebras connected with $\widetilde{sl}(r, C)$, $r > 2$, may have more than one set of discrete symmetries, depending on the grading. What differential-difference equations do they satisfy? For example, if τ were to depend on two discrete variables m, n, would the "lattice" equations satisfied by $\tau(m, n, t_k)$ have any relevance in statistical mechanics?

The idea that it is important to allow the τ function to depend on discrete variables originated with Jimbo and Miwa [125]. The reader interested in reading about some of the latest developments should consult volume 2 of the new Springer series which reports the proceedings of workshops held at the Mathematical Sciences Research Institute at the University of California, Berkeley. The title of that particular workshop is "Vertex Operators in Mathematics and Physics".

Exercises 5j.

1. Show that the successive application of the Zakharov–Shabat "dressing" procedure is a group action.

Answer. Let $V_1 = \{(V_0 g V_0^{-1})_-\}^{-1} V_0$ and consider

$$V_2 = \{(V_1 h V_1^{-1})_-\} V_1$$
$$= \{(V_0 g V_0^{-1})_1^{-1} V_0 h V_0^{-1} (V_0 g V_0^{-1})_-\}_-^{-1} (V_0 g V_0^{-1})_-^{-1} V_0.$$

The first factor on the left in the large bracket already belongs to $\bar{K}$ and thus can be removed on the left; after the inverse operation, however, it is transposed to the right and cancels $(V_0 g V_0^{-1})_-^{-1}$. We then have

$$V_2 = \{V_0 h V_0^{-1} (V_0 g V_0^{-1})_-\}_-^{-1} V_0.$$

But since $(k^{-1} n)_- = (k^{-1} nn')_-$, we can multiply under the second minus subscript by $(V_0 g V_0^{-1})_+$. Hence

$$\{V_0 h V_0^{-1} (V_0 g V_0^{-1})_-\}_-^{-1} = (V_0 h g V_0^{-1})_-^{-1}.$$

2. Recall in (5.90), we expressed the phase space element $Q = \lim_{k \to \infty} Q^{(k)}/\lambda^k$, with $Q^{(k)}$ given by (5.57) in Exercise 5c(3), as

$$\lambda Q = \hat{V}(-iH\zeta)\hat{V}^{-1} = \hat{U}X\lambda\hat{U}^{-1}. \tag{5.233}$$

In particular, notice that in the principal grading of 5h, which gives rise to the decomposition used in Exercise 5c(3), $X\lambda$ belongs to K^* and $\hat{U}$, which has asymptotic expansion I + terms of weight (-1) or less, is the exponential of an element in K. $\hat{U}$ can be expressed in terms of the KdV τ-function.

$$\hat{U} = \frac{1}{2\tau} \begin{pmatrix} \tau_+ + \tau_- & \frac{i}{\zeta}(\tau_- - \tau_+) \\ -i\zeta(\tau_- - \tau_+) + \frac{\partial}{\partial x}(\tau_+ + \tau_-) & \tau_+ + \tau_- + \frac{i}{\zeta}\frac{\partial}{\partial x}(\tau_- - \tau_+) \end{pmatrix}$$

where $\tau_\pm = \tau(t_k \pm i/(2k-1)\zeta^{2k-1})$, $k = 1, 2, \ldots$. Now interpret equation (5.233), which arises naturally when we take the algebra as a phase space and the flows as curves in this space, as telling us what happens to the τ function. First, in parallel with the last subsection of Section 5e, show that (5.233) contains the Hirota equations for the KdV family. Second, using (5.233) and the Zakharov–Shabat "dressing" scheme explained earlier, show that the formula of adding one soliton to the vacuum state can be re-expressed as $\tau \to \exp \beta Y(\zeta) \cdot \tau$ where $\beta = \exp(-2\eta x_0)$, $\zeta = i\eta$ and $Y(\zeta)$ is the vertex operator (4.124).

5k. The "sine-Gordon" flows. In Section 5c, we introduced the Lax equations for the positive time (t_k, $k \geq 0$) flows (5.52),

$$Q_{t_k} = [Q^{(k)}, Q]$$

where Q was $\lim_{j \to \infty} (1/\zeta^j) Q^{(j)}$, and

$$Q^{(j)} = \zeta^j \left(Q_0 + \frac{Q_1}{\zeta} + \cdots + \frac{Q_j}{\zeta^j} \right).$$

These Lax equations are the integrability conditions for the set

$$V_{t_k} = Q^{(k)} V, \qquad k \geq 0. \tag{5.234}$$

But, it is known [23] that the sine-Gordon flows come from including new equations in (5.233) corresponding to $k < 0$. For example, with $k = -1$,

$$Q^{(-1)} = \frac{1}{\zeta} Q_{-1},$$

the compatibility of (5.234) with $k = 1$ and -1, gives

$$Q^{(1)}_{t_{-1}} - Q^{(-1)}_x + [Q^{(1)}, Q^{(-1)}] = 0.$$

Equating powers of ζ, one obtains

$$Q_{0t_{-1}} = 0, \quad Q_{1,t_{-1}} = -[Q_0, Q_{-1}], \quad Q_{-1x} = [Q_1, Q_{-1}]. \tag{5.235}$$

A little calculation shows that this is sine–Gordon. Let $Q_0 = -iH$, $Q_1 = qE + rF$, $Q_{-1} = h_{-1}H + e_{-1}E + f_{-1}F$. We find

$$q_{t_{-1}} = 2ie_{-1}, \tag{5.236a}$$

$$r_{t_{-1}} = -2if_{-1}, \tag{5.236b}$$

$$h_{-1x} = qf_{-1} - re_{-1}, \tag{5.236c}$$

$$e_{-1x} = -2qh_{-1}, \tag{5.236d}$$

$$f_{-1x} = 2rh_{-1}. \tag{5.236e}$$

Now look at the equations satisfied by the quadratic products,

$$h = -i(\bar\psi_1\psi_2 + \psi_1\bar\psi_2), \quad e = 2i\psi_1\bar\psi_1, \quad f = -2i\psi_2\bar\psi_2, \tag{5.237}$$

where ψ and $\bar{\psi}$ are the vector solutions of (5.234) defined in Section 5f(i), except that at $x \to +\infty$ they are normalized to have the asymptotic behavior

$$\psi \sim \begin{pmatrix} 0 \\ 1 \end{pmatrix} \exp \left(i \sum_{-\infty}^{\infty} \zeta^j t_j \right),$$

$$\bar{\psi} \sim \begin{pmatrix} 1 \\ 0 \end{pmatrix} \exp \left(-i \sum_{-\infty}^{\infty} \zeta^j t_j \right). \tag{5.238}$$

We find

$$h_x = qf - re,$$
$$e_x + 2i\zeta_e = -2qh, \tag{5.239}$$
$$f_x - 2i\zeta f = 2rh.$$

Notice that if we expand (5.239) about $\zeta = \infty$,

$$h = \sum \frac{h_r}{\zeta^r}, \quad e = \sum \frac{e_r}{\zeta^r}, \quad f = \sum \frac{f_r}{\zeta^r}, \tag{5.240}$$

we obtain exactly the h, e, f in

$$Q = \sum_0^\infty \frac{h_r H + e_r E + f_r F}{\zeta^r}.$$

Now, expand about $\zeta = 0$. Notice that the first terms satisfy exactly the same equations as h_{-1}, e_{-1}, f_{-1}. Hence (5.236a, b) are

$$q_{t_{-1}} = -4\psi_1 \bar{\psi}_1|_{\zeta=0}, \quad r_{t_{-1}} = -4\psi_2 \bar{\psi}_2|_{\zeta=0}. \tag{5.241}$$

Let us simplify things by taking $q = r = (u_x/2)$. Then $\psi_1 = \sinh(u/2)$, $\psi_2 = \cosh(u/2)$, $\bar{\psi}_1 = \cosh(u/2)$, $\bar{\psi}_2 = \sinh(u/2)$, (where we demand that $u \to 0$ at $x = \pm\infty$) whence (5.241) is simply the sinh–Gordon equation

$$u_{xt_{-1}} = -4 \sinh u. \tag{5.242}$$

Similarly, if we set $q = -r = -u_x/2$, we find (5.241) gives the sine-Gordon equation. In this case we may allow u to be any integer multiple of π at $\pm\infty$. We can continue. It is not hard to show that if we take

$$V_{t_{-2}} = Q^{(-2)} V = \left(\frac{1}{\zeta^2} Q_{-1} + \frac{1}{\zeta} Q_{-2} \right) V, \tag{5.243}$$

then

$$q_{t_{-2}} = 2i \frac{\partial}{\partial \zeta} \psi_1 \bar{\psi}_1 \Big|_{\zeta=0}, \quad r_{t_{-2}} = 2i \frac{\partial}{\partial \zeta} \psi_2 \bar{\psi}_2 \Big|_{\zeta=0}. \tag{5.244}$$

The cross-differentiation of (5.243) and (5.233) with $k = 1$ is

$$Q_{0t_{-2}} = 0, \tag{5.245a}$$
$$Q_{1t_{-2}} = -[Q_0, Q_{-2}], \tag{5.245b}$$
$$Q_{-1x} = [Q_1, Q_{-1}], \tag{5.245c}$$
$$Q_{-2x} = [Q_1, Q_{-2}] + [Q_0, Q_{-1}]. \tag{5.245d}$$

Note that (5.245c) is the same as (5.235c) and that the elements h_{-2}, e_{-2}, f_{-2} of $Q_{-2} = h_{-2}H + e_{-2}E + f_{-2}F$ satisfy the same equations as $\partial h/\partial \zeta, \partial e/\partial \zeta, \partial f/\partial \zeta$ at $\zeta = 0$.

As these calculations suggest, the Lax equation

$$Q_{t_k} = [Q^{(k)}, Q], \tag{5.246}$$

$$Q^{(k)} = \begin{cases} \zeta^k\left(Q_0 + \dfrac{Q_1}{\zeta} + \cdots + \dfrac{Q_k}{\zeta^k}\right), & k \geq 0, \\[2mm] \zeta^k(Q_{-1} + \zeta Q_{-2} + \cdots + \zeta^{-k-1}Q_k), & k < 0 \end{cases} \tag{5.247}$$

and

$$Q = \lim_{j \to \infty} \frac{Q^{(j)}}{\zeta^j}. \tag{5.248}$$

If we take $j \to +\infty$ in (5.248), we get the familiar element

$$Q = \sum_0^\infty \frac{Q_j}{\zeta^j}, \qquad Q_0 = -iH, \tag{5.249}$$

the general element of the phase space $N^* = K^\perp$ (in the first Hamiltonian structure) or $-iH + K^\perp$ (in the second Hamiltonian structure). On the other hand, for $j \to -\infty$, we have

$$Q = \sum_0^\infty Q_{-1-j}\zeta^i, \tag{5.250}$$

which is the general element in $K^* = N^\perp$ in the second Hamiltonian structure.

Now recall that the Lax equation is solved formally by

$$Q = V(-iH)V^{-1} = \begin{pmatrix} -i(\bar{\psi}_1\psi_2 + \psi_1\bar{\psi}_2) & 2i\psi_1\bar{\psi}_1 \\ -2i\psi_2\bar{\psi}_2 & i(\psi_1\bar{\psi}_2 + \psi_1\bar{\psi}_2) \end{pmatrix} \tag{5.251}$$

where

$$V = \begin{pmatrix} \bar{\psi}_1 & \psi_1 \\ \bar{\psi}_2 & \psi_2 \end{pmatrix} \tag{5.252a}$$

with asymptotic behavior

$$V \sim \exp\left(-i \sum_{-\infty}^\infty \zeta^i t_i H\right) \tag{5.252b}$$

as $\zeta \to \infty$. (Notice the summation in the exponent includes the negative powers and times.) Now it is relatively easy to see that the asymptotic expansion of (5.251) about $\zeta = \infty$ is exactly (5.249).

Exercises 5k.

1. Show that if

$$L = \frac{1}{2i}\begin{bmatrix} -\dfrac{\partial}{\partial x} - 2q\displaystyle\int_x^\infty dy\, r\cdot & -2q\displaystyle\int_x^\infty dy\, q\cdot \\[3mm] 2r\displaystyle\int_x^\infty dy\, r\cdot & \dfrac{\partial}{\partial x} + 2r\displaystyle\int_x^\infty dy\, q\cdot \end{bmatrix}, \tag{5.253}$$

then for $e = 2i\psi_1\bar\psi_1$, $f = -2i\psi_2\bar\psi_2$, (recall $e_1 = q$, $f_1 = r$),

$$(L - \zeta)\begin{pmatrix} e \\ -f \end{pmatrix} = -\begin{pmatrix} e_1 \\ -f_1 \end{pmatrix}.$$ (5.254)

Now, we know from [23] that the flows (5.246) for $k \geq 0$ can be written

$$\begin{pmatrix} e_1 \\ f_1 \end{pmatrix}_{t_n} = -2iL^n\begin{pmatrix} e_1 \\ -f_1 \end{pmatrix}.$$ (5.255)

But from (2.254),

$$\begin{pmatrix} e \\ -f \end{pmatrix} = \frac{1}{\zeta}\begin{pmatrix} e_1 \\ -f_1 \end{pmatrix} + \sum_1^\infty \frac{1}{\zeta^{n+1}} L^n\begin{pmatrix} e_1 \\ -f_1 \end{pmatrix}$$

$$= \frac{1}{\zeta}\begin{pmatrix} e_1 \\ -f_1 \end{pmatrix} + \sum_1^\infty \frac{1}{\zeta^{n+1}} \frac{i}{2} \cdot \begin{pmatrix} e_{1,t_n} \\ f_{1,t_n} \end{pmatrix}.$$ (5.256)

But from (5.55), we know $e_{1,t_n} = -2ie_{n+1}$, $f_{1,t_n} = 2if_{n+1}$ and thus the right-hand sides are precisely the e and f defined in Section 5c. The quantity $h = -i(\psi_1\bar\psi_2 + \bar\psi_1\psi_2)$ is given by $h^2 + ef = -1$ exactly as defined in Section 5c. Thus the asymptotic expansion of (5.251) is the familiar $Q = \sum_0^\infty Q_r \zeta^{-r}$.

Next, expand Q as given by (5.251) about $\zeta = 0$. We obtain $\sum_0^\infty Q_{-1-j}\zeta^j$, the partial sums of which when multiplied by ζ^{-i} are the coefficient matrices $Q^{(-i)}$ in (5.233).

How does all this connect with what we have done in the last section? Recall that the positive flows arose by reducing the simple flow

$$\mu^. = 0, \qquad g^. = \frac{\delta\Phi}{\delta\mu} g,$$ (5.257)

on $T^*\bar G$ first by the symmetries $\bar K$ in a Marsden–Weinstein reduction to $\bar N \times (-iH\zeta + K^\perp)$ and then by a trivial Poisson reduction to $-iH\zeta + K^\perp$. On the reduced phase space, the flow was given by (5.210),

$$Q(t_j) = k(t_j)(-iH)k^{-1}(t_j)$$

and I showed you that $k(t_j)$ was the left factor $\hat V$ in the asymptotic expansion of V about $\zeta = \infty$; i.e.

$$V \sim \hat V \exp\left(-i\sum_{-\infty}^\infty \zeta^i t_j H\right).$$ (5.258)

It was also the inverse of the left factor in the factorization

$$g = \exp\left(-i\sum_{-\infty}^\infty \zeta^i t_j H\right) g_0 = k^{-1}n.$$

Suppose, now, that instead of factoring $\bar G$ and $\bar K\bar N$ ($g = k^{-1}n$), we factor it instead as $\bar N\bar K(g = n^{-1}k)$. Then in exactly the same way as we did in Section 5j, we will find that the element Q on the reduced phase space $-iH + N^\perp$ is given by

$$Q(t_j) = \mathrm{Ad}^*_{n^{-1}}(-iH) = n(-iH)n^{-1}, \tag{5.259}$$

where $n(t_j)$ is the left factor in the expansion of V about $\zeta = 0$. The right factor is $\exp(-i\sum_{-\infty}^{\infty} \zeta^k t_k H)$ which commutes with $-iH$.

Therefore, the positive t_k flows are found by reducing the big phase space $T^*\bar{G}$ with respect to the left action of the symmetries $\bar{K}$ (the Marsden–Weinstein reduction) followed by the trivial Poisson reduction with respect to the right action of the symmetries $\bar{N}$. The negative t_k flows are found by the dual procedure, namely by factoring $\bar{G} = \bar{N}\bar{K}$.

Remark. I want to stress that in identifying $k(t_j)$ with $\hat{V}(t_j)$, we mean that $\hat{V}$ is taken as the left factor in the formal asymptotic expansion of V at $\zeta = \infty$ and not the function $\hat{V}(t_j, \zeta) = V(t_j, \zeta) \exp(i\sum_{-\infty}^{\infty} \zeta^i t_i H)$. Similarly, the inverse of the left factor in the dual factorization $n(t_j)$ is identified with the formal expansion of $\hat{V}$ about $\zeta = 0$. Because $\zeta = 0$ is an irregular singular point when we include the negative flows, this expansion need not be uniform in all sectors of $\zeta = 0$. If we know the full analytic structure of $\hat{V}$ as function of ζ, then we could relate k and n through the function $\hat{V}$. But, from the algebraic viewpoint, we do not know this and we must consider the asymptotic expansions of $\hat{V}$ about $\zeta = \infty$ and $\zeta = 0$ as unrelated. k^{-1} is simply the left factor of $g = e^{-i\theta H} g_0$ when g is factored as $k^{-1}n'$ and n is the left factor of g factored as $n^{-1}k'$. I use primes on the right factors in order to stress that the $n(k)$ in the latter (former) factorization is not the right factor $n'(k')$ in the former (latter). Nevertheless, the interpretation of k and n in terms of $\hat{V}$, considered as a function of ζ, is still useful.

We will now use a direct method to show that two sets of flows corresponding to the positive and negative times commute.

Consider the decompositions

$$G = K + N \tag{5.260a}$$

and

$$G^* \cong G = K^{\perp} + N^{\perp}, \tag{5.260b}$$

where with the inner product $\langle X, Y \rangle = \mathrm{Tr}\,(XY)_0$, $Q = \sum_0^{\infty} Q_r \zeta^{-r} \in K^{\perp}$ and $\hat{Q} = \sum_1^{\infty} Q_{-r} \zeta^r \in N^{\perp}$. Note that here $\hat{Q}$ is ζ times the expansion of $Q = V(-iH)\hat{V}$ about $\zeta = 0$. Now, we know that for $k \geq 0$,

$$Q_{t_k} = -[\pi_N \nabla\Phi_k(Q), Q], \tag{5.261a}$$

$$\hat{Q}_{t_k} = -[\pi_N \nabla\Phi_k(Q), \hat{Q}], \tag{5.261b}$$

and for $k < 0$

$$Q_{t_k} = -[\pi_k \nabla\Phi_{k-1}(\hat{Q}), Q], \tag{5.262a}$$

$$\hat{Q}_{t_k} = -[\pi_k \nabla\Phi_{k-1}(\hat{Q}), \hat{Q}]. \tag{5.262b}$$

Recall

$$\nabla\Phi_k(k) = -\tfrac{1}{2}\langle S^k X, X \rangle \tag{5.263a}$$

where $S^k X = \zeta^k \sum_{-\infty}^{\infty} X_r \zeta^r$, and

$$\nabla \Phi_k(X) = -S^k X. \tag{5.263b}$$

As before $\Phi_k(X)$ is ad-invariant on G, the loop algebra of $sl(2, C)$.

2. Check that if we multiply $\hat{Q} = \sum_1^{\infty} Q_{-r} \zeta^r$ by ζ^{k-1} for $k < 0$ and only keep those terms for which $r + k - 1 \le -1$, we obtain $Q^{(k)} = \zeta^k (Q_{-1} + \cdots + Q_{-k} \zeta^{-k-1})$.

Let us define, on $G^* \simeq G$, the Hamiltonian function

$$\psi_k(X) = \Phi_k(\pi_{K^\perp} X) \tag{5.264a}$$

and

$$x_l(X) = \Phi_{l-1}(\pi_{N^\perp} X) \tag{5.264b}$$

where $X = Q + \hat{Q}$, $Q \in K^\perp$, $\hat{Q} \in N^\perp$. Then (5.262) can be written

$$X_{t_k} = -[\nabla \psi_k(X), X], \qquad k \ge 0 \tag{5.265a}$$

and (5.262) can be written

$$X_{t_k} = -[\nabla x_l(X), X], \qquad l < 0. \tag{5.265b}$$

To prove (5.265a), we want to show that

$$\nabla \psi_k(X) = \pi_N \phi_k(\pi_{K^\perp} X). \tag{5.266}$$

Proof. By definition

$$\langle \nabla \psi_k(X), X' \rangle = \frac{d}{dt} \psi_k(X + tX')|_{t=0}$$

$$= \frac{d}{dt} \Phi_k(\pi_{K^\perp} X + t\pi_{K^\perp} X')|_{t=0}$$

$$= \langle \nabla \Phi_k(\pi_{K^\perp} X), \pi_{K^\perp} X' \rangle$$

$$= \langle \pi_N \nabla \Phi_k(\pi_{K^\perp} X), \pi_{K^\perp} X' \rangle$$

because the inner product of an element in K with one in $K^\perp$ in zero,

$$= \langle \pi_N \nabla \Phi_k(\pi_{K^\perp} X), X' \rangle$$

because $\pi_{K^\perp} X' = X' - \pi_{N^\perp} X'$ and the inner product of an element in N and an element in $N^\perp$ is zero. Hence (5.266).

With a similar calculation, one can prove that the flows generated by the Hamiltonians $\psi_k(X)$, $k \ge 0$, and $x_l(X)$, $l < 0$ are in involution with respect to the Poisson bracket

$$\{\psi_k, \chi_l\}(X) = -\langle X, [\nabla \psi_k(X), \nabla \chi_l(X)] \rangle.$$

3. (i) Prove that $\{\psi_k, \chi_l\}(X) = 0$ (reference [38, VI]).
(ii) Show from (5.236) that

$$h_{-1} = \frac{i}{2} \frac{\partial^2 \ln \tau}{\partial t_1 \, \partial t_{-1}}.$$

Note that $e_{-1} = (i/2)e_{1,t_{-1}}$, $f_{-1} = -(i/2)f_{1,t_{-1}}$. Show that, in general,

$$h_{-s} = \frac{i}{2}\frac{\partial^2 \ln \tau}{\partial t_1 \partial t_{-s}}$$

and

$$e_{-s} = \frac{i}{2}e_{1,t_{-s}}, \qquad f_{-s} = -\frac{i}{2}f_{1,t_{-s}}.$$

(iii) What are the Hirota equations for the negative time flows?

51. The extension of $\widetilde{sl}(2,C)$ to $\hat{A}_1^{(1)}$. Let me first remind you of the basic ideas of this chapter. We take the loop algebra $G = \{X = \sum_{+\infty}^{-N} X_j \zeta^{-j}, X_j \in sl(2, C)\}$ of $\widetilde{sl}(2, C)$ on which we define a Killing form $\langle X, Y \rangle_0 = \sum_{j+k=0} \mathrm{Tr}\, X_j Y_k$. G is decomposed into two subalgebras, $G = K + N$ and via the Killing form on G, the dual N^* of N is identified with the orthogonal complement $K^\perp$ of K. $K^\perp$ has a natural Poisson structure and the Hamiltonian vector fields generated by a function Φ on $K^\perp$ can be written down. There is a special class of functions, the ad-invariant functions Φ_k which are the coefficients of ζ^{-k} in the expansion of $-\frac{1}{2}\mathrm{Tr}\, X^2$, which have special relevance. Because of ad-invariance, the Hamiltonian vector fields

$$Q_{t_k} = -\pi_{K^\perp}[\pi_N \nabla\Phi_k(Q), Q], \qquad Q \in K^\perp \tag{5.267}$$

take on Lax form,

$$Q_{t_k} = -[\pi_N \nabla\Phi_k(Q), Q], \qquad Q \in K^\perp. \tag{5.268}$$

We have found it convenient in Sections 5i,j to extend the phase space from $K^\perp$ to $\varepsilon + K^\perp$ where $\varepsilon \in K^* = N^\perp$ is a distinguished element in the dual of the algebra of the symmetry group $\bar{K}$. The set $\varepsilon + K^\perp$ also has a Poisson bracket for any $\varepsilon \in G$ and for those ε which belong to the orthogonal complements of $[K, K]$ and $[N, N]$ (the latter is obvious since by implication $\varepsilon \in K^\perp$ and therefore $\varepsilon \in N^\perp$), we have the set of commuting flows

$$(\varepsilon + Q)_{t_k} = -[\pi_N \nabla\Phi_{k-1}(\varepsilon + Q), \varepsilon + Q]. \tag{5.269}$$

Any element of the form $(aH + bE + cF)\zeta$ is acceptable as ε; for the homogeneous grading we choose to take $\varepsilon = -iH\zeta$. The reader now should check that (5.269) is exactly the same as (5.268) when we make the consistent choice $h_0 = -i$, $e_0 = f_0 = 0$ in (5.268) and think of the Q in (5.269) as $\sum_1^\infty Q_r \zeta^{r-1}$, $Q_r = h_r H + e_r E + f_r F$. For the alternative grading we used in the third example in Section 5c, we take $\varepsilon = -F\zeta + E$ (which has weight 1).

Why is it necessary to include any new elements in the algebra? Unfortunately, at this time, I do not really know. What is apparent, however, is that the theory is somewhat incomplete. No natural Lie-theoretic way of introducing the τ-function has been found. Moreover, several formulae, such as (5.63) for the flux tensor, seem to beg the introduction of the operator $\zeta\, d/d\zeta$. Further, since the complete integrability of the Lax equations means that we can reduce the strongly coupled system (5.268) by a canonical transformation to a set of uncoupled harmonic oscillators, one might expect that the Heisenberg algebra should make an appearance. But there is no subalgebra of $\widetilde{sl}(2, C)$ which is Heisenberg. An extra element, a center Z, is required.

Let us then consider

$$\hat{G} = G + cZ + dD, \qquad D = \zeta \frac{d}{d\zeta}. \tag{5.270}$$

We have to specify the commutator and inner product rules associated with the new elements. They are $(X, Y \in G)$

$$[Z, \text{anything}] = 0, \tag{5.271a}$$

$$[D, X] = \zeta \frac{dX}{d\zeta} = \sum_{\infty}^{-N} -jX_j\zeta^{-j}, \tag{5.271b}$$

$$[X, Y] = [X, Y]^\sim + \langle [D, X], Y \rangle_0 Z. \tag{5.271c}$$

In (5.271c), $[X, Y]^\sim$ refers to the commutator under the old $\widetilde{sl}(2, C)$ rules; for example $[H, E]^\sim = 2E$. The new term is $\langle DX, Y \rangle_0 Z$. The new nontrivial inner product rule is

$$\langle Z, D \rangle = 1 \tag{5.272}$$

which is required in order that the parallelogram volume law

$$\langle \hat{X}, [\hat{Y}, \hat{Z}] \rangle = \langle \hat{Z}, [\hat{X}, \hat{Y}] \rangle = \langle \hat{Y}, [\hat{Z}, \hat{X}] \rangle \tag{5.273}$$

is satisfied when $\hat{X}, \hat{Y}, \hat{Z} \in \hat{G}$. For example, check $\hat{X} = D$, $\hat{Y} = \hat{Z} = Y \in G$.

Let us look at some examples. Let $X \in G$ and calculate $[\nabla \Phi_k(X), X]$. Again $\nabla \Phi_k(X) = -S^k X = -\sum_{-\infty}^{N} X_{-j}\zeta^{j+k}$, $X_{-j} \in sl(2, C)$ but note that, because of the extra terms proportional to the center in (5.271c), the commutator is not zero. Therefore the functions $\Phi_k(X)$ are no longer ad-invariant and we lose one of the starting points of the original theory. Nevertheless one can check by following through all the details that, if

$$\hat{Q} = \sum_{0}^{\infty} (h_r H + e_r E + f_r F)\zeta^{-r} + cZ + dD,$$

the Lax equations $Q_{t_k} = [Q^{(k)}, Q]^\sim$ still hold and that c and d are constants.

Next let us calculate the commutator of the gradients Φ_{k-1}, Φ_{j-1}, the Hamiltonians for the t_k and t_j flows in $\widetilde{sl}(2, C)$. We calculate the gradients at the distinguished point $\varepsilon = -iH\zeta$ of K^*; i.e. $\nabla \Phi_{k-1}(-iH\zeta) = -iH\zeta^k$. We find,

$$[\nabla \Phi_{k-1}(-iH\zeta), \nabla \Phi_{j-1}(-iH\zeta)] = [-iH\zeta^k, -iH\zeta^j] = -2k\zeta^{j+k}Z.$$

The sequence $\{\nabla \Phi_{k-1}(-iH\zeta)\}_{k=-\infty}^{\infty}$ therefore generates a Heisenberg subalgebra of $A_1^{(1)}$ and we can introduce the representation

$$-iH\zeta^k \to \frac{\partial}{\partial t_k}, \qquad k > 0,$$

$$-iH\zeta^j \to 2jt_{-j}, \qquad j < 0.$$

The reader should also check that the sequence $\{\nabla\Phi_{k-1}(\varepsilon)\}$, where ε is the distinguished element $\varepsilon = -F\zeta + E$ of the alternative grading, also generates a Heisenberg subalgebra. The fact that the governing algebra $A_1^{(1)}$ has Heisenberg subalgebras is reassuring because we know that the algebra of Poisson brackets of the scattering data is Heisenberg. The reader will recall that the basic idea of inverse scattering is to transform the old coordinates $q, r, q_x, r_x, \ldots$ etc. (in this discussion we consider x distinguished) into new ones, the action-angle coordinates which, when $r = -q^*$, are

$$\mathbf{p} = \left(\frac{1}{\pi}\ln aa^*(\xi), \; \xi \text{ real}, \; 2i\zeta_k, \; 2i\zeta_k^*, \; k = 1, \ldots, N\right),$$

$$\mathbf{q} = (\ln b(\xi), \ln b_k, \ln b_k^*).$$

In [70] it is shown that with respect to the bracket

$$\{F, G\} = \int_{-\infty}^{\infty} \left(\frac{\delta F}{\delta q}\frac{\delta G}{\delta r} - \frac{\delta F}{\delta r}\frac{\delta G}{\delta q}\right) dx,$$

one finds $\{p_i, q_j\} = \delta_{ij}$, $\{p_i, p_j\} = \{q_i, q_j\} = 0$ and, for ξ, ξ' real, $\{p(\xi), q(\xi')\} = \delta(\xi - \xi')$, $\{p(\xi), p(\xi')\} = 0$, $\{q(\xi), q(\xi')\} = 0$. I do not yet know how to identify this Heisenberg subalgebra with the one generated by the sequence $\{-iH\zeta^k\}_{-\infty}^{\infty}$ but believe that one is the manifestation of the other.

In addition, I want to draw your attention to a number of circumstances in which the derivative element $\zeta \, d/d\zeta$ is important. We note that in the third definition of the τ-function, the formula (5.91b), is

$$\langle \hat{V}^{-1}D\hat{V}, -i\zeta^j H\rangle_0 = \frac{\partial}{\partial t_j}\ln \tau.$$

Also, we have already noted that the formulae for the flux tensor F_{jk} can best be written

$$F_{jk} = \langle [D, \pi_N \nabla\Phi_k(Q)], \zeta^i Q\rangle_0$$

where $Q = \sum_0^{\infty} Q_r \zeta^{-r}$, $Q_0 = -iH$ and $D = \zeta^d/d\zeta$. The reader will recognize this coefficient as the extra term proportional to the center Z which appears in the commutator

$$[\pi_N \nabla\Phi_k(Q), \zeta^i Q]$$

which would describe the time evolution of the element $\zeta^i Q$ under the t_k flow if $\zeta^i Q$ belonged to $K^\perp$. In general, of course, it does not. However, on a purely formal level, if we write the time dependence of $\zeta^i Q + c_j Z + d_j D$, then

$$\frac{\partial}{\partial t_k}(\zeta^i Q + c_j Z + d_j D) = [Q^{(k)}, \zeta^i Q]^{\sim} + F_{jk}Z. \qquad (5.274)$$

This would imply that d_j is constant and that c_j is the gradient of $\ln \tau$, i.e. $\partial \ln \tau/\partial t_j$. Note, in particular, that for $j = 0$, in which case formula (5.274) actually holds, $\partial \ln \tau/\partial t_0 = 0$. This is because the dependence of all quantities e_r, f_r

on t_0 is exponential; $e_1(t_0, t_1, \ldots) = e_1(t_1, \ldots)e^{-2it_0}$, $f_1(t_0, t_1, \ldots) = f_1(t_1, \ldots)e^{2it_0}$. Then, since h_r is computed from the ζ^{-r} component of the equation $h^2 + ef = -1$, each h_r, and therefore τ, is t_0 independent.

Finally I want you to observe how the operator $\zeta\, d/d\zeta$, together with its product by powers of ζ, is important in generating exactly integrable nonautonomous equations. These equations are also a natural hierarchy of $\hat{A}_1^{(1)}$ and are quite different from the ones we have already seen. In certain limiting cases, they are the equations one would obtain by looking for the solutions of the former hierarchy which possess the scaling invariance property. The reader should recall the discussion in Section 5f(iii). One of the simplest examples of these new flows is found by considering the integrability condition of the equation pair,

$$V_x = Q^{(1)}V, \tag{5.275a}$$

$$V_t + \zeta V_\zeta = (Q^{(3)} + xQ^{(1)})V. \tag{5.275b}$$

Cross-differentiating, we obtain

$$Q_t^{(1)} + \zeta\frac{d}{d\zeta}Q^{(1)} - Q_x^{(3)} - (xQ^{(1)})_x + [Q^{(1)}, Q^{(3)}]^{\sim} = 0$$

where $[\cdot, \cdot]^{\sim}$ is the old $\widetilde{sl}(2, C)$ matrix commutator. The term $\zeta(d/d\zeta)Q^{(1)}$ cancels the term $1 \cdot Q_0$ from $(xQ^{(1)})_x$ and leaves us with

$$Q_{1t} - (xQ_1)_x - Q_x^{(3)} + [-i\zeta H + Q_1, Q^{(3)}] = 0.$$

This equation is satisfied precisely for our original choice of $Q^{(3)} = -iH\zeta^3 + Q_1\zeta^2 + Q_2\zeta + Q_3$, $Q_r = h_r H + e_r E + f_r F$. Equating the coefficient of ζ^0, we obtain the evolution equations

$$q_t - (xq)_x + \tfrac{1}{4}(q_{xxx} - 6qrq_x) = 0, \tag{5.276a}$$

$$r_t - (xr)_x + \tfrac{1}{4}(r_{xxx} - 6qrr_x) = 0. \tag{5.276b}$$

There are several features worth noting about these equations. First, as already mentioned, they reflect the scaling symmetries inherent in the AKNS hierarchy, some details of which we discussed in Section 5f(iii). It is easy to check that if we had written the RHS of (5.275a) as $3t_3Q^{(3)}$ instead of $Q^{(3)}$, the third terms in (5.276a,b) would be $-3t_3q_{t_3}$ and $-3t_3r_{t_3}$ respectively, and the equations tell us that q and r are functions of $X = (x/(3t_3)^{1/3})$ and $T = xe^t$. If we ask that q and r are independent of t, we recover again the self-similar solutions of the same type to those already discussed in Section 5f(iii). The second feature of interest is the nature of the conservation laws. Without the terms $(xq)_x$ and $(xr)_x$ they are given by

$$\frac{\partial}{\partial t}F_{1j} = \frac{\partial}{\partial t_1}F_{3j}, \qquad j = 1, 2, \ldots;$$

for example:

$$\frac{\partial}{\partial t}F_{11} = -\frac{\partial}{\partial t}qr = \frac{\partial}{\partial x}F_{31} = \frac{\partial}{\partial x}\frac{1}{4}(rq_{xx} + qr_{xx} - r_xq_x - 3q^2r^2).$$

Now, what is $-r(xq)_x - q(xr)_x$? It is simply the x derivative of $(-xqr + \partial \ln \tau/\partial x)$ and so

$$-\frac{\partial qr}{\partial t} = \frac{\partial}{\partial x}\left\{\frac{1}{4}(rq_{xx}+qr_{xx}-r_xq_x-3q^2r^2)-xqr+\frac{\partial}{\partial x}\ln \tau\right\}$$

or

$$-\frac{\partial F_{11}}{\partial t} = \frac{\partial}{\partial x}\left(F_{13}+xF_{11}+\frac{\partial \ln \tau}{\partial x}\right).$$

It would appear that the gradient of $\ln \tau$ again wants to make an appearance. Let us explore further. Consider $F_{12} = -2ih_3 = i/2(rq_x - r_xq)$.

$$\frac{\partial F_{12}}{\partial t} = \frac{\partial F_{32}}{\partial x} + \frac{i}{2}(r(xq)_{xx}+q_x(xr)_x-q(xr)_{xx}-r_x(xq)_x)$$

$$= \frac{\partial F_{32}}{\partial x} + \frac{i}{2}(x(rq_x-r_xq)_x+3(rq_x-r_xq))$$

$$= \frac{\partial F_{32}}{\partial x} + \frac{\partial}{\partial x}\,xF_{12}+2\frac{\partial^2 \ln \tau}{\partial x\,\partial t_2}$$

$$= \frac{\partial}{\partial x}\left(F_{32}+xF_{12}+2\frac{\partial \ln \tau}{\partial t_2}\right).$$

But if we are going to introduce t_2 and think of it as an independent variable, we must ensure that the t_2 flow commutes with the t flow. In order to guarantee this, we have to add $2t_2Q^{(2)}$ to the RHS of (5.274) which does not change the compatibility of (5.275a,b); it simply is equivalent to choosing $h^2+ef = -1+c_1\zeta^{-1}+\cdots$, with a suitably chosen c_1. Consider then,

$$V_x = Q^{(1)}V, \tag{5.277a}$$

$$V_{t_2} = Q^{(2)}V, \tag{5.277b}$$

$$V_t + \zeta V_\zeta = (Q^{(3)}+2t_2Q^{(2)}+xQ^{(1)})V. \tag{5.277c}$$

Cross-differentiate (5.277a,c) and find, after a little calculation,

$$Q_{1t} - (xQ_1)_x - 2t_2Q_{1t_2}+[Q_1, Q_3]=0, \tag{5.278}$$

where we have used the compatibility

$$Q^{(1)}_{t_2}-Q^{(2)}_{t_1}+[Q^{(1)}, Q^{(2)}] \tag{5.279}$$

of (5.277a,b). From (5.277b,c) we have

$$Q^{(2)}_t - iH\zeta^2 + Q_1\zeta - Q^{(3)}_{t_2} - 2t_2Q^{(2)}_{t_2} - 2Q^{(2)} - xQ^{(1)}_{t_2}$$
$$+[Q^{(2)}, Q^{(3)}+xQ^{(1)}]=0.$$

Recall that $Q^{(1)}_{t_2} = Q^{(2)}_{t_1}-[Q^{(1)}, Q^{(2)}]$, and $Q^{(3)}_{t_2}+[Q^{(3)}, Q^{(2)}] = Q^{(2)}_{t_3}$. Then we have

$$Q^{(2)}_t - Q_1\zeta - 2t_2Q^{(2)}_{t_2} - xQ^{(2)}_x - Q^{(2)}_{t_3} = 0.$$

Equating powers of ζ, we obtain

$$\zeta: \quad Q_{1t} - (xQ_1)_x - 2t_2 Q_{1t_2} - Q_{1t_3} = 0,$$
$$\zeta^0: \quad Q_{2t} - xQ_{2x} - 2(t_2 Q_2)_{t_2} - Q_{2t_3} = 0.$$

The reader should check that the last equations are indeed compatible.

In summary, we note:

(i) The inclusion of the new basis elements $\zeta \, d/d\zeta$ and Z in the phase space introduces a new class of flows which are nonautonomous.

(ii) It would appear that, in addition to the phase space variables (h_r, e_r, f_r), one should also include as *dependent* variables t_j and $\partial \ln \tau / \partial t_j$, $j = 1, 2, \ldots$, the latter as coefficients of the centre.

(iii) The new flows have also an infinite set of local conservation laws and symmetries as long as we think of $\partial \ln \tau / \partial t_j$ as a new local variable. For example, since $\partial \ln \tau / \partial t_j = -2i \int^{x=t_1} h_{j+1} \, dx$, we may think of the new variable as the lower limit in the integration. The symmetries seem to be the ones suggested by Chen, Lee and Lin [122] although I have not checked this. I note that the algebra which their symmetries σ_m satisfy is $[\sigma_m, \sigma_n] = (m-n)\sigma_{m+n-1}$. Observe that this is precisely the Virasoro algebra satisfied by the sequence $\sigma_n = -\zeta^n \, d/d\zeta$.

(iv) It would be my guess that the element

$$\pi_N \zeta^i Q + \frac{\partial \ln \tau}{\partial t_j} Z + \left(-\zeta^i \frac{d}{d\zeta} \right)$$

is very important in the whole theory.

And so, I leave you with a lot of open ends. I hope you find the hints and suggestions of this section as tantalizing as I do and, moreover, that you do something about it. Good luck!

Note. I do not want to leave the reader with the impression after reading Section 5b that the Wahlquist–Estabrook procedure is a foolproof way of discovering the Lax pair formulation of integrable systems. Much depends on being able to identify the dependence of P in (5.3) on the phase space coordinates $q, q_x, q_{xx}, \ldots$ etc. Indeed I know of several examples of finite dimensional integrable systems for which all the constants of the motion are known and for which no Lax pair has yet been found. For example, I challenge the reader to consider the stationary equation for the t_5 flow in the KdV family, $q_{xxxx} + 5q_x^2 + 10qq_{xx} + 10q^3 = 0$, which we know is Hamiltonian with canonical coordinates

$$p_1 = q_x, \quad q_1 = \tfrac{1}{4}(q_{xx} + 3q^2), \quad p_2 = \tfrac{1}{4}q_{xxx} + qq_x, \quad q_2 = q$$

and Hamiltonian

$$H = p_1 p_2 + \tfrac{1}{4}p_1^2 - 2q_1^2 q_2 + \tfrac{1}{2}q_1 q_2^4$$

with second constant

$$G = -\tfrac{1}{2}(p_2 + \tfrac{1}{2}p_1 q_2)^2 + \tfrac{1}{4}q_1 p_1^2 - 2q_1^2 q_2 + \tfrac{1}{2}q_1 q_2^3$$

in involution. H generates the x flow, G the t_3 flow. Given the equation and the above information, can you show via Wahlquist–Estabrook that the Lax equation is $Q_x = [Q^{(1)}, Q]$, the compatibility condition for $yV = QV$, $V_x = Q^{(1)}V$ where

$$Q^{(1)} = -i\zeta H + qE - F, \qquad Q = (i\zeta B - \tfrac{1}{2}B_x)H + (-\tfrac{1}{2}B_{xx} + i\zeta B_x - qB)E + BF,$$

$$B = -\zeta^4 + \frac{q}{2}\zeta^2 - \frac{1}{8}(q_{xx} + 3q^2) \quad ?$$

Another example for which I do not yet know the answer is generated by

$$H = \tfrac{1}{2}p_1^2 + \tfrac{1}{2}p_2^2 + q_1^2 + 2q_2^3$$

with second constant

$$G = q_1^4 + 4q_1^2q_2^2 - 4p_1^2q_2 + 4p_1p_2q_1.$$

The difficulty is that one does not have much help in calculating the dependence of Q, $Q^{(1)}$ on the coordinates in which the problem is originally given.

However, I do not want to sound too gloomy because the scheme does work some of the time (especially when one couples the method with information gleaned from the Painlevé test) and, when it does, has the great advantage of pointing to the coordinates in which the constants of the motion separate. For example, in the first example quoted, let

$$B = -(\zeta^2 - \mu_1)(\zeta^2 - \mu_2)$$

and find

$$H = -8\frac{\mu_1^5 - \mu_2^5}{\mu_1 - \mu_2} + 2(\mu_{1x}^2 - \mu_{2x}^2)(\mu_2 - \mu_1),$$

which separates into

$$-\mu_1^5 - \frac{H}{8}\mu_1 - \frac{1}{4}(\mu_1 - \mu_2)^2\mu_{1x}^2 = -\mu_2^5 - \frac{H}{8}\mu_2 - \frac{1}{4}(\mu_2 - \mu_1)^2\mu_{2x}^2 = \frac{G}{8},$$

which is (3.168) and, as we have shown, integrable via the Abel map.

It is therefore reasonable to ask: Does every Hamiltonian system integrable in the sense of Liouville (N constants of the motion in involution) have an equivalent Lax pair formulation and, if so, how does one construct it?

References

(*Please see Key, pp. 243–244, for alphabetical listing.*)

[1] J. SCOTT RUSSELL, *Report on waves*, Rept. Fourteenth Meeting of the British Association for the Advancement of Science, John Murray, London, 1844, pp. 311–390 + 57 plates.

[2] ———, *The Modern System of Naval Architecture*, Vol. 1, Day and Son, London, 1865, p. 208.

[3] J. BOUSSINESQ, *Théorie des ondes et des remous qui se propagent le long d'un canal rectangulaire horizontal, en communiquant au liquide contenu dans ce canal des vitesses sensiblement pareilles de la surface au fond*, J. Math. Pures Appl., 17(2) (1872), pp. 55–108.

[4] D. J. KORTEWEG AND G. DEVRIES, *On the change of form of long waves advancing in a rectangular canal, and on a new type of long stationary waves*, Phil. Mag., 39 (1895), pp. 422–443.

[5] A. FERMI, J. PASTA AND S. ULAM, *Studies of nonlinear problems*, I. Los Alamos Report LA 1940, 1955; in Nonlinear Wave Motion, A. C. Newell, ed., Lectures in Applied Mathematics, 15, American Mathematical Society, Providence. RI, 1974, pp. 143–196.

[6] N. J. ZABUSKY AND M. D. KRUSKAL, *Interaction of "solitons" in a collisionless plasma and the recurrence of initial states*, Phys. Rev. Lett., 15 (1965), pp. 240–243.

[7] M. D. KRUSKAL, *Asymptotology in numerical computation: Progress and plans on the Fermi–Pasta–Ulam problem*, Proc. IBM Scientific Computing Symposium on Large-Scale Problems in Physics, IBM Data Processing Division, White Plains, NY, 1965, pp. 43–62.

[8] ———, *The Korteweg–deVries equation and related evolution equations*, in Nonlinear Wave Motion, A. C. Newell, ed., Lectures in Applied Mathematics, 15, American Mathematical Society, Providence, RI, 1974, pp. 61–83.

[9] M. D. KRUSKAL AND N. J. ZABUSKY, *Progress on the Fermi–Pasta–Ulam non-linear string problem*, Princeton Plasma Physics Laboratory Annual Rept. MATT-Q-21, 1963, Princeton, NJ, pp. 301–308.

[10] N. J. ZABUSKY, *Computational synergetics and mathematical innovation*, J. Comp. Phys., 43 (1981), pp. 195–249.

[11] R. M. MIURA, *Korteweg–deVries equation and generalizations. I. A remarkable explicit nonlinear transformation*, J. Math. Phys., 9 (1968), pp. 1202–1204.

[12] C. S. GARDNER, J. M. GREENE, M. D. KRUSKAL, AND R. M. MIURA, *Method for solving the Korteweg–deVries equation*, Phys. Rev. Lett., 19 (1967), pp. 1095–1097.
———, *The Korteweg–deVries equation and generalizations. VI. Methods for exact solution*, Comm. Pure Appl. Math., 27 (1974), pp. 97–133.

[13] C. S. GARDNER, *The Korteweg–deVries equation and generalizations IV. The Korteweg–deVries equation as a Hamiltonian system*, J. Math. Phys., 12 (1971), pp. 1548–1551.
V. E. ZAKHAROV AND L. D. FADDEEV, *The Korteweg–deVries equation: a completely integrable Hamiltonian system*, Funct. Anal. Appl., 5 (1971), pp. 280–287.

[14] P. D. LAX, *Integrals of nonlinear equations of evolution and solitary waves*, Comm. Pure Appl. Math., 21 (1968), pp. 467–490.

[15] S. L. MCCALL AND E. L. HAHN, *Self-induced transparency by pulsed coherent light*, Phys. Rev. Lett., 18 (1967), pp. 908–911.
———, *Self-induced transparency*, Phys. Rev., 183 (1969), pp. 457–485.

[16] L. P. EISENHART, *A Treatise on the Differential Geometry of Curves and Surfaces*, Ginn & Co, 1909, reprinted by Dover, New York, 1960.

[17] A. SEEGER, H. DONTH AND A. KOCHENDORFER, *Theorie der Versetzungen in eindimensionalem Atomreihen*, Z. Phys., 134 (1953), pp. 173–193.

[18] J. K. PERRING AND T. H. R. SKYRME, *A model unified theory*, Nucl. Phys., 31 (1962), pp. 550–555.

[19] A. C. SCOTT, F. Y. F. CHU AND D. W. MCLAUGHLIN, *The soliton: A new concept in applied science*, Proc. IEEE, 61 (1973), pp. 1443–1483.

[20] G. L. LAMB, JR., *Analytical descriptions of ultra-short optical pulse propagation in a resonant medium*, Rev. Mod. Phys., 43 (1971), pp. 99–129.

[21] V. E. ZAKHAROV AND A. B. SHABAT, *Exact theory of two-dimensional self focusing and one-dimensional self-modulation of waves in nonlinear media*, Soviet Phys. JETP, 34 (1972), pp. 62–69.

[22] M. WADATI, *The exact solution of the modified Korteweg–deVries equation*, J. Phys. Soc. Japan, 32 (1972), pp. 1681ff.

[23] M. J. ABLOWITZ, D. J. KAUP, A. C. NEWELL AND H. SEGUR, *Method for solving the sine-Gordon equation*, Phys. Lett., 30 (1973), pp. 1262–1264.

———, *Nonlinear-evolution equations of physical significance*, Phys. Rev. Lett., 31 (1973), pp. 125–127.

———, *The inverse scattering transform–Fourier analysis for nonlinear problems*, Stud. Appl. Math., 53 (1974), pp. 249–315.

[24] H. FLASCHKA, *The Toda lattice. I. Existence of integrals*, Phys. Rev. B., 9 (1974), pp. 1924–1925.

———, *On the Toda lattice. II. Inverse scattering solution*, Progr. Theor. Phys., 51 (1974), pp. 703–716.

[25] H. P. MCKEAN AND P. VAN MOERBEKE, *The spectrum of Hill's equation*, Invent. Math., 30 (1975), pp. 217ff.

[26] S. P. NOVIKOV, *The periodic problem for the Korteweg-deVries equation*, Funct. Anal. Appl., 8 (1974), pp. 236–246.

[27] A. R. ITS AND V. B. MATVEEV, *The periodic Korteweg–deVries equation*, Funct. Anal. Appl., 9 (1975), pp. 67ff.

[28] I. M. KRICHEVER AND S. P. NOVIKOV, Mathematical Physics Reviews, Vol. 1, S. P. Novikov, ed., (Publ. Soviety Scientific Reviews), pp. 5–23.

[29] H. P. MCKEAN AND E. TRUBOWITZ, *Hill's operator and hyperelliptic function theory in the presence of infinitely many branch points*, Comm. Pure Appl. Math., 29 (1976), pp. 143–226.

[30] J. L. BURCHNALL AND T. W. CHAUNDY, *Commutative ordinary differential operators I, II*, Proc. London Math. Soc., 21 (1922), pp. 420–440; Proc. Royal Soc. Lond. Ser. A, 118 (1928), pp. 557–583.

[31] S. V. MANAKOV, *The inverse scattering transform for the time dependent Schrödinger equation and Kadomtsev–Petviashvili equation*, Physica D, 3 (1981), pp. 420–427.

[32] A. S. FOKAS AND M. J. ABLOWITZ, *On the inverse scattering and direct linearizing transforms for the Kadomtsev–Petviashvili equation*, to be published.

H. SEGUR, *Comments on inverse scattering for the Kadomtsev–Petviashvili equation*, Proc. La Jolla Workshop, Dec. 1981.

[33] M. F. ATIYAH, N. J. HITCHIN, V. G. DRINFELD AND YU. I. MANIN, *Construction of instantons*, Phys. Lett., A, 65 (1978), p. 285.

[34] R. HIROTA, *Direct methods in soliton theory*, in Topics in Current Physics, 17, R. Bullough and P. Caudrey, eds., Springer-Verlag, New York, 1980, pp. 157–175.

[35] M. J. ABLOWITZ, A. RAMANI AND H. SEGUR, *Nonlinear evolution equations and ordinary differential equations of Painlevé type*, Lett. Nuovo Cimento, 23 (1978), pp. 333–338.

———, *A connection between nonlinear evolution equations and ordinary differential equations of p-type I, II*, J. Math. Phys., 21 (1980), pp. 715–721; pp. 1006–1015.

[36] H. FLASCHKA AND A. C. NEWELL, *Monodromy and spectrum preserving deformations*, Comm. Math. Phys., 76 (1980), pp. 65–116.

[37] H. D. WAHLQUIST AND F. B. ESTABROOK, *Bäcklund transformation for solutions of the Korteweg–deVries equation*, Phys. Rev. Lett., 31 (1973), pp. 1386–1390.

———, *Prolongation structures of nonlinear evolution equations*, J. Math. Phys., 16 (1975), pp. 1–7.

[38] H. FLASCHKA, A. C. NEWELL AND T. RATIU, *Kac–Moody algebras and soliton equations,* II, *Lax equations associated with* $A_1^{(1)}$, Physica D, 9 (1983), pp. 300–323.
———, III, *Stationary equations associated with* $A_1^{(1)}$, Physica D, 9 (1983), pp. 324–331, Papers I & IV to follow.

[39] E. DATE, M. JIMBO, M. KASHIWARA AND T. MIWA, *Transformation groups for soliton equations,* Proc. RIMS Symposium on Nonlinear Integrable Systems—Classical and Quantum Theory, M. Jimbo and T. Miwa, eds., World Scientific Press, 1983.

[40] G. SEGAL AND G. WILSON, *Loop groups and equations of KdV type,* to be published.

[41] J. W. MILES, *The Korteweg–deVries equation: A historical essay,* J. Fluid Mech., 106 (1981), pp. 131–147.

[42] ———, *On the Korteweg–deVries equation for a gradually changing channel,* J. Fluid Mech., 91 (1979), pp. 181–190.

[43] C. J. KNICKERBOCKER AND A. C. NEWELL, *Shelves and the Korteweg–deVries equation,* J. Fluid Mech., 98 (1980), pp. 803–818.
———, *Internal solitary waves near a turning point,* Phys. Lett., 75A (1980), pp. 326–330.
———, *Reflections from a solitary wave in a channel of varying depth,* J. Fluid Mech., to appear.

[44] R. S. JOHNSON, *On the development of a solitary waves moving over an uneven bottom,* Proc. Camb. Phil. Soc., 83 (1973), pp. 183.

[45] D. J. KAUP AND A. C. NEWELL, *Solitons as particles, oscillators and in slowly changing media: a singular perturbation theory,* Proc. Roy. Soc. London A, 361 (1978), pp. 413–446.

[46] V. I. KARPMAN AND E. M. MASLOV, *Structure of tails produced under the action of perturbations on solitons,* Sov. Phys. JETP, 48 (1978), pp. 252ff.

[47] V. E. ZAKHAROV AND A. B. SHABAT, *A scheme for integrating the nonlinear equations of mathematical physics by the method of the inverse scattering problem,* Funct. Anal. Appl., 8 (1974), pp. 226–235.

[48] B. B. KADOMTSEV AND V. I. PETVIASHVILI, *On the stability of solitary waves in weakly dispersing media,* Sov. Phys. Doklady, 15 (1970), pp. 539–541.

[49] C. H. SU, *An evolution equation for a stratified flow having two characteristics coalesced,* Preprint, Div. Appl. Math., Brown Univ., Providence, RI, 1983; Advances in Nonlinear Waves, Cambridge Univ. Press, to appear in special volume.

[50] M. TODA, *Theory of Nonlinear Lattices,* Solid State Sciences 20, Springer-Verlag, New York, 1981.

[51] T. B. BENJAMIN AND J. F. FEIR, *The disintegration of wave trains on deep water,* J. Fluid Mech., 27 (1967), pp. 417–430.

[52] A. C. NEWELL, *Nonlinear tunneling,* J. Math. Phys., 19 (1977), pp. 1–26.

[53] D. J. BENNEY AND A. C. NEWELL, *The propagation of nonlinear wave envelopes,* J. Math. and Phys. (now Stud. Appl. Math)., 46 (1967), pp. 133–139.

[54] H. HASIMOTO AND H. ONO, *Nonlinear modulation of gravity waves,* J. Phys. Soc. Japan, 33 (1972), pp. 805–811.
A. DAVEY, *The propagation of a weakly nonlinear wave,* J. Fluid Mech., 53 (1972), p. 769ff.

[55] G. B. WHITHAM, *Linear and Nonlinear Waves,* Wiley-Interscience, New York, 1974.

[56] D. J. BENNEY AND G. J. ROSKES, *Wave instabilities,* Stud. Appl. Math., 48 (1969), pp. 377–385.

[57] A. DAVEY AND K. STEWARTSON, *On three-dimensional packets of surface waves,* Proc. Roy. Soc. London A, 338 (1974), pp. 101–110.

[58] V. E. ZAKHAROV, *Collapse of Langmuir waves,* Sov. Phys. JETP, 35 (1972), pp. 908–914.

[59] B. M. LAKE, H. C. YUEN, H. RUNGALDIER AND W. E. FERGUSON, *Nonlinear deep water waves: theory and experiment, Part 2,* J. Fluid Mech., 83 (1977), pp. 49–74.

[60] H. C. YUEN AND W. FERGUSON, *Fermi–Pasta–Ulam recurrence in the two-space dimensional nonlinear Schrodinger equation,* Phys. Fluids, 21 (1978), pp. 2116–2118.

[61] V. E. ZAKHAROV AND V. S. SYNAKH, *The nature of the self-focusing singularity,* Sov. Phys. JETP, 41 (1976), pp. 465–468.

[62] M. J. ABLOWITZ AND D. J. BENNEY (1976), *The evolution of multi-phase modes for nonlinear dispersive waves*, Stud. Appl. Math., 49 (1970), pp. 225–238.

[63] M. J. ABLOWITZ, *Applications of slowly varying nonlinear dispersive wave theories*, Stud. Appl. Math., 50 (1971), pp. 329–344.

[64] M. C. CROSS AND A. C. NEWELL, *Convection patterns in large aspect ratio systems*, Physica D, 10 (1984), pp. 299–329.

[65] H. FLASCHKA, G. FOREST AND D. W. MCLAUGHLIN, *Multiphase averaging and the inverse spectral solution of KdV*, Comm. Pure Appl. Math., 33 (1979), pp. 739–784.

[66] P. D. LAX AND D. LEVERMORE, *The zero dispersion limit for the Korteweg–deVries equation*, Proc. Natl. Acad. Sci., 76 (1979), pp. 3602–3606.

[67] M. V. G. KRISHNA, Ph.D. thesis, Clarkson College of Technology, Potsdam, NY, 1974.

[68] G. L. LAMB, *Analytical descriptions of ultrashort optical pulse propagation in a resonant medium*, Rev. Mod. Phys., 43 (1971), pp. 99–124.

[69] ———, *Elements of Soliton Theory*, John Wiley, New York, 1980.

[70] H. FLASCHKA AND A. C. NEWELL, *Integrable systems of nonlinear evolution equations*, in Dynamical Systems, Theory and Application, J. Moser, ed., Lecture Notes in Physics 38, Springer-Verlag, New York, 1973, pp. 355–440.

[71] A. R. BISHOP, J. A. KRUMHANSL AND S. E. TRULLINGER, *Solitons in condensed matter: A paradigm*, Physica D, 1 (1980), pp. 1–44.

[72] D. J. KAUP, A. RIEMAN AND A. BERS, *Space-time evolution of nonlinear three wave interactions I, Interaction in a homogeneous medium*, Rev. Mod. Phys., 51 (1979), pp. 275–309.

[73] D. J. KAUP, *A perturbation theory for inverse scattering transforms*, SIAM J. Appl. Math., 31 (1976), pp. 121–123.

[74] ———, *The solution of the general initial value problem for the full three-dimensional three-wave resonant interaction*, Physica D, 3 (1981), pp. 374–395.

[75] A. C. NEWELL, *The inverse scattering transform*, in Topics in Current Physics 17, R. Bullough and P. Caudrey, eds., Springer-Verlag, New York, 1980, pp. 177–242.

[76] M. J. ABLOWITZ AND R. HABERMAN, *Resonantly coupled nonlinear evolution equations*, J. Math. Phys., 16 (1975), pp. 2301–2305.

[77] A. C. NEWELL, *The general structure of integrable evolution equations*, Proc. Roy. Soc. A, 365 (1979), pp. 283–311.

[78] D. J. KAUP AND A. C. NEWELL, *An exact solution for a derivative nonlinear Schrödinger equation*, J. Math. Phys., 19 (1978), pp. 798–801.

[79] D. J. KAUP AND A. C. NEWELL, *On the Coleman correspondence and the solution of the massive Thirring model*, Lett. Nuovo Cimento, 20 (1977), pp. 325–331.

[80] P. DEIFT AND E. TRUBOWITZ, *Inverse scattering on the line*, Comm. Pure Appl. Math., 32 (1979), pp. 121–151.

[81] I. M. GEL'FAND AND B. M. LEVITAN, *On the determination of a differential equation from its spectral function*, Amer. Math. Soc. Transl., Ser. 2, 1, (1955), pp. 259–309.

[82] S. LEIBOVICH AND G. D. RANDALL, *Amplification and decay of long nonlinear waves*, J. Fluid Mech., 58 (1973), pp. 481–493.

[83] I. M. KRICHEVER AND S. P. NOVIKOV, *Holomorphic bundles and nonlinear equations*, Physica D, 3 (1981), pp. 267–293.

[84] R. ABRAHAM AND J. E. MARSDEN, *Foundations of Mechanics*, 2nd edition, Benjamin-Cummings, New York, 1978.

[85] S. P. NOVIKOV, *A method for solving the periodic problem for the KdV equation and its generalizations*, in Topics in Current Physics 17, R. Bullough and P. Caudrey, eds., Springer-Verlag, New York, 1980, pp. 325–338.

[86] H. FLASCHKA, *Construction of conservation laws for Lax equations. Comments on a paper by G. Wilson*, Quart. J. Math. Oxford, 34 (1983), pp. 61–65.

[87] H. GOLDSTEIN, *Classical Mechanics*, Addison-Wesley, Reading, MA, 1950.

[88] J. MARSDEN AND A. WEINSTEIN, *Reduction of symplectic manifolds with symmetry*, Rep. Math. Phys., 5, (1974), pp. 121–130.

[89] R. HIROTA, *Exact solution of the Korteweg-deVries equation for multiple collisions of solitons*, Phys. Rev. Lett., 27 (1972), pp. 1192–1194.

———, *Exact solution of the sine-Gordon equation for multiple collisions of solitons*, J. Phys. Soc. Japan, 33 (1972), pp. 1459–1463.

———, *Exact envelope-soliton solutions of a nonlinear wave equation*, J. Math. Phys., 14 (1973), pp. 805–809.

———, *Exact N-soliton solution of a nonlinear lumped network equation*, J. Phys. Soc. Japan, 35 (1973), pp. 289–294.

———, *Exact N-soliton solution of the wave equation of long waves in shallow water and in nonlinear lattices*, J. Math. Phys., 14 (1973), pp. 810–814.

———, *Exact three-soliton solution of the two-dimensional sine-Gordon equation*, J. Phys. Soc. Japan, 35 (1973) p. 1566.

———, *A new form of Bäcklund transformation and its relation to the inverse scattering problem*, Prog. Theor. Phys., 52 (1974), pp. 1498–1512.

———, *Direct methods of finding exact solutions of nonlinear evolution equations*, in Bäcklund Transformations, R. M. Miura, ed., Lecture Notes in Mathematics 515, Springer-Verlag, New York, 1976.

———, *Nonlinear partial difference equations I. A difference analogue of the Korteweg-deVries equation*, J. Phys. Soc. Japan, 43 (1977), pp. 1429–1433.

———, *Nonlinear partial difference equations II. Discrete time Toda equation*, J. Phys. Soc., Japan, 43 (1972), pp. 2074–2078.

———, *Nonlinear partial difference equations III. Discrete sine-Gordon equations*, J. Phys. Soc. Japan, 43 (1977), pp. 2079–2089.

———, *Nonlinear partial difference equations IV. Bäcklund transformation for the discrete Toda equation*, J. Phys. Soc. Japan, 45 (1978), pp. 321–332.

———, *Nonlinear partial difference equations V. Nonlinear equations reducible to linear equations*, J. Phys. Soc. Japan, 46 (1979), pp. 321–319.

[90] L. SCHLESINGER, J. Reine Angew., Math., 141 (1912), pp. 96–145.

[91] R. GARNIER, *Sur une classe de systemes differentials abéliens deduits de la théorie des équations linéares*, Rend. Cir. Mat. Patermo, 43 (1919), pp. 155–191.

[92] E. L. INCE, *Ordinary Differential Equations*, 1927, reprinted by Dover, New York, 1956.

[93] M. J. ABLOWITZ AND H. SEGUR, *Exact linearization of a Painlevé transcendent*, Phys. Rev. Lett., 38 (1977), pp. 1103–1106.

[94] J. WEISS, M. TABOR AND G. CARNEVALE, *The Painlevé property for partial differential equations*, J. Math. Phys., 24 (1983), pp. 522–526.

[95] H. RUND, *Variational problems and Bäcklund transformations associated with the sine-Gordon and Korteweg–deVries equations and their extensions*, in Bäcklund Transformations, R. M. Miura, ed., Lecture Notes in Mathematics 515, Springer-Verlag, New York, 1974, pp. 199–226.

[96] A. C. NEWELL, *The interrelation between Bäcklund transformations and the inverse scattering transform*, in Bäcklund Transformations, R. M. Miura, ed., Lecture Notes in Mathematics 515, Springer-Verlag, New York, 1974, pp. 227–240.

H. FLASCHKA AND D. W. MCLAUGHLIN. *Some comments on Bäcklund transformations*, Lecture Notes in Mathematics 515, R. M. Miura, ed., Springer-Verlag, New York, 1974, pp. 251–295.

L. D. FADDEEV, *The inverse problem in the quantum theory of scattering*, J. Math. Phys., 4 (1963), pp. 72–104.

[97] D. J. KAUP, *The Wahlquist–Estabrook method with examples of applications*, Physica D, 1 (1980), pp. 391–411.

[98] J. MOSER, *Dynamical systems, finitely many mass points on the line under the influence of an exponential potential – an integrable system*, in Dynamical Systems, Theory and Applications, J. Moser, ed., Lecture Notes in Physics 38, Springer-Verlag, New York, 1975.

[99] B. KOSTANT, *The solution to a generalized Toda lattice and representation theory*, Adv. Math. 34, 3 (1979), pp. 195–338.

[100] D. Kazhdan, B. Kostant and S. Steinberg, *Hamiltonian group actions and dynamical systems of Calogero type*, Comm. Pure Appl. Math., 31 (1978), pp. 481–507.

[101] I. B. Frenkel and V. G. Kac, *Basic representations of affine Lie algebras and dual resonance models*, Invent. Math., 62 (1980), pp. 23–66.

[102] J. Lepowsky and R. L. Wilson, *Construction of the affine Lie algebra $A_1^{(1)}$*, Comm. Math. Phys., 62 (1978), pp. 43–63.

[103] M. Jimbo, T. Miwa, Y. Mori and M. Sato, *Density matrix of an impenetrable box gas and the fifth Painlevé transcendent*, Physica D, 1 (1980), pp. 80–139.
——, *Holonomic quantum fields*, in Lecture Notes in Physics 116, Springer-Verlag, New York, 1979; pp. 119ff.
M. Sato, T. Miwa and M. Jimbo, *Aspects of holonomic quantum fields*, Lecture Notes in Physics 126, Springer-Verlag, New York, 1979, pp. 429ff.

[104] K. Sawada and T. Kotera, *A method for finding n-soliton solutions of the KdV equation and KdV-like equations*, Prog. Theor. Phys., 51 (1974), pp. 1355ff.

[105] V. Arn'old, *Mathematical Methods of Classical Mechanics*, Graduate Texts in Mathematics 60, Springer-Verlag, New York, 1978.

[106] A. Reyman and M. Semenov-Tian-Shansky, *Reduction of Hamiltonian systems, affine Lie algebras and Lax equations. II*, Invent Math., 63 (1981), pp. 423–432.
——, *Current algebras and nonlinear partial differential equations*, Doklady. Acad. Sci USSR, 251 (1980), pp. 1310ff.

[107] J. Moser, *Dynamical Systems*, CIME Lectures, Bressarone, Italy, Program in Mathematics #8, Birkhauser, Boston.

[108] V. E. Zakharov and A. B. Shabat, *Integration of the nonlinear equations of mathematical physics by the method of the inverse scattering problem*, Funct. Anal. Appl., 13, 3 (1979), pp. 13–22.
A. V. Mikhailov, *The reduction problem and the inverse scattering method*, in Solitons, Topics in Current Physics 17, R. Bullough and P. Caudrey, eds., Springer-Verlag, New York, 1980, pp. 243–285.

[109] B. McCoy and T. T. Wu, preprint.

[110] J. M. Greene and I. C. Percival, *Hamiltonian maps in the complex plane*, Physica D, 3 (1981), pp. 530ff.

[111] H. Segur, *Lectures given at the International School of Physics (Enrico Fermi)*, Varenna, Italy, unpublished, 1980.

[112] A. S. Davydov, *The role of solitons in the energy and electron transfer in one-dimensional molecular systems*, Physica D, 3 (1981), pp. 1–22.
J. M. Hyman, D. W. McLaughlin and A. C. Scott, *On Davydov's alpha-helix solitons*, Physica D, 3 (1981), pp. 23–44.

[113] R. Bullough and P. Caudrey, eds., *Solitons*, Topics in Current Physics 17, Springer-Verlag, New York, 1980.

[114] M. J. Ablowitz and H. Segur, *Solitons and the Inverse Scattering Transform*, SIAM Studies in Applied Mathematics 4, Society for Industrial and Applied Mathematics, Philadelphia, 1981.

[115] K. Lonngren and A. C. Scott, eds., *Solitons in Action*, Academic Press, New York, 1978.

[116] F. Calogero and A. Degasperis, *Solitons and the Spectral Transform* I, North-Holland, Amsterdam, 1982.

[117] W. Eckhaus and A. Van Harten, *The Inverse Scattering Transformation and the Theory of Solitons: An Introduction*, Mathematical Studies 50, North-Holland, Amsterdam, 1981.

[118] H. Flaschka and D. W. McLaughlin, eds., *The Theory and Application of Solitons*, Rocky Mountain J. Math. vol. 8, issues 1, 2, 1978.

[119] F. Calogero, ed., *Nonlinear Evolution Equations Solvable by the Spectral Transform* Research Notes in Mathematics 26, London, Pitman, 1978.

[120] R. Miura, *The Korteweg–deVries equation, a survey of results*, SIAM Rev, 18 (1976), pp. 412–459.

[121] G. S. Emmerson, *J. S. Russell, A Biography*, John Murray, London, 1971.

[122] H. H. CHEN, Y. C. LEE AND J. E. LIN, *On a new hierarchy of symmetries for the integrable nonlinear evolution equations*, preprint.

[123] H. L. SWINNEY AND J. P. GOLLUB, eds., *Hydrodynamic Instabilities and the Transition to Turbulence*, Topics in Applied Physics 45, Springer-Verlag, New York, 1981.

[124] *Nonlinear and Turbulent Processes*, Proceedings 2nd International Workshop, Kiev, 1983, Gordon and Breach, New York, 1984.

[125] M. JIMBO AND T. MIWA, *Monodromy preserving deformations of linear ordinary differential equations with rational coefficients* II, Physica D, 2 (1981), pp. 407ff.
———, III, Physica D, 4 (1981), pp. 26ff.

[126] T. BROOKE BENJAMIN, *Lectures on Nonlinear Wave Motion*, Lectures in Applied Mathematics 15, American Mathematical Society, Providence, Rhode Island, 1974, pp. 3–48.

[127] A. C. NEWELL, *Envelope Equations*, Lectures in Applied Mathematics 15, American Mathematical Society, Providence, RI, 1974, pp. 157–163. *Bifurcation and nonlinear focusing*, Pattern Formation and Pattern Recognition, H. Haben, ed., Springer Series on Synergetics, Vol. 5, Springer-Verlag, New York, 1979, pp. 244–265.

KEY

D. J. Kaup, [23], [45], [72], [73], [74], [78], [79], [97]
D. Kazhdan, [100]
C. J. Knickerbocker, [43]
A. Kochendorfer, [17]
D. J. Korteweg, [4]
B. Kostant, [99], [100]
T. Kotera, [104]
I. M. Krichever, [28], [83]
M. V. G. Krishna, [67]
J. A. Krumhansl, [71]
M. D. Kruskal, [6], [7], [8], [9], [12]

B. M. Lake, [59]
G. L. Lamb, Jr., [20], [68], [69]
P. D. Lax, [14], [66]
Y. C. Lee, [122]
S. Leibovich, [82]
J. Lepowsky, [102]
D. Levermore, [66]
B. M. Levitan, [81]
J. E. Lin, [122]
K. Lonngren, [115]

S. V. Manakov, [31]
Yu. I. Manin, [33]
J. E. Marsden, [84], [88]
E. M. Maslov, [46]
V. B. Matveev, [27]
S. L. McCall, [15]
B. McCoy, [109]
H. P. McKean, [25], [29]
D. W. McLaughlin, [19], [65], [96], [112], [118]
A. V. Mikhailov, [108]
J. W. Miles, [41], [42]
R. M. Miura, [11], [12], [120]
T. Miwa, [39], [103], [125]
Y. Mori, [103]
J. Moser, [98], [107]

A. C. Newell, [23], [36], [38], [43], [45], [52], [53], [64], [70], [75], [77], [78], [79], [96], [127]
S. P. Novikov, [26], [28], [83], [85]

H. Ono, [54]

J. Pasta, [5]
I. C. Percival, [110]
J. K. Perring, [18]
V. I. Petviashvili, [48]

A. Ramani, [35]
G. D. Randall, [82]
T. Ratiu, [38]
A. Reyman, [106]
A. Rieman, [72]
G. J. Roskes, [56]
H. Rund, [95]
H. Rungaldier, [59]
J. Scott Russell, [1], [2]

M. Sato, [103]
K. Sawada, [104]
L. Schlesinger, [90]
A. C. Scott, [19], [112], [115]
A. Seeger, [17]
G. Segal, [40]
H. Segur, [23], [32], [35], [93], [111], [114]
M. Semenov-Tian-Shansky, [106]
A. B. Shabat, [21], [47], [108]
T. H. R. Skyrme, [18]
S. Steinberg, [100]
K. Stewartson, [57]
C. H. Su, [49]
H. L. Swinney, [123]
V. S. Synakh, [61]

M. Tabor, [94]
M. Toda, [50]
E. Trubowitz, [29], [80]
S. E. Trullinger, [71]

S. Ulam, [5]

A. Van Harten, [117]
P. Van Moerbeke, [25]

M. Wadati, [22]
H. D. Wahlquist, [37]
A. Weinstein, [88]
J. Weiss, [94]
G. B. Whitham, [55]
G. Wilson, [40]
R. L. Wilson, [102]
T. T. Wu, [109]

H. C. Yuen, [59], [60]

N. J. Zabusky, [6], [9], [10]
V. E. Zakharov, [13], [21], [47], [58], [61], [108]

CBMS-NSF REGIONAL CONFERENCE SERIES
IN APPLIED MATHEMATICS

A series of lectures on topics of current research interest in applied mathematics under the direction of the Conference Board of the Mathematical Sciences, supported by the National Science Foundation and published by SIAM.

GARRETT BIRKHOFF, *The Numerical Solution of Elliptic Equations*

D. V. LINDLEY, *Bayesian Statistics, A Review*

R. S. VARGA, *Functional Analysis and Approximation Theory in Numerical Analysis*

R. R BAHADUR, *Some Limit Theorems in Statistics*

PATRICK BILLINGSLEY, *Weak Convergence of Measures: Applications in Probability*

J. L. LIONS, *Some Aspects of the Optimal Control of Distributed Parameter Systems*

ROGER PENROSE, *Techniques of Differential Topology in Relativity*

HERMAN CHERNOFF, *Sequential Analysis and Optimal Design*

J. DURBIN, *Distribution Theory for Tests Based on the Sample Distribution Function*

SOL I. RUBINOW, *Mathematical Problems in the Biological Sciences*

P. D. LAX, *Hyperbolic Systems of Conservation Laws and the Mathematical Theory of Shock Waves*

I. J. SCHOENBERG, *Cardinal Spline Interpolation*

IVAN SINGER, *The Theory of Best Approximation and Functional Analysis*

WERNER C. RHEINBOLDT, *Methods of Solving Systems of Nonlinear Equations*

HANS F. WEINBERGER, *Variational Methods for Eigenvalue Approximation*

R. TYRRELL ROCKAFELLAR, *Conjugate Duality and Optimization*

SIR JAMES LIGHTHILL, *Mathematical Biofluiddynamics*

GERARD SALTON, *Theory of Indexing*

CATHLEEN S. MORAWETZ, *Notes on Time Decay and Scattering for Some Hyperbolic Problems*

F. HOPPENSTEADT, *Mathematical Theories of Populations: Demographics, Genetics and Epidemics*

RICHARD ASKEY, *Orthogonal Polynomials and Special Functions*

L. E. PAYNE, *Improperly Posed Problems in Partial Differential Equations*

S. ROSEN, *Lectures on the Measurement and Evaluation of the Performance of Computing Systems*

HERBERT B. KELLER, *Numerical Solution of Two Point Boundary Value Problems*

J. P. LASALLE, *The Stability of Dynamical Systems—Z.* ARTSTEIN, *Appendix A: Limiting Equations and Stability of Nonautonomous Ordinary Differential Equations*

D. GOTTLIEB and S. A. ORSZAG, *Numerical Analysis of Spectral Methods: Theory and Applications*

PETER J. HUBER, *Robust Statistical Procedures*

HERBERT SOLOMON, *Geometric Probability*

FRED S. ROBERTS, *Graph Theory and Its Applications to Problems of Society*

JURIS HARTMANIS, *Feasible Computations and Provable Complexity Properties*

ZOHAR MANNA, *Lectures on the Logic of Computer Programming*

ELLIS L. JOHNSON, *Integer Programming: Facets, Subadditivity, and Duality for Group and Semi-Group Problems*

SHMUEL WINOGRAD, *Arithmetic Complexity of Computations*

J. F. C. KINGMAN, *Mathematics of Genetic Diversity*

(continued)

MORTON E. GURTIN, *Topics in Finite Elasticity*

THOMAS G KURTZ, *Approximation of Population Processes*

JERROLD E. MARSDEN, *Lectures on Geometric Methods in Mathematical Physics*

BRADLEY EFRON, *The Jackknife, the Bootstrap, and Other Resampling Plans*

M. WOODROOFE, *Nonlinear Renewal Theory in Sequential Analysis*

D. H. SATTINGER, *Branching in the Presence of Symmetry*

R. TEMAM, *Navier–Stokes Equations and Nonlinear Functional Analysis*

MIKLÓS CSÖRGŐ, *Quantile Processes with Statistical Applications*

J. D. BUCKMASTER and G. S. S. LUDFORD, *Lectures on Mathematical Combustion*

R. E. TARJAN, *Data Structures and Network Algorithms*

PAUL WALTMAN, *Competition Models in Population Biology*

S. R. S. VARADHAN, *Large Deviations and Applications*

KIYOSI ITÔ, *Foundations of Stochastic Differential Equations in Infinite Dimensional Spaces*

ALAN C. NEWELL, *Solitons in Mathematics and Physics*